Modeling with Mathematics:
A Bridge to Algebra II

Modeling with Mathematics:
A Bridge to Algebra II

PROJECT LEADERSHIP

Jo Ann Wheeler
REGION 4 ESC, HOUSTON, TX

Solomon Garfunkel
COMAP INC., LEXINGTON, MA

David Eschberger
REGION 4 ESC, HOUSTON, TX

Roland Cheyney
COMAP INC., LEXINGTON, MA

LEAD AUTHORS

Gary Cosenza
REGION 4 ESC, HOUSTON, TX

Paul Gray
REGION 4 ESC, HOUSTON, TX

Julie Horn
REGION 4 ESC, HOUSTON, TX

AUTHORS

Sharon Benson
REGION 4 ESC, HOUSTON, TX

Roland Cheyney
COMAP, LEXINGTON, MA

David Eschberger
REGION 4 ESC, HOUSTON, TX

Jo Ann Wheeler
REGION 4 ESC, HOUSTON, TX

W. H. Freeman and Company
New York
www.whfreeman.com

Published and distributed by
W. H. Freeman and Company
41 Madison Avenue, New York NY 10010

Library of Congress Control Number: 2006921261

ISBN 07167-0780-2 (EAN 9780716707806)

Printed in the United States of America
Second Printing 2007

PUBLISHER: Craig Bleyer

MARKETING MANAGER: Cindi WeissGoldner

DIRECTOR OF HIGH SCHOOL SALES AND MARKETING: Mike Saltzman

SENIOR MEDIA EDITOR: Roland Cheyney

ASSOCIATE EDITOR: Brendan Cady

PRODUCTION MANAGER: George Ward

PRODUCTION DEPARTMENT: Daiva Kiliulis, Tim McLean, George Ward, Pauline Wright

PHOTO RESEARCHERS: COMAP Production

ILLUSTRATIONS: Lianne Dunn, George Ward

COVER AND TEXT DESIGN: Daiva Kiliulis

COVER IMAGES: Shutterstock

MANUFACTURING: RR Donnelley

PROJECT LEADERS:

Jo Ann Wheeler, Region 4 ESC, Houston, TX

David Eschberger, Region 4 ESC, Houston, TX

Solomon Garfunkel, COMAP, Lexington, MA

Roland Cheyney, COMAP, Lexington, MA

LEAD AUTHORS:

Gary Cosenza, Region 4 ESC, Houston, TX

Paul Gray, Region 4 ESC, Houston, TX

Julie Horn, Region 4 ESC, Houston, TX

AUTHORS:

Sharon Benson, Region 4 ESC, Houston, TX

Roland Cheyney, COMAP, Lexington, MA

David Eschberger, Region 4 ESC, Houston, TX

Jo Ann Wheeler, Region 4 ESC, Houston, TX

REVIEWER:

Sandra Nite, Texas A&M University,
College Station, TX

Writers who contributed material used from
COMAP's *Mathematical Models with Applications*
and Region 4's *Mathematical Models with
Applications/Algebra II Alignment*:

Marsha Davis, Eastern Connecticut State
University, Williamantic, CT

Juliann Doris, Seguin High School, Seguin, TX

Anne Konz, Cypress-Fairbanks Independent
School District, Cypress, TX

Jerry Lege, COMAP, Inc., Lexington, MA

Jennifer May, Independent Consultant

Paul Mlakar, St. Mark's School, Dallas, TX

Sandra Nite, Texas A&M University,
College Station, TX

Richard Parr, Rice University, Houston, TX

Gary Simundza, Wentworth Institute of
Technology, Boston, MA

Ann Worley, Spring Branch Independent School
District, Houston, TX

Region 4 and COMAP would like to thank the
following teachers and school districts for
piloting *Modeling with Mathematics: A Bridge to
Algebra II* and providing essential feedback:

Kim Armstrong
Lindale High School
Lindale Independent School District
Lindale, Texas

Toni Ericson
Hallsville High School
Hallsville Independent School District
Hallsville, Texas

Jason Hendricks
Madison High School
North East Independent School District
San Antonio, Texas

Sara Kamphaus
Summit High School
Mansfield Independent School District
Mansfield, Texas

Mary Ann Knight
Mansfield High School
Mansfield Independent School District
Mansfield, Texas

Connie Koehn
Elsik High School
Alief Independent School District
Houston, Texas

Garnette Lamm
Madison High School
North East Independent School District
San Antonio, Texas

Caroline Martin
Timberview High School
Mansfield Independent School District
Mansfield, Texas

Michael McCabe
Grapevine High School
Grapevine-Colleyville Independent School
District
Grapevine, Texas

Robert McFarland
Aldine Senior High School
Aldine Independent School District
Houston, Texas

Jerry McHugh
MacArthur High School
North East Independent School District
San Antonio, Texas

Monica Merchant
Aldine Senior High School
Aldine Independent School District
Houston, Texas

Theresa Patton
Elsik High School
Alief Independent School District
Houston, Texas

Brenda Porter
Reagan High School
North East Independent School District
San Antonio, Texas

DeAnna Ramirez
Mansfield High School
Mansfield Independent School District
Mansfield, Texas

Clint Reynolds
Van High School
Van Independent School District
Van, Texas

Laura Salazar
Reagan High School
North East Independent School District
San Antonio, Texas

Michael Seibert
Grapevine High School
Grapevine-Colleyville Independent School
District
Grapevine, Texas

Glenn T. Smith
Aldine Senior High School
Aldine Independent School District
Houston, Texas

Sheryl Smith
Taylor High School
Alief Independent School District
Houston, Texas

Diane Tobin
Aldine Senior High School
Aldine Independent School District
Houston, Texas

Maria Tolentino
Hastings High School
Alief Independent School District
Houston, Texas

Ron Van Raemdonck
Aldine Senior High School
Aldine Independent School District
Houston, Texas

Michael Williams
Gilmer High School
Gilmer Independent School District
Gilmer, Texas

Yangki Wojcik
Reagan High School
North East Independent School District
San Antonio, Texas

Aubrey Wright
Colleyville Heritage High School
Grapevine-Colleyville Independent School
District
Colleyville, Texas

Randy Zelahy
Aldine Senior High School
Aldine Independent School District
Houston, Texas

Dear Student,

You have probably seen descriptions of how people use mathematics in their careers and lives in other math textbooks. The message of these little sidebars is simple, trust us … if you keep taking math courses you will have great career options and know more about our world. We agree!

But when we set out to make *Modeling with Mathematics: A Bridge to Algebra II* we asked a question: What if we do more? Could we not just tell you that mathematics is useful, but show you too? Why not create hands-on activities in science, business, design, and other fields and have you learn and use the mathematics in these real-world settings? This text uses a tool called mathematical modeling to explore these and other fields.

Mathematical modeling is used to clarify and solve a wide range of real-world problems. It is a tool used by engineers when they design a new style of MP3 player and by architects when they design a skyscraper. The best part about mathematical modeling is that it can simplify a tricky problem and help you focus on the things that really matter. The hard part about it is that it makes you read more, think more, and write more. We also think it makes learning math more interesting. The Introduction will tell you a little bit more about mathematical modeling.

In a few places in this text, you will see A Look Ahead section. These sections are a chance for you to see a little of the content you will do when you take Algebra II.

Some of you may not be confident about your math skills. Some of you may even feel that you can't take any more math courses. The goal of *Modeling with Mathematics: A Bridge to Algebra II* is to help you get back on track. It is a chance to look back at some of the math you may not have really understood and to look forward to a great experience in Algebra II.

Sincerely,

Jo Ann Wheeler
Region 4 ESC, Houston, TX

David Eschberger
Region 4 ESC, Houston, TX

Solomon Garfunkel
COMAP, Lexington, MA

Roland Cheyney
COMAP, Lexington, MA

Table of Contents

CHAPTER 3 The World of Business: Systems of Equations 130–193

CHAPTER 4   Art: Transformations, Symmetry, and Proportions   194–255

CHAPTER 5   Motion: Quadratic Functions   256–343

CHAPTER 7 Money and You: Mathematics and Finance 414–493

The Modeling Process

INTRODUCTION

The process of starting with a situation or problem and gaining understanding about that situation through the use of mathematics is known as **mathematical modeling**.

Mathematical modeling attempts to describe real-world relationships in mathematical terms. Mathematical descriptions are useful because mathematics provides ways to obtain solutions. There are several ways to represent real-world relationships.

Mathematical modelers use a number of representations in their work. Among them are:

❖ verbal descriptions

❖ graphs, such as the graph of a straight line

❖ formulas, such as the formula for the area of a square

❖ tables, such as a table of values of length and width that give a rectangle with an area of 24 square inches

Length (l)	Width (w)
2	12
4	6
6	4
8	3
12	2

❖ drawings, such as a scale drawing of a room in a house

❖ diagrams, such as an arrow diagram or flow chart.

The mathematical modeling process can be described in steps.

Step 1. Identify the problem.

Read and ask questions about the situation. Identify a problem that you want to solve.

Step 2. Simplify the situation.

Select the features of the situation that you think are most important. These are your assumptions that you will use to build a model. Note the features that you will ignore at first.

Step 3. Build the model and solve the problem.

Describe relationships among the parts of the problem in mathematical terms. Then find a mathematical solution to the problem. In this step you might do all or some of the following:

❖ define variables

❖ write equations

❖ draw shapes

❖ measure objects

❖ gather data and organize into tables

❖ make graphs.

Step 4. Evaluate and revise the model.

Go back to the original situation and see if the results of the mathematical model make sense. If so, use the model until new information becomes available or assumptions change. If not, reconsider the assumptions you made in step 2 and revise them.

You might go through a revision process several times before you have a good mathematical model. One helpful principle that guides all modelers is *keep it simple*. All models ignore something, and first models usually ignore several things.

A way of visualizing the mathematical modeling process is shown in **Figure 1**.

FIGURE 1.
The modeling process.

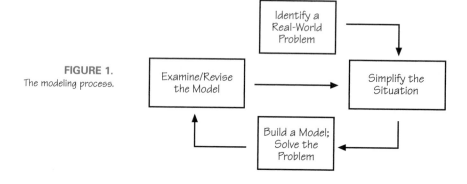

Mathematical modeling is often used to describe the relationship between two or more real-world variables. (A variable is a quantity that changes.) Discovering and describing a relationship among variables allows modelers to predict values of one of the variables from values of the others.

Some examples of real-world situations where the modeling process might be used include:

* predicting the growth of a population based on variables such as the quantity of food available, disease rates, or birth rate

* predicting the profit for a company based on the costs of producing a product or the level of demand for the product

* predicting the effect on the economy of factors such as the unemployment rate or the value of exported goods

* predicting the effect of widening a road on traffic volume

* describing the relationship between the volume of blood flowing through an artery and the pulse rate of a person.

In this course you will use mathematics to create models that explore and explain a wide range of real-world issues. You will also use your imagination to create these models, then use mathematics to test them and see if they make sense. In order to create and test your models tools are needed, such as paper and pencil problem solving, using calculators, and doing research at your library and on the Internet. Most importantly you will need to find good ways to show your model to your teacher and to other students.

This may seem a very different way of learning math, but in this course you will see how mathematical modeling can help you succeed, both in school and in many careers. So give mathematical modeling a chance, you may find it changes how you feel about math.

1

CHAPTER

Science: Modeling with Direct and Inverse Variation

Extra! Extra! Scientists Use Mathematical Modeling!

From ancient times, people have tried to make sense of the world around them. Nomadic Native Americans made observations about the behavior of the buffalo herds in the central plains of North America. These observations led them to make conjectures about the movement of the herd, which was essential to the tribe's survival. If they successfully followed the herd, then food was plentiful. If the herd left them behind, then they starved.

Some civilizations began to use numbers to describe relationships. In ancient Egypt, people began writing down numerals to keep count of things. They began to write simple equations to perform calculations.

These numeric observations prompted early scientists to want to learn more. They noticed that certain quantities are related, such as the length of day and the location of certain stars, and began to use mathematics to describe these relationships. These observations were the early stages of what we now call mathematical modeling.

In this chapter, you will explore some of the work of three notable scientists who used mathematics to describe relationships that they noticed in the natural world.

Archimedes (**Figure 1.1**) was a Greek scientist who lived in Syracuse, Sicily, during the third century BC. Archimedes wrote many books about geometry, mathematics, and physical science and was a good friend of King Hiero of Syracuse. Archimedes was instrumental in defending his home city against the Roman siege in 212 BC, before his death that same year.

FIGURE 1.1.
Archimedes.

Archimedes is perhaps best known for what is today called Archimedes' Principle. This idea describes the forces that interact between a fluid, such as water, and an object that is submerged in that fluid.

FIGURE 1.2.
Robert Hooke.

Robert Hooke (**Figure 1.2**) was an English scientist who lived in London during the 1600s. This was an exciting time to be a scientist in England. Like most of Western Europe, England was enjoying the Age of Enlightenment, a time when scientists and philosophers explored new ideas about the world around them. After a long period of silence, they rediscovered the work of the ancient Greeks and Romans. As technology had become quite advanced for the day, scientists now had many new tools at their disposal, and they began to look for new ways to use them.

Robert Hooke began his work with springs as he searched for a way to build a better clock. Ships had been sailing from Europe to North America and Asia for over 200 years in Hooke's day. Yet, they had difficulty telling time in order to determine their longitude. European clocks at this time used a pendulum that swung back and forth, keeping steady time. However, on long sea voyages, waves disrupted the pendulum. So, Hooke began to explore the use of springs to make clocks tick. Along the way, he discovered a relationship between the length that a spring stretches and the amount of force required to make it stretch to that length. This relationship today is known as Hooke's Law.

Robert Boyle (**Figure 1.3**) was also an English scientist in London during the 1600s. In fact, Robert Hooke was one of his students. Boyle studied religion, chemistry, and physics. Like many of his colleagues, he wanted to take the mystery out of science, and he relied heavily on mathematics to do so.

FIGURE 1.3.
Robert Boyle.

Boyle was keenly interested in the properties of air, which were still largely misunderstood. Boyle's predecessors in Italy had found that air has weight and exerts a force on the ground called air pressure. Boyle learned through a series of experiments that not only did air have weight but also that fire cannot burn without air.

Boyle also studied the relationship between air pressure and volume. In 1662, he wrote in *Touching the Spring of the Air and Its Effects* about his experiments with a vacuum chamber to determine the relationship between pressure and volume of a gas. His findings are today known in chemistry as Boyle's Law.

Scientists continue to use mathematical modeling to describe natural events and to make predictions. Astrophysicists use mathematical modeling to chart the paths of stars and solar systems. Aeronautical engineers use mathematical modeling to build better airplanes and space vehicles. Social scientists use mathematical modeling to make predictions about populations and natural resource management.

Throughout this book, you will study how different people use mathematical modeling to make decisions about everyday life. You will also see how mathematics appears in unlikely places and is used by people from a variety of backgrounds.

In Chapter 1 you will explore some important ideas in science using mathematical modeling.

Archimedes and the Crown

In the third century BC lived a famous mathematician and scientist named Archimedes. Archimedes is famous for many discoveries including an irrigation device, the law of the lever, the formula for the volume of a sphere, and possibly even the odometer. One of his discoveries is the Archimedes Principle, which describes the concept of buoyancy. Vitruvius, a Roman architect, tells a famous story about Archimedes and King Hiero of Syracuse.

King Hiero hired a craftsman to make a crown of gold. The king measured out the exact amount of gold for the craftsman to use. The craftsman later delivered to the king a beautiful crown. The crown weighed the same as the measured amount of gold. Rumors began floating around Syracuse that the crown was not made of pure gold. It was suggested that the craftsman had replaced part of the gold with an equal mass of silver. King Hiero asked Archimedes to prove or disprove the rumors without damaging the crown in any way. (Today this is called nondestructive testing.) As Archimedes began to sit down in a bath pondering the problem, he noticed the water level in the bath rising as he submerged more and more of his body into the bath. Realizing he had found a way to solve the problem he ran down the street, still naked, shouting, "Eureka!" ("I have found it!") The craftsman admitted he was guilty of stealing part of the gold and replacing it with silver.

1. What do you think Archimedes' solution might have been?

2. Gold has greater mass than silver. If a gold crown and a silver crown have the same weight, would they have the same volume? Why?

3. How can the volume of an object be determined by placing it in water?

4. Suppose your teacher places 15 pennies in a film canister and 20 pennies in another. Which canister has greater mass?

5. Which canister, the one with 15 pennies or the one with 20 pennies, would displace more water when submerged?

6. Does mass or volume cause displacement?

Your teacher will do a demonstration with film canisters, pennies, and a graduated cylinder. Answer the following questions as she/he does the demonstration.

7. What did you observe when your teacher placed the film canister into the graduated cylinder?

8. What did you observe when your teacher placed 20 pennies into the film canister?

9. Which canister, one with 15 pennies or one with 20 pennies, has greater mass?

10. Calculate the volume of a film canister.

11. What do you observe about the relationship between the calculated volume of the film canister and the amount of water displaced when the canister is submerged?

Displacement Investigation

In this section you will explore displacement. You have a graduated cylinder filled with a given amount of water. You will record three things: the number of film canisters in the water, the total displacement of the canisters (the change in the volume of water), and the total volume contained in the cylinder.

✓ CHECK THIS!

Displacement is the amount of water moved when an object is placed in the water.

1. Fill a large graduated cylinder to the 500 mL mark. In a table, record the amount of displacement when no film canisters have been placed in the cylinder. Also record the total volume contained in the cylinder.

2. Place a film canister with the minimum number of pennies to make it sink into the water in the large graduated cylinder (**Figure 1.5**). Read and record in your table (similar to the table in **Figure 1.4**) the volume of water displaced by 1 film canister and the total volume contained in the cylinder. Repeat this process adding 1 additional canister at a time. Continue until you submerge all 6 of your canisters.

Total Number of Submerged Canisters	Total Volume of Water Displaced (mL)	Total Volume Contained in the Graduated Cylinder (mL)

FIGURE 1.4.
Displacement data.

3. If you place 20 pennies in the film canister, does it displace more water than a film canister with 15 pennies?

FIGURE 1.5.
Displacement illustration.

Your table has three columns. The first step of the mathematical modeling process is selecting a problem. In the rest of this section you will examine the relationship between displacement and the number of submerged canisters. You will examine the relationship between total volume and the number of submerged canisters in a later section.

4. Make a scatterplot of the pairs (displacement, number of submerged canisters) using the data in your table.

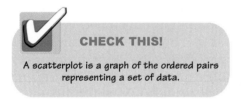

CHECK THIS!

A scatterplot is a graph of the ordered pairs representing a set of data.

5. Describe the pattern of the points that you observe in your scatterplot.

6. Predict the total volume of water displaced if you submerge 8 canisters.

7. Predict the number of canisters that you need to submerge in order to displace 500 mL.

A Question of Variation

In the previous section, you collected and analyzed data regarding the relationship between volume of water displaced by canisters and the number of canisters submerged in water in a graduated cylinder.

You have already collected data and answered some questions about what you found. Now you will look at some important ideas about collecting data and building models based on that data.

The number of film canisters in the water and the total volume of water displaced are variables because their values change. A **variable** is a quantity that changes. Letters of the alphabet can be used to label variables. For example, c is a natural choice for the total number of film canisters in the water, and w is a good choice for the total volume of water displaced. The variable that you want to use to make predictions in your investigation is called the input or **independent** variable. The variable that you want to predict is called the output or **dependent** variable. Another way to think about this is that the dependent variable responds to changes in the independent variable or that changes in the independent variable explain changes in the dependent variable. In an experiment such as the one you just did, the independent variable is the one that you control.

When making a table of data in an experiment, it is customary to place the values of the independent variable in the first (left) column.

There are other variables in this situation, but in order to keep the modeling simple you will consider only two of them: the total number of canisters in the water (c) and the volume of water displaced (w). In modeling, variables have **units** that describe the kind of quantity the variable uses. For c, the units are numbers of canisters, for w, the units are milliliters

1. In the Displacement Investigation in Section 1.2, which variable is the independent variable? Why? Which variable is the dependent variable? Why?

2. The set of possible values of the independent variable is called the **domain of the situation**. What is a reasonable domain in this situation? Why?

3. The set of possible values of the dependent variable is called the **range of the situation**. What is a reasonable range for this situation? Why?

4. What patterns do you observe in your collected data?

5. One important aspect of mathematical modeling is the way the variables change in relation to each other. In the table where you recorded the number of submerged film canisters and the total displacement of the volume, find the value of the total displacement divided by the total number of film canisters for each row, $\frac{w}{c}$ (as in **Figure 1.6**).

Total Number of Submerged Canisters	Total Volume of Water Displaced (mL)	$\frac{w}{c} = \dfrac{\text{Total volume of water displaced}}{\text{Total number of film canisters}}$

FIGURE 1.6.
More displacement data.

6. A **proportional relationship** is one in which the ratio $\frac{\text{dependent variable}}{\text{independent variable}}$ is constant. The constant is called the **constant of proportionality**. In a proportional relationship, this constant is also a **rate of change**. A **rate of change** is a ratio that compares the change in the output variable to the change in the input variable. What do you observe in your table?

7. Real-world data are seldom perfect. Is the constant of proportionality the same or does it vary in your table?

One way to visualize a relationship between two variables is with a graph of enough data to get a sense of the pattern they form. In mathematics, a pair of perpendicular **axes** (number lines) creates a **coordinate plane**, in which a pair of numbers locates a point. The two numbers, a value of the independent variable and the corresponding value of the dependent variable, are called the **coordinates** of the point. In a graph, values of the independent variable are located along the horizontal axis (traditionally called the x-axis), and values of the dependent variable are located along the vertical axis (traditionally called the y-axis).

A **scatterplot** (**Figure 1.7**) is a graph of pairs (x, y) where the x-values are the independent values from your data and the y-values are the corresponding dependent values from your data. You created a scatterplot with the number of submerged film canisters on the x-axis and the total displacement of water on the y-axis.

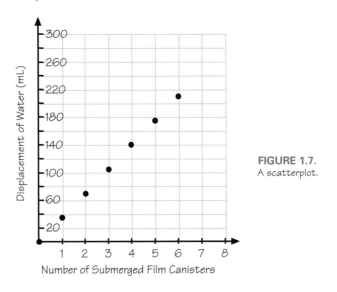

FIGURE 1.7.
A scatterplot.

This graph is called **discrete** because it shows distinct (separated) points. It does not make sense to connect the points since you cannot have a part of a canister. Since you cannot have a fraction of a canister, the domain is restricted to whole numbers. However, to use a scatterplot to make predictions, it is convenient to connect the points with a straight line, which creates a type of **continuous** graph. However, doing so changes the domain shown in the graph from '*some*' values to '*all*' values. Even though it makes mathematical sense, drawing a continuous graph can cause confusion when working with real-world situations.

8. Does it make sense to ask how much water 1.5 film canisters displace? Why or why not?

In a proportional relationship:

❖ The continuous graph is a straight line that passes through the origin;

❖ The ratio $\frac{\text{dependent variable}}{\text{independent variable}}$ is constant (or nearly so) for all data.

A proportional relationship can be described with an equation of the form $y = kx$, which is called a **direct variation function**. When a function is used to describe a relationship among variables, it is sometimes said that the function models the relationship.

9. Do your data demonstrate a proportional relationship between displacement and number of canisters? How do you know?

10. Describe how the constant of proportionality is displayed in the graph.

11. Based on your table or graph, what is the volume of 1 film canister in cubic centimeters?

Two quantities have a **positive relationship** (sometimes called a positive correlation) when the dependent variable increases as the independent variable increases.

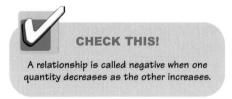

CHECK THIS!

A relationship is called negative when one quantity decreases as the other increases.

12. Is there a positive relationship between the total displacement of the water and the number of canisters submerged? Why or why not?

13. Use a graphing calculator to make a scatterplot of your data from the Displacement Investigation in Section 1.2. Your teacher will give you instructions on how to do this.

Assignment

For their science project, Judy and Jose plant a kudzu vine seed in order to investigate how quickly it grows. Every four days after the seed sprouts, they measure and record the height of their plant. Their results are shown in **Figure 1.8.**

Day Number	Height (cm)
0	0
4	3
8	6
12	9
16	12
20	15

FIGURE 1.8.
Kudzu data.

1. For this situation, what is the dependent variable and what is the independent variable?

2. State a reasonable domain and range for this situation.

3. Make a scatterplot for these data.

4. Is there a proportional relationship between height and day number? Why or why not?

5. What is the constant of proportionality for this situation?

6. Which kind of graph, continuous or discrete, is more appropriate for this situation? Why?

7. What is an appropriate direct variation function to model the relationship between height and day number?

Alexia and Alex collected data by placing marbles in a graduated cylinder partially filled with water and measuring the volume of displaced water. Their data are shown in the scatterplot in **Figure 1.9.**

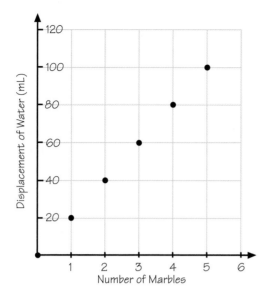

FIGURE 1.9.
Marble data.

8. Make a table for the data displayed in the scatterplot.

9. For this situation, what is the dependent variable and what is the independent variable?

10. State a reasonable domain and range for this situation.

11. Is this a proportional relationship? Why or why not?

12. What is the constant of proportionality for this situation?

13. Which kind of graph, continuous or discrete, is more appropriate for this situation?

14. What is an appropriate direct variation function to model the relationship between displacement and number of marbles?

15. Use the volume of 1 marble to find the approximate radius of a marble.

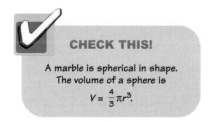

CHECK THIS!

A marble is spherical in shape. The volume of a sphere is

$$V = \frac{4}{3}\pi r^3.$$

Displacement Investigation Round 2

In Section 1.2 you collected data and recorded the data in a table. You made a scatterplot of the relationship between displacement and number of submerged canisters, and found that the data demonstrated a proportional relationship. The direct variation model that you made involves two variables: the number of submerged film canisters (c) and the amount of water displaced (w).

If you change the variables you examine in a situation, your model may change.

In Section 1.2 you also recorded the total volume of the graduated cylinder. For this next investigation, continue to use c for the number of submerged canisters, your independent variable. Use v for the total volume contained in the graduated cylinder, your dependent variable.

1. Make a table showing the number of submerged canisters and the total volume of the graduated cylinder.

2. Make a scatterplot of the data in your table.

3. What are the domain and range for this situation?

4. When the independent variable has a value of 0, what is the value of the dependent variable?

5. Describe the relationship displayed in your scatterplot.

6. Do the variables appear to be positively related?

7. Do you think there is a proportional relationship between the total volume of the graduated cylinder and the number of submerged canisters?

8. What is the value of the dependent variable when the independent variable is 0?

9. What does your answer to the Question 8 value mean in this situation?

10. What do you predict the total volume contained in the cylinder will be if you submerge 7 film canisters?

11. How many canisters do you need to submerge for the total volume to be 900 mL?

12. When the points in a scatterplot lie along a straight line, the relationship between the two variables is called **linear**. Not all linear relationships are proportional. Do you think the relationship between total volume and number of canisters is linear?

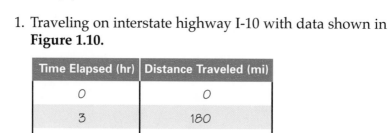

Assignment

Tell whether the relationship in each situation is proportional or not. Justify your answer.

1. Traveling on interstate highway I-10 with data shown in **Figure 1.10.**

Time Elapsed (hr)	Distance Traveled (mi)
0	0
3	180
6	360
9	540

FIGURE 1.10.
Highway travel.

2. Fare in a taxicab: it costs $2.00 to get in the cab and an additional $0.20 for each $\frac{1}{4}$ mile traveled.
   (Hint: Make a table!)

3. Temperature conversions shown in **Figure 1.11.**

Celsius Scale (°C)	Fahrenheit Scale (°F)
5	41
10	50
15	59
20	68

FIGURE 1.11.
Temperature conversions.

4. Money spent on many cans of coffee when the sale price is $1.89/can.

A bank savings plan book keeps track of the amount of money in the account at the end of each month. The data are recorded in **Figure 1.12.**

Number of Months	Account Balance
0	$220
1	$250
2	$280
3	$310
4	$340

FIGURE 1.12.
Savings account.

5. Explain how the balance can be something other than $0 after 0 months.

6. Make a scatterplot of the savings account data.

7. What are the domain and range for this situation?

8. When the independent variable has a value of 0, what is the value of the dependent variable?

9. Describe the type of relationship displayed in your scatterplot.

10. Do the variables appear to be positively related? Why or why not?

11. Is the relationship proportional? Why or why not?

12. What does the pair (0, 220) mean in this situation?

13. Is this set of data better represented by a continuous or discrete graph?

14. If the person deposits money into his savings account each month at the same rate, how much money is in the account after 10 months?

15. How many months does it take for this person to save $1000?

16. Make a scatterplot on the graphing calculator of the savings account data.

Mystery Metals

Tom and Tara's teacher gives them one of the metals shown in **Figure 1.13.**

Metal	Density $\left(\frac{g}{cm^3}\right)$
Aluminum	2.70
Zinc	7.13
Silver	10.49
Mercury	13.55
Gold	19.32

FIGURE 1.13.
Mystery metals.

Tom and Tara collect the following data (**Figure 1.14**) in the science lab on their metal. They use a graduated cylinder to find the volume and a balance to find the mass.

Volume (cm^3)	Mass (g)
10	135
20	270
30	410
50	680
100	1350

FIGURE 1.14.
Metal data.

1. Use three different methods (table, graph, and equation) to examine the relationship between the mass of the metal and its volume. Then use the results of your investigation to determine the volume of 500 grams of the metal. Justify your answer.

2. Which metal do you think Tom and Tara were given? Why?

SUMMARY

In these sections you learned that proportional relationships have certain characteristics.

1. Continuous graphs of proportional relationships are linear and contain the point $(0, 0)$.

2. The ratio $\frac{y}{x}$ is a constant in a proportional relationship and is also called the **constant of proportionality.**

3. Proportional relationships can be described by a function of the form $y = kx$, which is called a **direct variation function**.

You also saw that there are relationships that are linear, but not proportional.

This is Really a Stretch

In this section your teacher will conduct a class demonstration. Be sure to observe the demonstration carefully as you answer the following questions.

1. What do you observe when your teacher holds up the Slinky® with an empty film canister attached?

2. What observations can you make when your teacher holds up the Slinky with the attached film canister containing 5 pennies?

3. What observations can you make when your teacher holds up the Slinky with the attached film canister containing 10 pennies?

4. What observations can you make when your teacher holds up the Slinky with the attached film canister containing 15 pennies?

5. What patterns do you notice?

6. What are the input and output variables for this situation?

7. Do you think that the relationship between the amount of stretch of the Slinky and the number of pennies in the film canister is a proportional relationship? Why or why not?

8. Do you think that the amount of stretch of the Slinky and the number of pennies in the film canister are positively related? Why or why not?

9. What experiment could you do to further investigate the relationship between the stretch of a Slinky and the amount of weight attached to it?

Hooke's Law

Robert Hooke (1635–1703) was an English scientist. He grew up during the Age of Enlightenment, a time of much scientific discovery. Hooke spent his life studying astronomy, geometry, mathematics, mechanics, and physics. He worked with such famous colleagues as Robert Boyle and Isaac Newton and served as the chief assistant to Christopher Wren, the famous architect who designed the rebuilt London after the Great Fire of 1666.

Today Hooke is known for his work with springs. In the 1600s, most people kept time with pendulum clocks. These clocks use a large pendulum that swings with a regular period to drive a wheel and turn the clock's hands. They work well on land. However, on a ship the large waves of the ocean disrupt the pendulum. Since determining a ship's longitude requires knowing the time in Greenwich, England when the sun is directly overhead, the ship's navigator needs an accurate way to keep time.

Hooke recognized this problem. In 1658, he began experiments using springs instead of a pendulum to control the wheel of a clock. By 1660, he made two significant improvements on spring clock design. While refining these improvements, he made many observations about the stretch of springs. One of these is a relationship between the amount of force applied to a spring and the length the spring stretches as a result. This relationship is known as Hooke's Law.

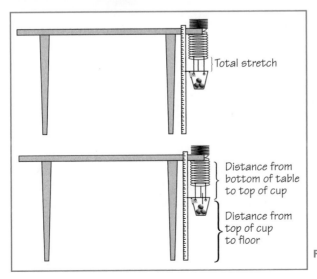

Total stretch

Distance from bottom of table to top of cup

Distance from top of cup to floor

HOOKE'S LAW EXPERIMENT

Place paper clips through three holes in the sides of a cup and hang the cup from the bottom of a Slinky. Suspend the Slinky and cup from a desk to allow the cup to move up and down freely. Tape a bent paper clip to the top of the cup to use as a pointer. Tape a meter stick vertically to the table (see **Figure 1.15**).

FIGURE 1.15. Slinky set-up.

In this experiment, measure the stretch that is explained by the marbles, but not the stretch that is explained by the cup or the weight of the Slinky. Place a mark or a piece of tape on the table or the meter stick at the level of the top of the empty cup and measure the stretch from this mark.

1. You will be adding one object at a time to the cup and measuring the total stretch of the Slinky as the objects are added. What is the independent variable? What is the dependent variable? What are a reasonable domain and range for this situation? Why?

2. You will also be measuring the distance from the bottom of the table to the top of the cup. What is the independent variable? What is the dependent variable? What are a reasonable domain and range for this situation? Why?

3. The third measurement that you will take is the distance from the top of the cup to the floor. What is the independent variable? What is the dependent variable? What are a reasonable domain and range for this situation? Why? Record and keep these data. You will need them in Section 1.9.

4. Sketch what you predict a scatterplot of total stretch versus number of objects placed in the cup would look like. Explain why you made the scatterplot the way you did.

CHECK THIS!

In scatterplots, the term "versus" (or the abbreviation "vs.") describes the position of the variables. The first variable (the dependent variable) is associated with the vertical axis and the second variable (the independent variable) with the horizontal axis.

5. Sketch what you predict a scatterplot of the distance from the bottom of the table to the top of the cup versus the number of objects placed in the cup would look like. Explain why you made the scatterplot the way you did.

6. Add objects to the cup one at a time. After each object is added, record the total stretch of the Slinky, the distance from the bottom of the table to the top of the cup, and the distance from the top of the cup to the floor (see **Figure 1.16**).

Number of Objects in the Cup	Total Stretch of the Slinky® (cm)	Distance from the Bottom of the Table to the Top of the Cup (cm)	Distance from the Top of the Cup to the Floor (cm)
0	0		
1			
2			
3			
4			
5			

FIGURE 1.16.
Slinky data.

7. Is the relationship between the total stretch and the number of objects placed in the cup a proportional relationship? Use your table to justify your answer.

8. Use the data in your table to create a scatterplot of total stretch versus number of objects placed in the cup. Record your window. Sketch and describe your graph.

9. Do you think that your scatterplot shows a positive relationship between the total stretch and the number of objects placed in the cup? Why?

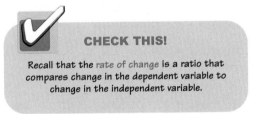

CHECK THIS!

Recall that the rate of change is a ratio that compares change in the dependent variable to change in the independent variable.

10. What characteristics of your graph verify that the relationships between total stretch and number of objects placed in the cup is or is not a proportional relationship?

Mathematicians use Δ, the Greek letter delta, for the change in a variable. Thus, Δs means "the change in the variable named s," or simply "the change in s." For example, between $s = 3$ and $s = 5$, $\Delta s = 2$.

You can calculate a rate of change between any two pairs of values in your table. To do so, divide the change in the dependent variable by the change in the independent variable. That is, find the ratio $\frac{\Delta s}{\Delta c}$, where s is the total stretch in the spring and c is the number of objects in the cup.

11. Find the rate of change between two pairs of values in your table. How does the rate compare to the constant of proportionality? What does the rate mean in this situation?

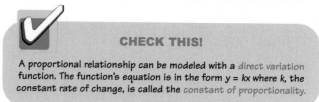

CHECK THIS!

A proportional relationship can be modeled with a *direct variation* function. The function's equation is in the form $y = kx$ where k, the constant rate of change, is called the *constant of proportionality*.

12. What is an appropriate function rule for a direct variation function that models the relationship between total stretch and the number of objects placed in the cup?

A **function rule** is a symbolic equation that describes a relationship between two variables. For example, if each value of the independent variable is triple the corresponding value of the dependent variable, then a function rule is $y = 3x$.

13. Graph your function rule over your scatterplot. Sketch your results. Does the graph show that your function models the relationship well?

14. Use your function rule to determine the total stretch of the Slinky when 12 objects are added to the cup.

15. How many objects should it take to make the Slinky stretch a total of 32 cm?

16. Is the relationship between the distance from the bottom of the table to the top of the cup to the number of objects placed in the cup a proportional relationship? What characteristics in your table justify your answer?

17. Use the data in your table to create a scatterplot of the distance from the bottom of the table to the top of the cup versus the number of objects placed in the cup. Record your window. Sketch and describe your graph.

18. What characteristics of your graph show that the relationship between the distance from the bottom of the table and the top of the cup versus the number of objects placed in the cup is or is not a proportional relationship?

19. Find the rate of change between two pairs of values in your table. Do the same for another two pairs. What is the meaning of the rate in this situation?

20. How does the rate of change for the relationship between the distance from the bottom of the table to the top of the cup and the number of objects placed in the cup compare to the rate of change for the relationship between total stretch and the number of objects placed in the cup? How are the graphs related?

Assignment

For Questions 1 and 2, decide whether the relationship between the two variables is or is not a proportional relationship. Explain how you made your decision.

1.
x	0	0.2	0.6	2.5	6.1
y	0	1.2	3.6	15.0	36.6

2.
t	0	1.6	2.4	3.0
d	0	3.2	4.2	5.1

For Questions 3 and 4, decide whether the function rule describes a direct variation function. Explain how you made your decision.

3. $x \cdot y = 12$

4. $y = 0.3875x$

For Questions 5 and 6, decide whether the graph depicts a proportional relationship between the two variables. Explain how you made your decision.

5. The graph shown in **Figure 1.17**

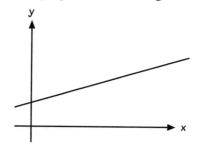

FIGURE 1.17.
A graph.

6. The graph shown in **Figure 1.18**

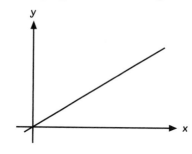

FIGURE 1.18.
A graph.

For Questions 7 and 8, let the relationship between x and y be a proportional relationship, and let $y = 12$ when $x = 5$.

7. What is the constant of proportionality?

8. If $x = 11$, what is the value of y?

Use the following information for Questions 9–17. Marbles were placed on a scale and weighed. The collected data are shown in **Figure 1.19.**

Number of Marbles (m)	Mass in Ounces (w)
1	1.5
2	3.0
4	6.0
8	12.0
10	15.0

FIGURE 1.19.
Mass of marbles.

9. What is the independent variable in this situation? Why?

10. What is the dependent variable in this situation? Why?

11. What are a reasonable domain and range for this situation?

12. Create a scatterplot of mass in ounces versus number of marbles.

13. Is this a proportional relationship? How do you know?

14. Find the rate of change between two pairs in the table. Do the same for two other pairs. What do you notice?

15. What is a function rule that models this relationship?

16. Predict the weight of 27 marbles.

17. Predict the number of marbles that have a mass of 54 ounces.

Variation on a Theme

In the previous section, you collected and analyzed data for the relationship between the amount of stretch in a Slinky and the number of marbles in a plastic cup.

You have already collected data and answered some questions about what you found. Now you will look at some important ideas about data collection and building models based on those data.

1. How did you decide which is the independent and which is the dependent variable for this situation?

2. How did you decide on a reasonable domain and range for this situation?

3. Why did you sketch your predicted scatterplots the way you did?

4. How did you use your table to determine if the relationship between total stretch and the number of objects in the cup is a proportional relationship? What happens when you divide w by n?

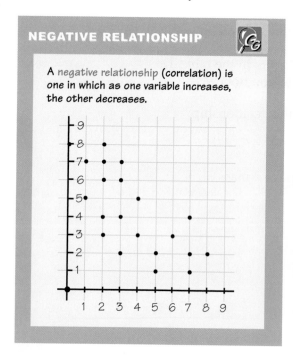

NEGATIVE RELATIONSHIP

A negative relationship (correlation) is one in which as one variable increases, the other decreases.

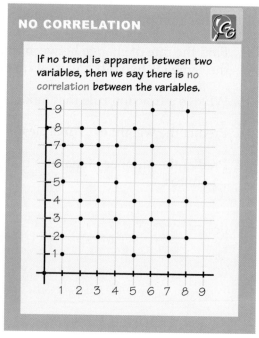

NO CORRELATION

If no trend is apparent between two variables, then we say there is no correlation between the variables.

5. Why do you think that there is a positive relationship between the total stretch of the Slinky and the number of objects in the cup?

6. What do you think is meant by a negative relationship?

7. What do you think is meant by no relationship?

8. How did you determine rates of change in your table?

To find the rate of change $\frac{\Delta w}{\Delta n}$ for any two pairs of values in the table, subtract the values of the dependent variable and divide by the corresponding difference in the values of the independent variable.

9. Find the rate of change for the pairs (2 objects, 4 cm) and (5 objects, 10 cm).

10. Explain how the rate of change for pairs of values in the table appears in the graph of this relationship.

11. In a proportional relationship, how can you use the rate of change and the values of the independent variable to find the values of the dependent variable?

12. Write a function rule to model the relationship between the total stretch of the Slinky, w, and the total number of objects in the container, n.

For a linear function including direct variation functions, the rate of change is also the slope of the line. In terms of x and y, the formula for the slope of the line through (x_1, y_1), (x_2, y_2) is $m = \frac{\Delta y}{\Delta x} = \frac{y_2 - y_1}{x_2 - x_1}$. In Question 12, you used your table to write a direct variation function rule, $w = 2n$, to relate the variables. In terms of x and y, the equation of the line is $y = 2x$.

13. Show how you can use the slope formula to determine the slope of the line that relates the total amount of stretch to the number of objects in the cup.

Another way to find the equation of a line through (x_1, y_1) and (x_2, y_2) is to use the **point-slope formula**: $y - y_1 = m(x - x_1)$.

14. Show how you can use the point-slope formula to determine the equation of the line that relates the total amount of stretch to the number of objects in the cup.

15. How can you predict the total stretch of the Slinky when 12 objects are added to the cup?

16. How can you predict the number of objects it takes to stretch the Slinky 32 cm?

Now, consider the situation where you compared the number of objects in the cup with the distance from the bottom of the table to the top of the cup. Data that one group gathered are in **Figure 1.20.** Use these data to answer the remaining questions.

Number of Objects in the Cup (n)	Distance (cm) from the Bottom of the Table to the Top of the Cup (d)
0	17
1	19
2	21
3	23
4	25
5	27

FIGURE 1.20.
More Slinky data.

17. How can you decide if this relationship is or is not proportional?

18. How can you find the slope of a function that models this relationship?

19. The **y-intercept** of a function's graph is the point where the graph crosses the y-axis. For the function that models this relationship, where do you find this point's coordinates in the table?

20. What is the meaning of the y-intercept in this situation?

Another representation for a linear function is called **slope-intercept form**: $y = mx + b$. This general form of a linear function expresses the dependent variable, y, in terms of the independent variable, x, the slope of the line (m), and the y-intercept of the line whose coordinates are $(0, b)$.

The slope-intercept form of a linear function is useful when you know the rate of change, or slope, and the starting point, or y-intercept, for a relationship that can be described with a linear function.

21. Use your answers to Questions 18 and 19 to write a function rule that can be used to find the distance between the bottom of the table and the top of the cup for n marbles.

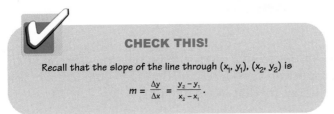

CHECK THIS!

Recall that the slope of the line through (x_1, y_1), (x_2, y_2) is

$$m = \frac{\Delta y}{\Delta x} = \frac{y_2 - y_1}{x_2 - x_1}.$$

22. How does the function relating the distance from the table to the cup and the number of objects in the cup (Question 21) compare to the function relating the total stretch of the Slinky and the number of objects in the cup (Question 14)?

Assignment

Predicting values from a proportional relationship can be done using a **proportion** (two fractions that are equal to each other), by writing an equation that looks like this:

$$\frac{12}{5} = \frac{y}{11}.$$

In this proportion 12 and 5 are corresponding values of y and x. The y-value that corresponds to the x-value 11 is not known.

1. Why must the two fractions be equal?

One algebraic method to solve a proportion involves "cross-multiplying" in the manner shown in **Figure 1.21**, and setting the two products equal to each other.

FIGURE 1.21.
Cross-multiplying.

2. In this case, what are the two products formed?

3. What equation do you get when you set the products equal to each other?

4. How do you solve that equation for y?

Another way to solve the equation $\frac{12}{5} = \frac{y}{11}$ is to multiply both sides of the equation by 11.

5. Why does that one step *solve* the equation?

6. In order to find the answer using this method, what arithmetic steps must you do? How does this compare to the arithmetic you did in Questions 2 through 4?

Solve each of the following proportions:

7. $\frac{4}{x} = \frac{15}{23}$

8. $\frac{7}{6} = \frac{y}{25}$

9. x and y form a proportional relationship. When $x = 13$, $y = 5$. Find x, when $y = 13$.

Write a direct variation function rule to model each of the following situations. Be sure to list the constant of proportionality and its units. Then use the rule to solve the problem.

10. On a map scale, 1 cm represents a distance of 25 miles. Two cities on the map are 3.2 cm apart. What actual distance separates the two cities?

11. On a typing test, Jane typed 98 words correctly in 2 minutes. Assuming she can type for longer periods of time without losing her concentration, how many minutes would it take her to type an essay that contains 343 words?

12. Hanging 3 washers on a spring stretches it a total of 4.5 cm. If 13 washers are placed on it instead, how far will the spring stretch? (Assume that the spring can be stretched several feet without problems.)

A local theme park charges $15 for admission. The amount of revenue generated by admission sales can be found using the function rule $R = 15n$, where R is the total revenue and n is the number of admissions sold.

13. Complete the table in **Figure 1.22**, and then use that information to answer Questions 14–15.

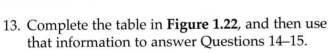

n	2	20		1000	
R			3000		30,000

FIGURE 1.22.
Theme park admissions.

14. For each of the five pairs in the table, calculate the ratio $\frac{R}{n}$. What do you notice?

15. Find $\frac{\Delta R}{\Delta n}$ (the rate of change of the revenue compared to the number of tickets sold) between the first and second entries of the table. What units does your answer have?

In Questions 16–18, find the slope of the line.

16. The line whose equation is $y = \frac{4}{3}x + 12$

17. The line that goes through the points $P(4, 9)$ and $Q(7, 21)$

18. The line that goes through the points $P(12, 5)$ and $Q(2, 8)$

This Floors Me

Recall that in Section 1.7 you took three measurements during the Hooke's Law experiment. You measured the total stretch of the Slinky, the distance from the bottom of table to the top of cup, and the distance from the top of the cup to the floor (**Figure 1.23**). In this section you will investigate the relationship between the number of objects placed in the cup and the distance from the top of the cup to the floor.

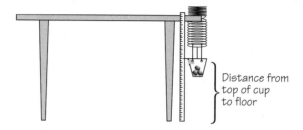

FIGURE 1.23.
Slinky stretch.

Distance from top of cup to floor

Copy your data from Section 1.7 into a table like **Figure 1.24.**

Number of Objects in the Cup	Distance from the Top of the Cup to the Floor (cm)
0	
1	
2	
3	
4	
5	

FIGURE 1.24.
Distance from cup to floor data.

1. In the relationship between the distance from the top of the cup to the floor and the number of objects in the cup, what is the independent variable? What is the dependent variable? What are a reasonable domain and range for this situation? Why?

2. Is the relationship between the distance from the top of the cup to the floor and the number of objects in the cup a proportional relationship? Use your table to justify your answer.

3. Suppose the starting position (empty cup) is 10 cm. How would this change your table?

4. Sketch what you predict a scatterplot of the distance from the top of the cup to the floor versus the number of objects in the cup would look like. Explain why you made the scatterplot the way you did.

5. Use the data from your table to create a scatterplot of the distance from the top of the cup to the floor versus the number of objects in the cup. Record your window, sketch and describe your graph.

6. Does your scatterplot show a positive, negative, or no correlation between the distance from the top of the cup to the floor and the number of objects in the cup? How do you know?

7. What characteristics of your graph verify that distance from the top of the cup to the floor versus the number of objects in the cup is or is not a proportional relationship?

8. Suppose the top of your cup is 10 cm higher. How would this change your graph?

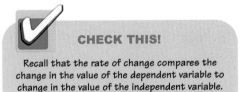

CHECK THIS!

Recall that the rate of change compares the change in the value of the dependent variable to change in the value of the independent variable.

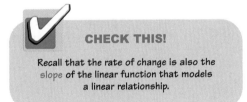

CHECK THIS!

Recall that the rate of change is also the slope of the linear function that models a linear relationship.

9. Find the rate of change between successive pairs of values in your table. To estimate the slope of a linear model for this relationship, find the average of these rates of change. What is the meaning of the slope in this situation?

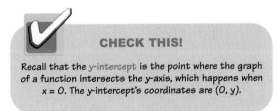

CHECK THIS!

Recall that the y-intercept is the point where the graph of a function intersects the y-axis, which happens when x = 0. The y-intercept's coordinates are (0, y).

10. What is the y-intercept of the graph of a linear function that models the relationship between the distance from the top of the cup to the floor and the number of objects in the cup? What is the meaning of the y-intercept in this situation?

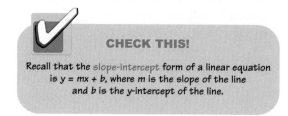

CHECK THIS!

Recall that the slope-intercept form of a linear equation is $y = mx + b$, where m is the slope of the line and b is the y-intercept of the line.

11. Write the slope-intercept form of a function rule that models the relationship between the distance from the top of the cup to the floor and the number of objects in the cup.

12. Suppose the top of your cup is 10 cm higher. Write the new function rule.

13. Graph your function rule on your scatterplot. Sketch your results. Do you think your function models the data well?

14. Predict the distance from the top of the cup to the floor when 9 objects are added to the cup.

15. Predict the smallest number of objects to place in the cup to make the bottom of the cup touch the floor.

Assignment

In Questions 1–3, an equation in slope-intercept form is provided. Use the equation to fill in the table. Then calculate Δx, Δy, and the ratio $\frac{\Delta y}{\Delta x}$ for successive pairs. (In these questions, x and y have no units.)

1. $y = 3x - 8$

x	y
4	
8	
12	
16	

2. $y = 1.2x + 15$

x	y
5	
7	
9	
11	

3. $y = 4.1x - 12$

x	y
13	
19	

Use the tables from Questions 1–3 to answer Questions 4–6.

4. In a table, do the values of the input variable (x) <u>have</u> to increase by a constant amount?

5. What is the smallest number of ordered pairs needed to calculate the ratio $\frac{\Delta y}{\Delta x}$?

6. What is the relationship between the rate of change ratio $\frac{\Delta y}{\Delta x}$ and the equation given?

7. The following functions were graphed with a computer drawing utility: $y = 1.2x + 5$, $y = 0.8x + 5$, $y = 2.3x - 5$, $y = -0.5x + 10$, and $y = -1.4x + 10$. The graphs are shown in **Figure 1.26.**

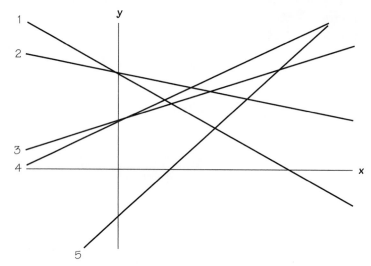

FIGURE 1.26.
Graph of 5 functions.

The order in which the lines were drawn is unknown, and the axes don't show a scale. Explain how you can identify the graph of each equation.

For Questions 8–11, find the equation of the line that goes through the two points given. Express your answer in slope-intercept form.

8. $P(-4, 12)$ and $Q(2, 3)$

9. $P(3, 8)$ and $Q(7, 15)$

10. $P(5, 12)$ and $Q(9, 14)$

11. $P(4, 14)$ and $Q(8, 4)$

Recall that the equation for a line in point-slope form looks like this: $y - k = m(x - h)$.

12. Graphs of direct variation functions always go through the origin $(0, 0)$. What does $y = mx$ look like if written in point-slope form?

13. If a function is written in slope-intercept form, like $y = 2x + 5$, you already know one point on its graph, the y-intercept. What are the coordinates of that point? What does the point-slope form of that same equation look like?

14. Suppose that Garth Brooks can sell out the Cotton Bowl, which has 25,704 seats, when the ticket price is set at $50, but a crowd of 16,000 is estimated when the admission is raised to $80. Assume that ticket sales drop off in a linear pattern as the price increases. Find a function rule to model the relationship between the ticket sales and ticket price.

Use the following information for Questions 15–17.

A concert promoter keeps track of advance ticket sales. Those sales can be used to estimate the attendance at the performance and to help determine how much help is needed in planning for the event. The data shown in **Figure 1.27** compares ticket sales 4 weeks in advance of several concerts with the actual attendance.

Advance Ticket Sales (A)	Actual Attendance (n)
5628	13,581
7043	15,902
8912	19,873
9117	21,683
9741	22,705

FIGURE 1.27.
Ticket sales.

15. Use the grid in **Figure 1.28** to make a scatterplot that shows the relationship between the actual attendance and the advance ticket sales.

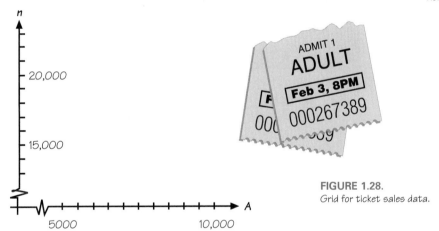

FIGURE 1.28.
Grid for ticket sales data.

16. When the pattern in a scatterplot is nearly linear, but not perfectly so, you can estimate a linear model that you can use to make fairly accurate predictions. Draw a line on the scatterplot you made in Question 15 that you think captures the trend of the data fairly well. Then find an equation for the line. Explain how you found the equation.

17. If a performance features a popular act, would you expect the data point to be above the line or below it? Explain.

Distance vs. Ramp Height

After investigating the relationship between the number of objects in a cup and the stretch of a Slinky, the students in the Math Club decided to roll a tennis ball down a ramp and measure how far it travels. To see if there is a relationship between the distance that a tennis ball rolls and the steepness of the ramp, they stacked books beneath one side of the ramp (**Figure 1.29**). As they changed the number of books, they rolled the ball and measured the distance of the roll.

FIGURE 1.29.
Changing ramp height.

The data from their experiment are in **Figure 1.30.**

Ramp Height in Books	Distance Traveled from the End of the Ramp (in)
1	27
2	53
3	84
4	110
5	129
6	160
7	180

FIGURE 1.30.
Ramp height data.

1. Is there a proportional relationship between the distance traveled and the number of books? Justify your answer.

2. Use three different methods (table, graph, and function rule) to predict the distance a tennis ball travels if the ramp is 10 books high.

SUMMARY

In this lesson you learned about different methods to get the value of m in $y = mx$, function rules that model proportional relationships.

❖ Find the constant of proportionality by dividing each y-value by its corresponding x-value.

❖ Find the rate of change.

You also learned about methods to write symbolic rules for functions that model non-proportional, linear relationships.

❖ From a table, find the rate of change (slope) and the y-intercept. Then use the slope-intercept formula, $y = mx + b$.

❖ If you know the coordinates of two points, (x_1, y_1), (x_2, y_2), use the slope formula, $m = \frac{\Delta y}{\Delta x} = \frac{y_2 - y_1}{x_2 - x_1}$, to find the slope. Then use the point-slope formula, $y - y_1 = m(x - x_1)$.

Two variables can have a positive relationship, a negative relationship, or no relationship at all.

Positive Relationship:

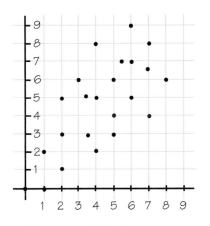

Negative Relationship:

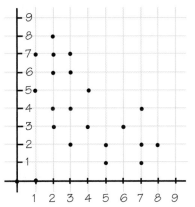

No Relationship:

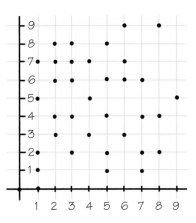

If one variable increases as the other variable increases, then there is a **positive relationship (correlation)** between the two variables.

If one variable decreases as the other variable increases, then there is a **negative relationship (correlation)** between the two variables.

If there is no apparent trend in the scatterplot, then there is **no relationship (correlation)** between the two variables.

Apparent Size

In this section you will mark the apparent size of a meter stick that your teacher has attached to a wall of the room. You will mark the apparent size on a pencil that you hold a short distance from your eyes.

1. Hold a pencil a few inches from your eyes. Align one end of the pencil with the top of the stick. Use your thumb to mark the bottom of the stick. Move the pencil farther from your eyes and repeat. What do you notice?

2. What are the variables in this situation?

3. Which is the independent variable? Why?

4. How are these variables related?

5. Again, mark the apparent size of the meter stick by holding the pencil a few inches from your eyes. After you have done so, keep the distance between your eyes and the pencil the same. This time move farther away from the meter stick and mark the apparent size. What do you notice?

6. What are the variables in this second situation?

7. Which is the independent variable? Why?

8. How are the variables related?

9. What experiment could you do to investigate the relationship between an object's apparent size and the distance that an observer is from the object?

The Farther You Go

In this section you will conduct an experiment or use data provided by your teacher to further examine a relationship that you discussed in Section 1.11. The variables are the distance that an observer stands from a meter stick taped to a wall and the apparent length of the meter stick.

To do the experiment, begin by taping a meter stick vertically to one of your room's shorter walls or use a meter stick that your teacher taped to the wall in advance. The middle of the ruler should be about eye level.

Use a tape measure to measure a distance of 8 feet from the wall or use an 8-foot mark made by your teacher. Stand with your toes on the mark and hold a ruler vertically at arm's length. Align the top of the ruler with the meter stick. Use your thumb to mark the place on the ruler that is aligned with the bottom of meter stick (see **Figure 1.31**). Record this apparent length of the meter stick.

FIGURE 1.31.
Apparent length of a meter stick.

1. Continue taking measurements, moving an additional 2 feet from the meter stick each time. Record the measurements in a table like the one in **Figure 1.32**.

Distance from Meter Stick (ft)	Apparent Length of Stick (in)
8	
10	
12	
14	
16	
18	
20	
22	
24	
26	
28	
30	

FIGURE 1.32.
Meter stick data.

2. Enter the data points from the table into a graphing calculator and make a scatterplot. Record your viewing window and sketch the graph.

3. What happens to apparent length as distance from the meter stick increases?

4. Can the distance from the meter stick equal 0? Why or why not?

5. Do you think a decreasing linear function would be a good model for the relationship between apparent length and distance from the meter stick? Why?

6. Add a column to your table for the products of the apparent length and the distance from the meter stick. Calculate the product of each pair. What do you notice?

7. How can you use your observation in Question 6 to write a function rule for the relationship between apparent size and distance from the meter stick?

8. Use your equation to predict the distance you should stand from the meter stick for its apparent size to be 2 inches.

9. Change your viewing window to include negative values of x and y. (For example, if your scatterplot had Xmin = 0 and Xmax = 35, change Xmin to –35.) How does the graph change? Do you see more or fewer data points?

10. Enter your function from Question 7 into your calculator and graph it over your scatterplot. Use the same window you used in Question 9. Sketch your graph.

11. The domain and range of the mathematical function you just graphed are different from the domain and range of the situation shown in your table. Explain the differences.

Assignment

For each table of values determine an appropriate viewing window, create a scatterplot, write a function rule, and graph your function rule over your scatterplot.

1.

x	y
−3	−2
−2	−3
−1	−6
1	6
2	3
3	2

2.

x	y
−3	−3
−2	−4.5
−1	−9
1	9
2	4.5
3	3

3.

x	y
−3	$-\frac{2}{3}$
−2	−1
−1	−2
1	2
2	1
3	$\frac{2}{3}$

4.

x	y
−3	−4
−2	−6
−1	−12
1	12
2	6
3	4

5. A group of senior citizens is planning a trip from Houston to Seattle. The cost to charter a private airplane is $10,000. The plane holds a maximum of 80 passengers.

Number of Passengers	Cost per Passenger
10	
20	
30	
40	
50	
60	
70	
80	

a) Fill in the table for the cost per passenger.

b) Which is the independent variable? Why?

c) Which is the dependent variable? Why?

d) What is a reasonable domain for this situation?

e) What is a reasonable range for this situation?

f) Indicate an appropriate viewing window, create a scatterplot, write a function rule, and graph your function rule over your scatterplot.

g) If 52 people buy tickets on the flight, what is the cost per person?

6. In some situations, it is not clear which variable is independent. For example, consider a rectangle that has a fixed area of 20 square centimeters. If the length is 10 centimeters, then the width must be 2 centimeters. But neither the length nor the width is truly independent of the other. So you can choose either to be the independent variable.

a) Use x and y to write a function rule that describes the relationship between the length and the width in this 20 square centimeter rectangle.

b) What are a reasonable domain and range for this situation?

c) Graph your function rule on a graphing calculator and record the viewing window you used.

d) If the length of the rectangle is 13 cm, what is the width?

One Goes Up, One Goes Down

In the last section you investigated the relationship between the distance from an object and its apparent size. You found a decreasing non-linear relationship.

1. What function rule describes the relationship between the distance from a meter stick and its apparent size?

2. What function rule describes the relationship between the length and width of a rectangle that has an area of 20 square units?

3. What do these function rules have in common? How do they differ?

4. Graph the functions on separate axes. Sketch your graphs.

5. What do the graphs have in common? How do they differ?

6. As the independent variable increases, what happens to the dependent variable in each situation? Does this change happen at a constant rate?

7. Think back to the apparent size situation in Section 1.12. The table in **Figure 1.33** contains some sample data that one student group collected.

Distance from Meter Stick (x)	Apparent Length of Stick (y), in cm	xy
8	24.5	
10	20.0	
12	16.3	
14	13.5	
16	12.3	
18	11.3	

FIGURE 1.33.
Sample data.

For each pair, multiply the apparent length of the stick by the distance from it. Record your results in a table like Figure 1.33.

8. What is the average of the products of the distance from the meter stick and its apparent length?

9. In a proportional relationship, $y = kx$, or $k = \frac{y}{x}$. This means that $\frac{y}{x}$ is a constant value. Add a fourth column to your table and use it to find $\frac{y}{x}$ for each of your data points. Is this ratio constant? Are these ratios close to each other?

10. Is the relationship between distance from the meter stick and apparent size a proportional relationship? How do you know?

11. Use your graphing calculator to make a scatterplot of apparent size versus distance from the stick. Sketch your graph and identify your window.

12. If $xy = 196.8$, solve this equation for y in terms of x. Graph your equation on this scatterplot. Based on your graph, how well do you think this function models the relationship between apparent size and distance from the meter stick? Explain.

SUMMARY

In some relationships, the product of the values of the independent variable and the corresponding values of the dependent variable is a constant. In other words, xy equals (or is close to) a constant, k. These relationships are called **inverse variation** or **indirect variation**. In these situations, the variables are said to be **inversely proportional**.

The general form of an inverse variation function rule is $y = \frac{k}{x}$. The parent function for inverse variation functions is $y = \frac{k}{x}$, and the graphs of this family of functions have two branches (see **Figures 1.34 and 1.35**).

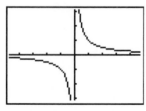

FIGURE 1.34. Graph of two branches of inverse functions.

FIGURE 1.35. Table of family of inverse functions.

When k is positive, one branch of the graph is in the first quadrant where small x-values yield large y-values and large x-values yield small y-values. This branch is commonly used to model inverse variation situations in the real world, such as the relationship between the apparent size of an object and distance from the object (see **Figure 1.36**).

Direct Variation	Inverse Variation
$\frac{y}{x}$ is a constant value, k	xy is a constant value, k
$y = kx$	$y = \frac{k}{x}$
x and y are positively related	x and y are negatively related
Linear graph	Non-linear graph

FIGURE 1.36. Traits of direct and inverse variation.

Assignment

For each inverse function, complete the table and graph the function using the ordered pairs from the table.

1. $y = \dfrac{1}{x}$

x	y
−4	
−2	
−1	
−0.5	
−0.25	
0.25	
0.5	
1	
2	
4	

2. $y = \dfrac{2}{x}$

x	y
−4	
−2	
−1	
−0.5	
−0.25	
0.25	
0.5	
1	
2	
4	

3. $y = \dfrac{5}{x}$

x	y
−4	
−2	
−1	
−0.5	
−0.25	
0.25	
0.5	
1	
2	
4	

4. $y = \dfrac{-1}{x}$

x	y
−4	
−2	
−1	
−0.5	
−0.25	
0.25	
0.5	
1	
2	
4	

5. Compare the graphs in Questions 1, 2, and 3. In what ways are they alike, and in what ways do they differ? How does the equation of each support your answers?

6. Compare the tables in Questions 1, 2, and 3. In what ways are they alike, and in what ways do they differ? How does the equation of each support your answers?

7. How do the graph and table in Question 4 differ from those in Questions 1, 2, and 3? How does the equation of each support your answers?

8. For each table, is the relationship an inverse variation, a direct variation, or neither?

a)

x	y
3	1
6	2
9	3
12	4
15	5

b)

x	y
2	12
7	32
11	48
13	56
14	60

c)

x	y
3	8
4	6
6	4
8	3
12	2

d)

x	y
1	5
2	$\frac{5}{2}$
3	$\frac{5}{3}$
4	$\frac{5}{4}$
5	1

9. Describe a method for determining if a direct variation function is a good model for a relationship in a table.

10. Describe a method for determining if an inverse variation function is a good model for a relationship in a table.

Boyle's Law

In 1662, Robert Boyle and an assistant found a way to increase the pressure on a quantity of air and measure both the volume of the air and the pressure exerted on it. They did these things by pouring mercury into a large U-shaped tube containing air. As more mercury was added, the air compressed. Boyle measured the volume of the air by measuring its height in the tube. He measured the pressure on the air by measuring the height of the mercury in the tube.

Some of Boyle's original data are shown in **Figure 1.36.**

Volume (height of air in inches)	Pressure (height of mercury in inches)
12	117.5625
16	87.875
20	70.6875
24	58.8125
28	50.3125
32	44.1875
36	39.3125
40	35.3125
44	31.9375
48	29.125

FIGURE 1.36.
Boyle's data.

1. Enter the data from the table into your graphing calculator and make a scatterplot. Sketch your graph and record your viewing window.

2. As volume increases, what happens to pressure?

3. Based on the scatterplot, do you think a direct variation function is a good model for the relationship between volume and pressure? Why?

4. Does the scatterplot appear to represent an inverse variation relationship? How do you know?

5. Add a third column to the table and label it product. Calculate and record the product for each pair.

6. Find the average product of volume and pressure.

7. Use the mean (average) product as the constant of variation and write an inverse variation function rule to model the relationship between pressure and volume for a gas.

8. Graph this function on your scatterplot. Based on the graph, does the function rule appear to model the relationship well? Explain.

Graphing a function rule on a scatterplot is one way to assess how well a function models a relationship. Another way is to compare numerical values. For example, you could use your rule to predict the pressure for a volume value of 12 and compare the prediction to the actual pressure that Boyle measured, 117.5625.

9. Use your function rule to calculate a predicted value for each volume value in your table. Compare the actual pressure values to the predicted ones. How well do you think the function models this relationship?

In the seventeenth century, most Europeans did not understand the nature of air. In 1644, Evangelista Torricelli, an Italian scientist and colleague of Galileo Galilei, correctly identified the concept of air pressure and built the first mercury barometer to measure it.

Robert Boyle, a prominent English chemist, set up a series of experiments with his research assistant, Robert Hooke (whom you met in Section 1.6), to further investigate properties of air. Among other things, Boyle and Hooke proved in the 1660s that air is necessary for sound, fire, and life.

The proof that fire needs air to exist and that sound cannot travel in a vacuum were radical at the time. Many scientists and philosophers insisted there was no such thing as a vacuum, or absence of air. Thus, the work of Boyle and Hooke was groundbreaking. Boyle's Law continues to be used as one of the fundamental principles of branches of science such as fluid dynamics and meteorology.

In 1662, Boyle published an appendix to his 1660 textbook *New Experiments Physio-Mechanicall, Touching the Spring of the Air and its Effects*. In this appendix, Boyle wrote about the mathematical relationship between the volume of air and its pressure.

10. Based on your work in this section and Boyle's ideas about the relationship between the volume of a gas and its pressure, make a conjecture about a function rule that could model this relationship. How can this function rule be generalized to model the pressure and volume in any data set? Explain your answer.

11. Boyle's Law is one example of inverse variation. Can you think of some others?

SUMMARY

Inverse variation can be used to describe many relationships in the natural world. In 1662, Robert Boyle established an inverse variation relationship between the pressure and volume of a gas when it is kept at a constant temperature. This relationship is known as Boyle's Law and is represented mathematically by the equation $P = \frac{k}{V}$, where P is the pressure, V is the volume, and k is the constant of variation.

 Assignment

1. A group of students decide to conduct an experiment to verify Boyle's law. They use a CBL and a pressure sensor attachment to vary the volume and record the resulting pressure. Their data are in **Figure 1.38.**

Volume (cubic cm)	6	8	10	12	14	16	18
Pressure (atmospheres)	3.94	3.00	2.41	2.00	1.67	1.44	1.26

FIGURE 1.38.
Student data.

a) Which is the independent variable? Why?

b) Which is the dependent variable? Why?

c) What is a reasonable domain for this situation?

d) What is a reasonable range for this situation?

e) Decide an appropriate viewing window, create a scatterplot, write a function rule, and graph this function rule on your scatterplot.

f) Predict the pressure that results from a volume of 2 cubic centimeters.

g) Predict the volume when the pressure is 1 atmosphere.

2. In a recent experiment, Javier used a 9-volt battery to test a circuit. He set various current levels and measured the resulting resistance. The data are in **Figure 1.39.**

Current (I)	Resistance (R)
2	4.5
5	1.8
6	1.5
10	0.9
12	0.75
15	0.6
20	0.45

FIGURE 1.39.
Javier's data.

a) Which is the independent variable? Why?

b) Which is the dependent variable? Why?

c) What is a reasonable domain for this situation?

d) What is a reasonable range for this situation?

e) Decide an appropriate viewing window, create a scatterplot, write a function rule, and graph your function rule on your scatterplot.

f) If the current is 7, predict the resistance.

g) Predict the current when the resistance is 1.75.

Using Inverse Variation

1. Pete's Construction Company has hired a team of engineers to study the company's policy for buying cement. Pete's foreman always orders 108 cubic feet of cement to pour a 12-foot wide driveway. The engineers conducted an experiment. They collected data by changing the length of a driveway, while keeping the width constant at 12 feet, and measuring the depth of cement after pouring 108 cubic feet of cement. Their data are in the table in **Figure 1.40**.

Width (feet)	Length (feet)	Depth (feet)	Volume of Cement (cubic feet)
12	1	9.00	108
12	2	4.50	108
12	3	3.00	108
12	4	2.25	108
12	5	1.80	108

FIGURE 1.40.
Pete's Construction data.

City code requires that driveways be 4 inches deep. What is the longest 12-foot wide driveway that can be constructed using the foreman's order? Justify your answer.

Modeling Project *Scope It Out*

When you look through a telescope, the amount you can see depends on several variables.

In this project, you will investigate the relationship between the size of a telescope's viewing circle and two other variables.

You will need several paper towel rolls, masking tape, a meter stick, a tape measure, and scissors.

Make a simple telescope by cutting a paper towel roll from one end to the other so that another paper towel tube can fit inside of it. Do the same with a third paper towel roll. Put the uncut tube inside one of the cut tubes. Then put these two inside the other cut tube. You can now vary the length of your scope from one tube to nearly three.

Next, tape a meter stick to a wall horizontally. Back off and look at the stick through your scope. If you align the left edge of your viewing circle with the left end of the meter stick, you can estimate the viewing circle's diameter. Now change the length of the scope. The diameter of the viewing circle should change.

There are several variables in this situation:
* The diameter of the viewing circle;
* The length of the scope;
* The distance you are standing from the meter stick.

In mathematical modeling, it can be difficult to analyze the relationship between several variables at once. So mathematical modelers use a simple trick: they keep all but two of the variables constant and examine just those two. To do this, they make changes in one variable and record the effect on the other. After they have figured out how these two variables are related, they do something similar with a different pair of variables.

So here is an approach to this situation:
* Keep the distance from the meter stick constant and examine the relationship between viewing circle diameter and telescope length;
* Keep telescope length constant and examine the relationship between viewing circle diameter and distance from the meter stick.

Conduct an investigation into these two relationships. Use the scope you have made to gather data. Prepare a report on your findings.

Practice Problems

1. A concert promoter is considering setting the ticket price at
   $30. **Figure 1.41** shows a table, function rule, and graph for
   revenue (R) based on number of tickets sold (n).

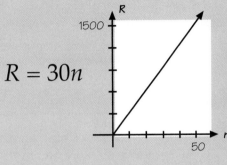

n	R
10	$300
20	$600
30	$900
40	$1200
50	$1500

$$R = 30n$$

FIGURE 1.41.
Data for Question 1.

Explain how to tell if the relationship between R and n is a
proportional relationship from the

a) table of values.

b) function rule.

c) graph.

2. For each linear relationship, find the slope (m). That is, find
   the constant rate of change $\frac{\Delta y}{\Delta x}$.

a) The table in **Figure 1.42**

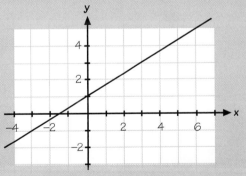

x	y
5	23
9	29
13	35
17	41
21	47

FIGURE 1.42.
Data for Question 2a.

FIGURE 1.43.
Graph for Question 2c.

b) The line going through the points P(5, 13) and Q(10, 10)

c) The line shown in **Figure 1.43**

3. Solve each of the following equations for the indicated variable:

a) $\frac{5}{12} = \frac{x}{18}$, for x.

b) $4.20n + 67000 = 76030$, for n.

c) $C = 2.45 \cdot n + 108000$, for n.

4. Two long-distance telephone carriers have quoted costs shown in **Figure 1.44:**

FIGURE 1.44.
Comparison of phone company rates.

Company	Monthly Service Charge	Rate (per min)
NCO	$5.00	$0.20
Jog	$3.00	$0.25

Write equations describing the total cost associated with each company plan for one month. Use t as the variable to represent the amount of time (number of minutes) spent on the phone.

5. Sam gathered data on the advertising costs and ticket sales from concerts at the Cotton Bowl. The data are shown in **Figure 1.45.** Sam is interested in the relationship between advertising costs and ticket sales.

Advertising Costs	Ticket Sales
$10,000	12,478
$15,000	16,395
$20,000	19,882
$25,000	22,043
$30,000	25,512
$40,000	25,704
$50,000	25,704

FIGURE 1.45.
Advertising costs
vs. ticket sales.

a) The pattern in the table shows increasing ticket sales as advertising costs go up. However, when the large sums are spent, there isn't an increase in ticket sales. What could explain this?

b) On a copy of the grid in **Figure 1.46**, make a scatterplot to show how the ticket sales are related to advertising costs.

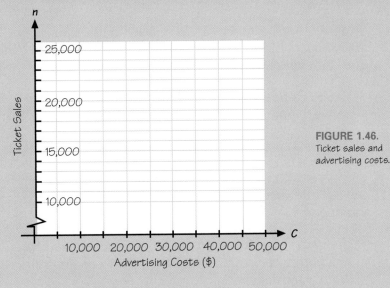

FIGURE 1.46.
Ticket sales and advertising costs.

c) The pattern for the first five data points is fairly linear. Draw a line on the scatterplot that you think captures this linear trend. Find the line's equation.

6. Data collected from ticket sales at the Cotton Bowl are shown in **Figure 1.47.**

Price (p)	Ticket Sales (n)
$10	25,000
$20	24,000
$30	22,000
$40	19,000
$50	16,000
$60	14,000
$70	10,500

FIGURE 1.47.
Price vs. ticket sales.

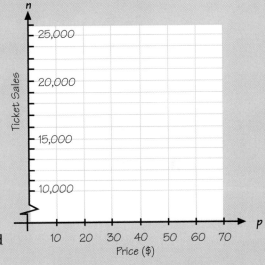

FIGURE 1.48.
A blank grid plot comparing price and ticket sales.

a) Make a scatterplot on a copy of the grid in **Figure 1.48.**

b) Draw a line on the graph that you think captures the trend. Find the line's equation.

c) On a graphing calculator, put the ticket price information in list L1 and the ticket sales information in list L2. Adjust the window settings to match the scatterplot you drew in a. What window settings did you use?

7. Suppose that a concert will have a sell-out attendance of 20,000 when the admission is $25, but the attendance drops 2000 for every $5 added to the ticket price.

   a) What is a function rule that models the relationship between attendance and admission price?

   b) What is a reasonable domain for this situation?

8. Direct variation functions have graphs that are straight lines, but not all functions with straight-line graphs are direct variations.

   a) Compare slope-intercept and point-slope forms of function rules for lines in general with function rules for direct variations.

   b) Compare the graphs of direction variations with the graphs of linear functions in general.

9. Data on various attendance projections for a concert are in **Figure 1.49**:

Ticket Price (p)	$10	$20	$30	$40	$50
Ticket Sales (n)	25,000	23,500	22,250	21,100	19,900

FIGURE 1.49. Projection of the ticket sales by ticket price.

   a) Draw a scatterplot on a copy of the grid in **Figure 1.50.**

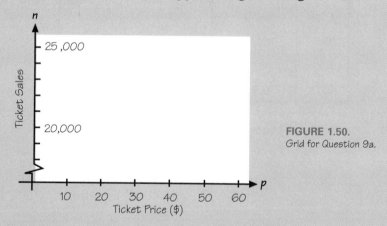

FIGURE 1.50.
Grid for Question 9a.

   b) Draw a line on your scatterplot that you think captures the trend of the data. Find the line's equation.

For each table in Questions 10 through 13, determine the constant of variation. Then match the table with a graph (a–d) and an equation (e–h) below.

10.

X	Y1
.5	32
1	16
2	8
4	4
8	2
16	1

$k =$ _____

graph _____

equation _____

11.

X	Y1
-16	-.25
-8	-.5
-4	-1
-2	-2
-1	-4
-.5	-8

$k =$ _____

graph _____

equation _____

12.

X	Y1
-16	-.5
-8	-1
-4	-2
-2	-4
-1	-8
-.5	-16

$k =$ _____

graph _____

equation _____

13.

X	Y1
-8	.5
-4	1
-2	2
-1	4
-.5	8
-.25	16

$k =$ _____

graph _____

equation _____

a)

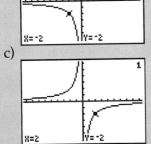

c)

b)

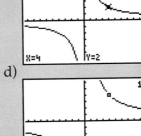

d)

e) $y = \dfrac{4}{x}$

f) $y = \dfrac{-4}{x}$

g) $y = \dfrac{16}{x}$

h) $y = \dfrac{8}{x}$

For Questions 14 and 15, fill in the missing values in the table and write the function rule.

14.

X	Y1
	ERROR
.25	
.5	48
1	12
2	6
	4

15.

X	Y1
-10	-10
-5	
	-100
0	
1	
5	20
	10

16. Students studying strengths of materials performed an experiment. They cut Styrofoam into beams of different lengths, all with the same width and thickness. Then they fastened each beam between two desktops and gradually placed one-ounce weights on it until it broke. Their data are shown in **Figure 1.51.**

Length of Beams (inches)	Breaking Weight (ounces)
4	28
5	22
6	19
7	16
8	14
9	13

FIGURE 1.51.
Data on length of beams vs. breaking weight.

a) Examine the relationship between beam length and breaking weight. What kind of relationship do you think it is?

b) Which is the independent variable in this situation? Why?

c) What function rule would you use to model this relationship? Explain.

17. The students in Question 16 conducted another experiment. This time they kept the length and thickness of the beams constant, but varied the width. Their data are shown in **Figure 1.52.**

Width of Beams (inches)	Breaking Weight (ounces)
0.5	6
1.0	13
1.5	19
2.0	22
2.5	29
3.0	37

FIGURE 1.52.
More data on width of beams vs. breaking weight.

Investigate the relationship between breaking weight and beam width. Give a function rule that models the relationship.

Glossary

KEY CONCEPTS

Axes: Number lines drawn at right angles used to locate points. Axes is the plural of axis.

Boyle's Law: A law that states that the pressure of a gas is inversely related to the volume of the gas.

Constant of proportionality: The constant ratio of y/x for (x, y) values of a direct variation function.

Continuous graph: A graph with the data points connected. For example, the relationship between the amount of hamburger purchased and the amount paid at a supermarket has a continuous graph. A shopper can order any amount of hamburger and pays more as the amount of hamburger increases.

Coordinate plane: A system for finding the location of points (x, y) in two dimensions, using axes with scales.

Coordinates: A pair of numbers that describes the location of a point in a coordinate plane. The first number is the x-coordinate, the second number is the y-coordinate.

Δ: The Greek capital letter, delta, used to show "the change in." The notation Δy means the change in the y variable.

Dependent variable: The output variable. In modeling, the dependent variable is the one you want to predict.

Direct variation function: A function with an equation of the form $y = kx$. Direct variation functions are used to describe proportional relationships.

Discrete graph: A graph with the data points unconnected. For example, the relationship between the number of oranges purchased at a supermarket and the price paid has a discrete graph. You cannot order 4.5 oranges, so connecting lines between 4 and 5 does not make sense.

Domain: The set of values that the input variable in a mathematical function can have.

Domain of the situation (or model): The set of reasonable values of the independent variable for a given situation.

Function: A set of ordered pairs of real numbers in which no two pairs with the same first coordinates have different second coordinates.

Function rule: A symbolic equation that describes a relationship between two variables.

Hooke's Law: A law that states that the force applied to a spring is directly proportional to the stretch of the spring.

Independent variable: The input variable. In modeling, the input variable is the one that is used to make predictions.

Inverse variation or indirect variation: A relationship in which one variable decreases as the other increases and the product of corresponding values of the variables is a constant. In other words, xy equals a constant, k.

Inversely proportional: Variables in an inverse variation relationship are said to be inversely proportional. That is, as one variable increases the other decreases, and the product of corresponding values of the variables is constant.

Linear relationship: When the points in a scatterplot lie along a straight line, the relationship between the two variables is called linear.

Mass: In science, mass is the amount of matter in an object. It is common to use weight and mass as the same.

Negative relationship: Two variables have a negative relationship (sometimes called a negative correlation) when the dependent variable decreases as the independent variable increases.

Point-slope formula: A method for finding the equation of a line. The formula is $y - y_1 = m(x - x_1)$.

Positive relationship: Two quantities have a positive relationship (sometimes called a positive correlation) when the dependent variable increases as the independent variable increases.

Proportional (relationship): A relationship is proportional if its continuous graph is a straight line that passes through the origin and the ratio $\frac{y}{x}$ is constant (or nearly so) for all data.

Proportions: Two fractions that are equal to each other.

Range: The values that the output variable in a mathematical function can have.

Range of the situation (or model): The set of reasonable values of the dependent variable for a given situation.

Rate of change: A ratio that compares the change in the output variable to the change in the input variable.

Scatterplot: A graph of pairs (x, y) where the x-values are the independent values from your data and the y-values are the corresponding dependent values from your data.

Slope: A measure of the steepness of a line: the ratio of the vertical change to the horizontal change between two points on a line, or "rise over run."

Slope-intercept form: A form represented by the equation $y = mx + b$. This general form of a linear function expresses the dependent variable, y, in terms of the independent variable, x, the slope of the line (m), and the y-intercept of the line whose coordinates are $(0, b)$.

Units: Units describe the kind of quantity or information that a variable uses. Inches, centimeters, and milliliters are all units.

Variable: A quantity that changes.

Versus (vs.): When used in the phrase y versus x, it describes a scatter plot in which y is the dependent variable (vertical axis) and x is the independent variable (horizontal axis).

Volume: Generally, refers to the size of an object. That is, volume is a measure of the amount of space that a three-dimensional object occupies.

y-intercept: The point at which a graph crosses the y-axis. The y-intercept is the term b in the slope-intercept form for the equation of a line.

Bones: Linear Functions and Predictions

2

CHAPTER

The Disappearance of Amelia Earhart

Amelia Earhart is one of the most famous pilots of all time. She was the first woman to fly solo across the Atlantic Ocean and later across the Pacific. On June 1, 1937 she set off on a flight around the world along with her navigator. On July 2, their plane vanished between New Guinea and Howland Island. The U.S. Navy searched but did not find a trace of them. To this day their fate is unknown.

A group trying to solve this mystery heard about bones that were found on a Pacific island in 1940. At the time a Dr. D. W. Hoodless examined the bones. He concluded that they were from a male about 5 feet 5 inches tall. Some of his measurements are given in **Figure 2.1**.

Bones	Length (cm)
Humerus	32.4
Tibia	37.2
Radius	24.5

FIGURE 2.1.
Bones analyzed by Dr. Hoodless.

Statements in his report raised doubt about his knowledge of the human skeleton. Lacking the bones, Dr. Karen Burns and Dr. Richard Jantz studied the measurements left by Dr. Hoodless. Using data from a data bank at the University of Tennessee they built models to predict height, gender, and ethnic background. They concluded that the bones were from a white female about 5 feet 7 inches tall, which fits Earhart. Could the bones have been hers? This question can only be answered if the bones are recovered.

In this chapter you will be asked to think like a forensic anthropologist. Given measurements of a set of bones, you will investigate the clues about a dead person. You will also collect data and build models to predict height and gender.

CHECK THIS!

A forensic anthropologist is a scientist who uses information about the human body and its bones to try to identify a dead person based on bones and teeth. This is one of many kinds of work that a forensic anthropologist does.

Mysterious Findings

Archaeologists and forensic scientists often use mathematical models to help investigate human remains found at historical sites and crime scenes. In this section you will explore models that describe the relationship between the length of a person's head and the person's height.

From time to time, bones are found in rugged areas. A hiker in Arizona's Superstition Mountains found a skull, eight long bones, and many bone fragments. He called the local police who sent a team to investigate. The team recorded facts about the bones, including their size. **Figure 2.2** shows some data that are similar to what the team recorded.

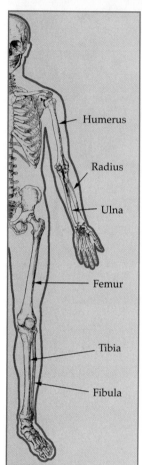

A human skeleton.

Bone Type	Number Found	Length (cm)
Femur	3	41.5, 41.4, 50.8
Tibia	1	41.6
Ulna	2	22.9, 29.0
Radius	1	21.6
Humerus	1	35.6
Complete Skull (including jaw)	1	23.0
Fragments	More than 10	From 3.0 to 5.0 cm

FIGURE 2.2. Measurements of bones found at Superstition Mountains.

Use all the facts you have so far to answer these questions.

1. Based on the data in the table, what is the precision of the measurements?

2. Study the data in Figure 2.2. The team reports that these bones belong to at least two people. How do they know?

3. Which bones do you think belong to the same person? What assumptions did you make to get your answer? Explain your answer. (To make it easier to classify the bones, refer to the dead people as Skeleton 1, Skeleton 2, and so on.) You will need the answers to this question later on.

4. Do the lengths of the bones make you think they belonged to a male or a female? What did you use to get your answer? (Remember, Dr. Hoodless concluded that the bones found on the island belonged to a male, but Burns and Jantz disagreed with his findings.)

5. Do you think the dead people were children or adults? Defend your answer.

6. Guess the heights of the dead people. How good do you think your guesses are?

A Model for Estimating Height

One way to turn your guess of the dead peoples' heights into a scientific estimate is to explore the relationship between the length of certain bones and a person's height.

In this section you will gather some data to explore a possible relationship between head length and a person's height.

Within your group, measure the length of each person's head (from chin to the top of the head) and then measure their height. Record your data in a table like **Figure 2.3**. Be sure to give the units of measure you used at the top of the second and third columns.

Name	Head Length	Height

FIGURE 2.3. Group head-length and height data.

1. What patterns do you see in this data set?

2. We aren't sure yet that there is a relationship between head length and height, so why does it make sense to consider head length as the independent variable here?

3. What is a reasonable domain for this situation? Why?

4. What is a reasonable range for this situation? Why?

5. On a sheet of chart paper create a scatterplot of height versus head length. Sketch your graph.

TREND LINES:
LINES THAT APPROXIMATE DATA

The plot of height versus head length is fairly spread out, but it appears to have a linear form. This large amount of variability makes it hard to pick a line that you can use to make good predictions for heights. The following two methods will help you choose a line that approximates the pattern of the height versus head-length data from your class:

METHOD 1

1. Pick a point that appears to lie in the middle of the points displayed in your scatterplot. The point doesn't have to be a data point.

2. Using this point as one point on your line, adjust the slope of your line until you find a line that you think best describes the pattern of the data.

3. Find the equation of your line.

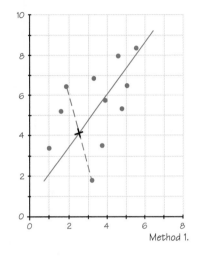

Method 1.

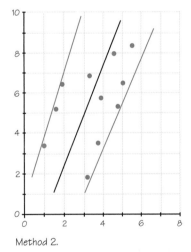

Method 2.

METHOD 2

1. Draw two lines in such a way that the points on your hand-drawn scatterplot are squeezed as tightly as possible between these lines. (The lines don't have to be parallel.)

2. Now draw one line halfway between the two lines that you have drawn.

3. Find the equation of this line.

6. Divide your group in half. Half of your group should use Method 1 and a meter stick to find a trend line for the graphed data. The other half should use Method 2 and two meter sticks to find a trend line. Draw both trend lines over the scatterplot and sketch your graph.

7. Write the equation of your trend line.

8. Find the ratio of head length to height by dividing each person's height by his or her head length. Record your results in a table like **Figure 2.4**.

Name	Height / Head Length

FIGURE 2.4. Ratio of head length to height.

Now you have a ratio of head length to height for each member of the class. You can start using these data to predict a person's height.

You need to find the best constant of proportionality (multiplier) that can be used to predict the height of the whole class.

One way to find the constant of proportionality is to use a measure of central tendency. You may have seen three before: mean, median, and mode. Let's review these and decide which one would be useful in making a prediction.

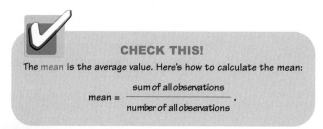

CHECK THIS!

The mean is the average value. Here's how to calculate the mean:

$$\text{mean} = \frac{\text{sum of all observations}}{\text{number of all observations}}.$$

9. What is the mean of the ratios of the head length to height?

CHECK THIS!

The median is the middle value. Here's how to calculate the median:

1. Arrange the data in order from smallest to largest.

2. If the number of data is odd, the median is the middle number. To find it, add 1 to the number of data then divide by 2. Start at the bottom of the ordered data and count up that number of observations.

3. If the number of data is even, the median is the average of the middle two numbers. To find it divide the number of data by 2. Again, start at the bottom and count up that number of observations. Add this number to the next number in the data set. Divide the sum by 2.

10. What is the median of the ratios of the head length to height?

CHECK THIS!

The mode is the most frequent data point in a data set.

11. What is the mode of the data set of ratios of the head length to height?

The question you are trying to answer is which measure of central tendency is the best constant of proportionality (multiplier).

In general, use the mode when you want to know what is the most common, as in the most common shoe size among American men. Use the mean when no values are much larger or much smaller than the rest of the data. Use median when a few relatively large or small values make the mean larger or smaller than most of the data.

Since all of the measures of central tendency are close in this case, it does not really matter which one we use. So, we will use the mean as the constant of proportionality (multiplier).

12. Use the constant of proportionality (multiplier) to write a function rule that could be used to predict a person's height given their head length.

13. Use a graphing calculator to create a scatterplot of height versus head length. Sketch your scatterplot and record your window.

14. Graph your function rule on the scatterplot. If needed, adjust the equation. Write your new equation and sketch your graph.

15. On the same scatterplot, graph the trend line that you found using the meter stick.

16. How does the trend line found using the meter stick compare to the function rule found with the constant of proportionality?

17. Does the relationship of height and head length appear to be a proportional relationship? Why or why not? Give at least two reasons to support your answer.

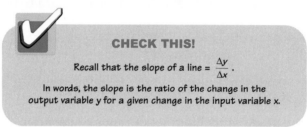

CHECK THIS!

Recall that the slope of a line = $\dfrac{\Delta y}{\Delta x}$.

In words, the slope is the ratio of the change in the output variable y for a given change in the input variable x.

18. What is the equation of a line that you think fits the data well? What is the slope of this line?

19. What meaning does the slope have in this context?

20. Using words, describe the relationship between the length of a person's head and the height of the person.

21. Predict the height of a person whose head length is 25 cm.

22. Predict the head length of a person whose height is 175 cm.

Assignment

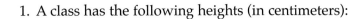

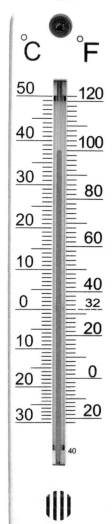

1. A class has the following heights (in centimeters):

   168, 178, 161, 155, 156, 155, 171, 188, 166, 161, 178, 180, 155, 167, 155, 157, 160, 158, 152, 152, 159, 155, 172, 188

   a) What is the mean height for this group of students?

   b) What is the median height?

   c) What is the mode of the heights?

2. The line plot in **Figure 2.5** shows the average daily temperature in a city for each day during the month of May. Each dot represents one day when the average daily temperature was that number of degrees Fahrenheit. For example, there were three days when the average daily temperature was 50°F.

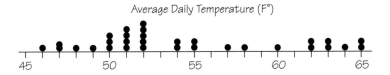

FIGURE 2.5. Line plot.

   a) What is the average (mean) daily temperature for the month of May? Explain how you found your answer.

   b) What is the median daily temperature? Explain how you found your answer.

   c) What is the mode daily temperature? Explain how you found your answer.

Use **Figure 2.6**, which shows the height of students in Ms. Jaczinko's tenth-grade class, for Questions 3–19.

Female		Male	
Name	**Height (cm)**	**Name**	**Height (cm)**
Alicia	157	Ahmed	173
Bia	166	Brian	177
Christi	164	Jesús	174
Chantalle	164	Davis	192
Coral-Mae	161	José	172
Juanita	164	Kelvin	180
Ji-Hyun	167	Lenny	174
Kim	162	Luis	175
Kristen	175	Mike	185
Maria	166	Pedro	185
Tianna	172	Sang	178
Teresa	176		

FIGURE 2.6.
Height data from Ms. Jaczinko's class.

3. What can you say about the precision of these measurements?

4. Use the table in **Figure 2.7** to find the mean, median, and mode for the male students, the female students, and all students in one class.

	Mean	**Median**	**Mode**
Male students			
Female students			
All students			

FIGURE 2.7. Measures of central tendency for Ms. Jaczinko's class.

5. Look over the data in Figure 2.6. What is the range in heights from the shortest student to the tallest?

6. A student joined Ms. Jaczinko's class. Is it reasonable to predict that the tenth-grader would be between 164 cm and 180 cm tall? Explain your answer.

7. What is the range of girls' heights from shortest to tallest?

8. If you know the new student in Ms. Jaczinko's class is a girl, what would you say about this person's height based on these data?

9. A girl who is 196 cm tall joins Ms. Jaczinko's class. How does this new girl's height compare with the other heights in the class?

10. What effect does this new height have on the mean, median, and mode of the girls' height data?

One way to predict a new student's height is to take the **mean** of all the heights.

11. Use the mean of the students' heights as a prediction of a new student's height. Record the mean student height.

12. If the new student is as short as the shortest student in the class, how far off is the prediction in Question 11?

13. If the new student is as tall as the tallest student in the class, how far off is the prediction in Question 11?

14. Do you think that the mean is a good prediction? Explain. Can you make a better one?

15. A new student, Melissa, joins the class. Does the new student, a girl, change your prediction? Find the mean for girls' heights. Use this average to predict Melissa's height.

16. If Melissa is as short as the shortest girl in Ms. Jaczinko's class, how far off is your prediction? What if Melissa is as tall as the tallest girl?

17. Do you think this is a better prediction than the one made in Question 11? Explain.

18. If the new student turns out to be Martin (a boy), not Melissa, can you choose a way to predict Martin's height? Give your prediction and your method.

19. If the new student's height is between the shortest boy and the tallest boy, what is the largest possible error for your prediction?

Height vs. Bone Lengths

In the last section, you collected and analyzed data on the relationship between a person's height and the length of their head from the bottom of the chin to the top of the skull. This relationship is important because it allows anthropologists who find skull bones to estimate the height of the person. Forensic scientists, such as those on crime-solving television shows, can also use this relationship to help find whose bones they were.

You have already collected data and answered questions about what you found. Now, let's look at some important ideas about data collection and building linear models based on those data.

1. When you measured the head length of your group members, what method did you use? How did you know that your method would yield good data?

2. How did you decide on a reasonable domain and range here?

3. How did you place your trend line?

4. How did you find the equation of your trend line?

5. When is it best to use mean as a measure of central tendency?

6. When is it best to use median as a measure of central tendency?

7. When is it best to use mode as a measure of central tendency?

8. What are the features of a direct variation function?

9. What effect does changing the value of the multiplier have on the graph of a direct variation function?

10. How might you use your model to make a rough prediction of the height of the person whose skull length is recorded in Figure 2.2?

    a) Recall that the skull measured 23.0 cm in length. Predict the height of the person in centimeters (cm). Describe the process you used to make your prediction.

    b) Does your prediction result in a reasonable height for a person? Explain.

11. How might you use your model to make a rough prediction of the head length of a person whose height was 175 centimeters?

    a) Predict the head length of the person in centimeters. Describe the process you used in making your prediction.

    b) Does your prediction result in a reasonable head length for a person? Explain.

12. What information do you think might give better estimates of the heights of the dead people whose bone lengths are recorded in Figure 2.2? How or where might you get this information?

MODELS OF THE FORM $y = mx$

- The graph of a function in the $y = mx$ family is a line that passes through the origin.

- The steepness and incline or decline of each of these lines is controlled by the value m, which is the slope of the line.

- These are also called direct variation functions, which are used to model proportional relationships.

SUMMARY

In Sections 2.1, 2.2, and 2.3 you explored the relationship between the length of certain bones and a person's height. You then used specific functions to predict a person's height from these bones.

❖ You also gathered data and applied a measure of central tendency—mean, median, or mode—to the data to find a multiplier (constant of proportionality) and a formula to predict height based on the length of the person's head.

❖ Since these models give rough estimates of a person's height, you will need to do more investigating to find a better model that gives better predictions. One way to improve your models is to collect data and use the data to test your model. Analyzing data forms the basis of the next section.

Assignment

1. Which graph(s) in **Figure 2.8** show(s) a trend line that appears to model the given data well? Explain your reasoning.

FIGURE 2.8.
Three graphs of trend lines.

Graph A   Graph B   Graph C

2. Hernando kept a record of his salon sales and tips for one afternoon as shown in **Figure 2.9.** Plot the data on a copy of the grid in Figure 2.9, number and label the graph appropriately. Use a straightedge to draw a trend line. Use this graph to answer the questions that follow.

Sale Amount ($)	Tip Amount ($)
10	2
15	3
20	2
25	4
30	4
35	6
40	5
45	7

FIGURE 2.9. A table of Hernando's tips with a grid to plot the data.

a) Use what you know about slope and y-intercept to write an equation for your trend line.

b) What does the slope mean here? Does this value make sense in this context?

c) What does the y-intercept mean? Does this value make sense in this context?

d) Predict Hernando's tip if his salon sale is $55. Justify your answer.

3. Nedra buys beads in bulk, by the pound, for her jewelry business. She has kept a record of her bead purchases (which are small) for the last 2 months. Use the data in **Figure 2.10** to graph a trend line and to answer the questions that follow.

Pounds of Beads Purchased	Total Amount Paid
1	1
1.25	1.50
2	2
2.25	2.5
2.75	3
3	3.5
3.75	4

FIGURE 2.10. Nedra's data.

a) Use your knowledge of slope and y-intercept to write an equation for your trend line in slope-intercept form.

b) What does the slope mean here? Does this value make sense in this context?

c) What does the y-intercept mean? Does this value make sense in this context?

d) How much would you expect Nedra to pay for 5 lbs of beads? Justify your answer.

4. Ten students' semester grades from Mr. Johnson's last year's class are given in **Figure 2.11.**

First Semester Grade (x)	55	62	64	71	75	78	82	85	95	97
Second Semester Grade (y)	50	54	64	65	60	70	77	72	88	88

FIGURE 2.11. Grades for 10 students.

a) Use these data to make a scatterplot. Find an appropriate trend line.

b) What does the slope mean here? Does this value make sense in this context?

c) What does the y-intercept mean? Does this value make sense in this context?

d) If you made a 72 the first semester, what is your predicted score for the second semester?

5. Psychologists are trying to find a relationship between test anxiety and success on a test. Examine the data in **Figure 2.12**, and answer the questions that follow.

X	0	7	14	14	15	20	20	21	23
Y	77	50	59	48	51	52	46	51	43

FIGURE 2.12.
Test anxiety data.

x = score on a test anxiety evaluation.
y = exam grade.

a) Looking at the overall trend in the table, as x-values increase do the y-values increase or decrease?

b) Use these data to create a scatterplot. Find the equation of a good trend line.

c) What does the slope mean here? Does this value make sense in this context?

d) If a person scores an 18 on the anxiety evaluation, what is the predicted exam grade?

e) What is the expected anxiety score if the exam grade is a 60?

I Think We Have Your Thighbones

In Section 2.2 you completed a process known as mathematical modeling, which is similar to using the Scientific Method, to solve problems. The first step in this process is to identify a problem that you need to solve. Collecting and analyzing data to develop a mathematical model and create a hypothesis is often the next step. The model is then used to make predictions. In this section you will analyze data that may have been collected by a famous doctor, develop a mathematical model of the data, and then use that model to make predictions.

Early models (from the late 1800s) that predicted height from the length of long bones were based on ratios. (See Figure 2.2 that shows some of the long bones.) For example, the ratio of the height H and femur length F is given by the formula $H/F = 3.72$.

1. Based on this model, estimate the heights of the people whose femur lengths are given in **Figure 2.13.**

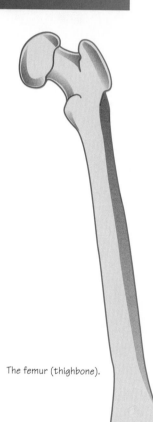

The femur (thighbone).

Bone Type	Number Found	Length (cm)
Femur	3	41.5, 41.4, 50.8
Tibia	1	41.6
Ulna	2	22.9, 29.0
Radius	1	21.6
Humerus	1	35.6
Complete Skull (including jaw)	1	23.0
Fragments	More than 10	From 3.0 to 5.0 cm

FIGURE 2.13. Measurements of bones found at Superstition Mountains.

2. For most adults, femurs range in size from 38 cm to 55 cm. According to this model, what is the range of heights of most adults? Are these reasonable heights for adults? (Recall that 2.54 cm = 1 in.)

3. What function rule would you enter into your graphing calculator in order to graph this model for the relationship between height and femur length?

The function rule that you wrote for Question 3 should be a member of the $y = mx$ family, as were the first models used to predict height from the lengths of long bones.

Dr. Mildred Trotter (1899–1991), a physical anthropologist, was well known for her work in the area of height prediction based on the lengths of long bones.

Dr. Mildred Trotter

During World War II, the armed services sometimes had problems identifying the remains of dead soldiers. Dr. Trotter was asked to help. She decided to measure each person's height and the length of each person's femur. Using these data she refined earlier models by adding a constant, thereby producing models of the form $y = mx + b$.

Data similar to the data she obtained are shown in **Figure 2.14.**

4. Why does it make sense to consider femur length as the independent variable?

5. What is a reasonable domain for this situation? Why?

6. What is a reasonable range for this situation? Why?

7. Use a graphing calculator to make a scatterplot of height versus femur length. Sketch your scatterplot and record your window.

8. Recall that Dr. Trotter's model is of the form $y = mx + b$. Find a function rule for the trend line for these data. So that you can compare your function rule to Dr. Trotter's model, be sure your rule is of the form $y = mx + b$.

9. Graph your function rule on the scatterplot. If necessary adjust the equation. Write your adjusted equation, sketch your graph, and record your window.

Femur Length (cm)	Height (cm)
45	165
53	190
46	168
38	150
54	188
52	185
39	155
51	186
41	157
42	160
47	178
40	160
55	195
50	175
43	165
48	175
44	166
49	178

FIGURE 2.14.
Adult height versus femur length.

Dr. Trotter models the relationship between the height measurements and femur measurements with the equation $H = 2.38F + 61.41$ where height, H, and femur length, F, are in centimeters.

10. Is your function rule similar to Dr. Trotter's model? How do you know?

11. How does the graph of the function rule for the early model ($y = 3.72x$) compare to the graph of the function rule for Dr. Trotter's model?

12. Would the early model or Dr. Trotter's model be the best way to predict a person's height given their femur length? Why?

13. Predict the height of an adult whose femur length is 42.5 centimeters.

14. If an adult is 170 centimeters tall, what is the approximate length of his or her femur?

SUMMARY

To build and test a model, you often need to collect data. In this section you did an analysis of collected data and made predictions. You also compared different models that are used to predict height. Most models used in this lesson are of the $y = mx$ family. Sometimes a model contains a non-zero constant such as those in the $y = mx + b$ family. These models often are used due to the variability of real-world data.

Assignment

1. Jackson's femur measures 39 cm and his brother's measures 40 cm. Based on Dr. Trotter's first formula, $H = 2.38F + 61.41$ where H is the person's height (cm) and F is the length of the femur (cm), predict the difference in the brothers' heights.

2. Femurs that belong to two women are found. One femur is one centimeter longer than the other. Predict the difference in their heights. Explain how you were able to get your answer even though the lengths of the two women's femurs were not given. In addition, tell how you could get your answer from Dr. Trotter's first formula.

3. If a man is 178 cm (about 5 ft 10 in) tall:

   a) Explain how you could use your graph to estimate the length of his femur. What is your estimate?

   b) Write a set of algebraic steps to solve Dr. Trotter's first formula, $H = 2.38F + 61.41$, for F. (A medical doctor might use such an equation to check that the length of a person's femur is normal for a person of that height.)

   c) Use your equation in b to predict the length of a man's femur if the man is 178 cm tall. Compare your answer to the femur length you estimated in a.

4. Another of Dr. Trotter's equations predicts height, H, from a known tibia length, T (this is called the second formula): $H = 2.52T + 78.62$ where H and T are measured in cm.

   a) The length of the tibia listed in Figure 2.13 is 41.6 cm. Use Dr. Trotter's second formula to predict the person's height. Is your answer a reasonable height for a person?

   b) Use your predicted height from 4a and your formula from 3b to predict the length of the femur of a person with the 41.6-cm tibia. Which femur length from Figure 2.13 is closest to your prediction?

c) Write a set of algebraic steps to solve Dr. Trotter's second formula, $H = 2.52T + 78.62$, for T. (A medical doctor might use such an equation to check that the length of a person's tibia is normal for a person of that height.)

5. In a third formula, Dr. Trotter used both the tibia and the femur to predict height:
$H = 1.30(F + T) + 63.29$. (All measurements are in cm.)

   a) A group of students measure the femur and tibia of a skeleton. They find that the femur is 42 cm long and the tibia is 43 cm long. Predict the height of the person using Dr. Trotter's third formula, $H = 1.30(F + T) + 63.29$.

   b) Compare the prediction in a with the predicted height using Dr. Trotter's equation, $H = 2.38F + 61.41$.

   c) Compare the predictions in a and b with the predicted height using Dr. Trotter's second formula, $H = 2.52T + 78.62$.

   d) You should have found a clear difference between your predictions in a–c. If a man is 175 cm tall (about 5 ft 9 in), based on Dr. Trotter's equations in b and c, would you expect his tibia or his femur to be longer and by how much?

Height vs. Forearm Length

Let's apply what you have learned thus far about building and using linear models to a new situation.

The table in **Figure 2.15** shows the heights and forearm lengths of students in Ms. Jaczinko's class.

Female			Male		
Name	Forearm Length (cm)	Height (cm)	Name	Forearm Length (cm)	Height (cm)
Alicia	24	157	Ahmed	26.5	173
Bia	24.5	166	Brian	27	177
Christi	27	164	Jesús	27	174
Chantalle	24	164	Davis	31	192
Coral-Mae	23	161	José	28	172
Juanita	27.5	164	Kelvin	29	180
Ji-Hyun	27	167	Lenny	27	174
Kim	26	162	Luis	28	175
Kristen	26	175	Mike	32	185
Maria	28.5	166	Pedro	30	185
Tianna	26.5	172	Sang	30	178
Teresa	25.5	176			

FIGURE 2.15. Heights and forearm data from Ms. Jaczinko's class.

1. Two new students will be added to the class next Monday. Each new student will get a made-to-fit class shirt. Ms. Jaczinko finds the sleeve lengths of shirts she orders by doubling the student's forearm length. Student A is 160 centimeters tall. Student B is 182 centimeters tall. Based on the class data, what sleeve lengths should Ms. Jaczinko order for the two students? Should she order only boys' shirts, only girls' shirts, or one of each? Justify your answer.

Chapter 2 Bones: Linear Functions and Predictions   **93**

Manatees and Powerboats

On the coast of Florida lives the manatee, a large and friendly marine mammal. However, the gentle Florida manatee does not live an easy life. It is one of the most endangered marine mammals in the United States. One major threat to the manatee is the large number of them killed each year by powerboats.

Should the Florida Department of Environmental Protection limit the number of registered boats in order to protect the manatee population? Before they decide, they will have to present a convincing argument to the public.

In a later section you must make a good case for whether or not to restrict powerboats. These are the main steps to present a good case to the authorities:

 ❖ Find a model that describes the relationship between manatee deaths and powerboat registrations.

 ❖ Show that the model does a good job of describing the data.

 ❖ Use the model to make your prediction.

Figure 2.16 contains data on the number of powerboats registered in Florida (in thousands) and the number of manatees killed.

1. Does there appear to be a relationship between powerboat registrations and manatee deaths? Explain why.

2. To make a recommendation about the number of powerboat registrations, you must find a model that represents the relationship between manatee deaths and powerboat registrations. To investigate this relationship, choose which variable—powerboat registrations or manatee deaths— should be the independent variable and which should be the dependent variable. Explain why.

Year	Powerboat Registrations (in thousands)	Manatees Killed
1977	447	13
1978	460	21
1979	481	24
1980	498	16
1981	513	24
1982	512	20
1983	526	15
1984	559	34
1985	585	33
1986	614	33
1987	645	39
1988	675	43
1989	711	50
1990	719	47

FIGURE 2.16. Powerboat registrations and manatee deaths in Florida, 1977–1990.

Assignment

1. The table in **Figure 2.17** lists the weights at birth for two groups of babies. The first group is babies whose mothers never smoked. The second group is babies whose mothers smoked at least ten cigarettes per day. From this small data set, does it appear that smoking affects a baby's birth weight? Explain your answer.

FIGURE 2.17.
Babies' birth weights.

Never Smoked	6.3	7.3	8.2	7.1	7.8	9.7	6.1	9.6	7.4	7.8	9.4	7.6
Smoked Ten or More Cigarettes per Day	6.3	6.4	4.2	9.4	7.1	5.9	6.8	8.2	7.8	5.9	5.4	6.3

2. Two groups of high school students were asked how much they typically spend on a date. The first group includes 12 students who did not exercise. Students in the second group exercised at least twice a week. **Figure 2.18** shows the results.

FIGURE 2.18.
Cost of a date (in dollars).

Does Not Exercise	10	5	20	4	20	20	15	0	8	40	8	15
Does Exercise	15	15	15	5	10	5	5	6	30	25	30	60

a) Find the mean, median, and mode for the Does Not Exercise group.

b) Find the mean, median, and mode for the Does Exercise group.

c) Based on this information, which of the two groups of students spends more on a date?

d) Make a scatterplot of the Does Not Exercise group vs. the Does Exercise group.

e) Is it valid to claim that there is a relationship between the exercise and non-exercise groups? Could you use the typical amount spent by someone in the Does Exercise group to predict how much a person in the Does Not Exercise group would spend? Explain.

CHECK THIS!

If the relationship between two variables is strong and has linear form, then the points in a scatterplot fall very close to a line. For weaker relationships, the data are more scattered.

A Scatterplot of Manatees Killed vs. Registrations

The last section established a problem: More manatees (an endangered species) are being killed as the number of powerboat registrations in the state of Florida rises. In this section, you will use the data comparing the number of manatee deaths and the number of powerboat registrations in order to build a mathematical model. You will use this model to make predictions in preparation for your report to the Florida Department of Environmental Protection on this issue.

1. Make a scatterplot of manatee deaths versus powerboat registrations on a sheet of graph or chart paper, and then make a scatterplot on your graphing calculator.

2. Are the two variables positively or negatively related? What does this mean?

CHECK THIS!

- If one variable increases as the other variable increases, then the two variables are positively related. This is sometimes called positive correlation.

- If one variable increases as the other variable decreases, then the two variables are negatively related. This is sometimes called negative correlation.

3. With your group, find an equation of a trend line that approximates the pattern of the data in your scatterplot. Draw the trend line on your scatterplot.

4. Use your model to predict the number of manatee deaths when 500,000 powerboats are registered. Does your prediction make sense?

5. Did all groups have the same prediction? Why or why not?

Assignment

In this assignment you will make scatterplots of some data sets, and in some cases fit a line to the scatterplot and make predictions. Pay particular attention to the features of the relationship between the variables.

Do the points in the scatterplot appear to be scattered close to the straight line? Then the scatterplot has linear form and it makes sense to describe it with a linear equation (a member of the $y = mx + b$ family).

Does the pattern made by the points move upward as you look from left to right? If so, the two variables are positively related (as one variable increases the other tends to increase). If the pattern drifts downward, then the two variables are negatively related (as one variable increases the other tends to decrease).

1. Linda heats her house with natural gas. She wonders how her gas consumption is related to the weather. The table in **Figure 2.19** shows the average outside temperature (in degrees Fahrenheit) each winter month and the average amount of natural gas Linda used (in hundreds of cubic feet) each day that month.

Month	Sep	Oct	Nov	Dec	Jan	Feb	Mar	Apr	May
Outdoor Temperature F	48	46	38	29	26	28	49	57	65
Gas Used Per Day (x100 ft³)	5.1	4.9	6.0	8.9	8.8	8.5	4.4	2.5	1.1

FIGURE 2.19. *Gas usage and temperature data.*

a) Make a scatterplot of these data. Which is the independent variable and which is the dependent variable? How did you decide?

b) Describe the features of the relationship between outside temperature and natural gas consumption. Why does the relationship have this direction?

c) Draw a trend line that you think best describes the pattern of these data. What is the equation of your line?

d) Use your equation from c to predict the gas used during a month when the average temperature is 60°F.

2. The 11 members of a college women's golf team play a practice round, then the next day play a round in competition on the same course. Their scores appear in **Figure 2.20.** (A golf score is the number of strokes required to complete the course, so low scores are better.)

Player	1	2	3	4	5	6	7	8	9	10	11
Practice	89	90	87	95	86	81	105	83	88	91	79
Competition	94	85	89	89	81	76	89	87	91	88	80

FIGURE 2.20. Golf scores.

a) Make a scatterplot of competition score versus the practice score.

b) Describe the relationship between practice and competition scores. Is there a positive or negative relationship? Explain why you would expect the scores to have a relationship like the one you observe.

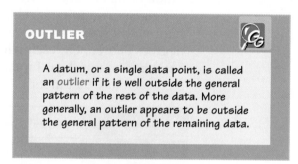

OUTLIER

A datum, or a single data point, is called an outlier if it is well outside the general pattern of the rest of the data. More generally, an outlier appears to be outside the general pattern of the remaining data.

c) One point falls clearly outside the overall pattern. Circle this point in your plot. A good golfer can have a bad round, or a weak golfer can have a good round. Can you tell from the given data whether the unusual point is produced by a good player or a poor player? What other data would you need to distinguish between the two possibilities?

d) You might expect a player to have about the same score on two rounds played on the same course. Draw on your graph the line that represents the same score on both days. Does this line fit the data well when you ignore the unusual point? If you don't like this line, draw a line that you would prefer to use to predict the competition score from the practice score.

e) Another golf team member shot a 95 in practice. Predict her score in competition.

3. **Figure 2.21** contains two scatterplots that might show the relationship between human height and ulna length (forearm-bone length). In both displays, a line describing the pattern of the data has been added.

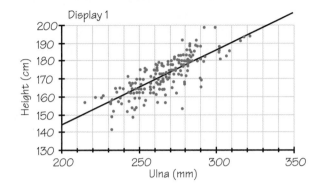

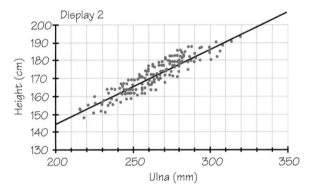

FIGURE 2.21. Two scatterplots showing height versus ulna length.

a) Both scatterplots have a linear form. Which of the two shows a stronger linear relationship? Explain.

b) A scientist used the model $y = 60.2 + 4.2x$ (where x is ulna length in cm and y is height in cm) to predict the height of a dead person whose ulna length of 29.0 cm is shown in Figure 2.21. What did the scientist predict as his or her height?

c) How close is the scientist's prediction if Display 1 shows the real data? How close is the scientist's prediction if Display 2 shows the real data? Which display better supports the scientist's prediction?

d) Describe the connection between a scatterplot and the predictions you can make using a trend line.

Residuals and Least-Squares Regression Lines

Previously, you learned methods for finding a trend line that describes a pattern of data. You used the equations of the functions you found to make predictions. In this section you will learn about a method used to find a line that approximates the pattern of data.

Whenever you make predictions based on data, you want to check the accuracy of your predictions. One way to evaluate your predictions is to calculate what are called the residual errors.

CHECK THIS!

Residual error: When a prediction is made from data, the difference between the actual value and the predicted value is called the residual error.

Residual error = Actual value − Predicted value.

You can represent the residuals graphically by drawing a vertical line from each point on the scatterplot to the line that represents your model as shown in **Figure 2.22**.

FIGURE 2.22.
Residuals for the plight of the manatee.

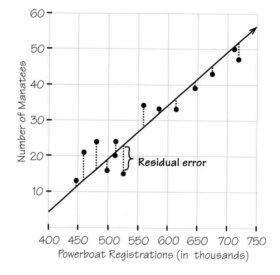

A good line should meet a few tests. One test is that the sum of the residual errors should be close to 0. The sum of the residual errors is often found with a calculator or computer software. At times, the graph allows you to see that the sum of the residuals is fairly close to 0, but often it is easier to use a table, such as **Figure 2.23,** where you can total the residuals.

Registrations (in thousands)	Manatees Killed	Predicted Number of Manatees Killed ($y = 0.15x - 56$)	Residual Error
447	13	11	2
460	21	13	7
481	24	16	8
498	16	19	–3
513	24	21	3
512	20	21	–1
526	15	23	–8
559	34	28	6
585	33	32	1
614	33	36	–3
645	39	41	–2
675	43	45	–2
711	50	51	–1
719	47	52	–5
Total			2

FIGURE 2.23. Residual table for manatee deaths.

As you can see in Figure 2.23, the sum of the residual errors can be close to zero even if most of the residual errors are not. This happens because some of the residuals errors are positive and others are negative. To give more emphasis to values that deviate from the overall pattern, the residual errors are squared and then totaled. The squares of the residuals are shown in **Figure 2.24**. The goal is then to find a line that makes the sum of the squared residual errors as small as possible.

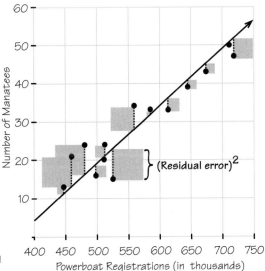

FIGURE 2.24. Squared residual errors for the plight of the manatee.

Chapter 2 Bones: Linear Functions and Predictions

To try and find the "line of best fit" by trial and error might take you a while. **Figure 2.25** shows some examples of equations that you might investigate as a trend line and the sum of squares for each line.

Equation	Sum of Squares
$y = 0.14x - 56$	745
$y = 0.13x - 50$	670
$y = 0.13x - 45$	228
$y = 0.12x - 45$	783
$y = 0.12x - 40$	246
$y = 0.125x - 41$	223

FIGURE 2.25.
Sum of the squared residual errors.

Your graphing calculator has built-in commands that calculate the equation of the **least-squares regression line**. Statisticians also call this line the **line of best fit**. To find the model that fits the data the best, they look for several criteria:

❖ The pattern of data in the scatterplot is randomly scattered close to the line.

❖ The sum of the residual errors is close to 0.

❖ The sum of the squared residual errors is as small as possible.

The general name for finding an equation that fits the data is **regression**. Since you are looking for a linear relationship, the technique used in fitting the "best" line is called **linear regression**.

1. Use your calculator (or spreadsheet) to calculate the values of the slope and y-intercept of the least-squares line. Write its equation.

2. What do the slope and y-intercept of the linear regression line tell you about boats and manatees?

3. Plot both the data and the least-squares line so that you can see both graphs in the same window. Does the least-squares line appear to fit the data well? Support your answer.

SUMMARY

After you collect data on two variables and display the data in a scatterplot, you may describe a relationship between the two variables by finding a line of best fit. Statisticians often use the least-squares regression line as the best model linear relationship. Criteria for a good regression line include the following:

❖ The line appears to go through the middle of the points in the scatterplot.

❖ The sum of the residual errors is close to 0.

❖ The sum of the squared residual errors is as small as possible.

Graphing calculators (and spreadsheets) can calculate the least-squares regression line for you. However, you should check your linear regression model to see if it fits the criteria above. If it does not fit, you should look for a different model to describe your data.

1. On graph paper, make a scatterplot of Ms. Jaczinko's class data (see **Figure 2.26**). Let forearm length be the independent variable and height be the dependent variable. Remember to label each axis and use an appropriate scale. To distinguish the boys from the girls, use two colors (or different shapes), one to represent the boys' data and the other to show the girls'.

FIGURE 2.26.
Height and forearm data from Ms. Jaczinko's class.

Female			Male		
Name	Forearm Length (cm)	Height (cm)	Name	Forearm Length (cm)	Height (cm)
Alicia	24	157	Ahmed	26.5	173
Bia	24.5	166	Brian	27	177
Christi	27	164	Jesús	27	174
Chantalle	24	164	Davis	31	192
Coral-Mae	23	161	José	28	172
Juanita	27.5	164	Kelvin	29	180
Ji-Hyun	27	167	Lenny	27	174
Kim	26	162	Luis	28	175
Kristen	26	175	Mike	32	185
Maria	28.5	166	Pedro	30	185
Tianna	26.5	172	Sang	30	178
Teresa	25.5	176			

2. Find a linear regression model to predict a girl's height from her forearm length. Find a second linear regression model to predict a boy's height from his forearm length.

3. Use your scatterplot and least-squares model to make these predictions.

   a) If a girl the same age as the students in Ms. Jaczinko's class has a forearm that measures between 25 cm and 27 cm, what would you predict for her height?

b) Predict the height of a tenth-grade boy with a 28.5-cm forearm. Explain your answer.

c) What would be the forearm length of a student who is 163 cm tall? Explain your answer.

The 16% Model

Next you will use the data from Ms. Jaczinko's class to examine and refine a model that represents the relationship between height and forearm length.

4. Archaeologists study ancient human life. They use general rules of proportions. So an archaeologist might use the rule of thumb that the forearm of a typical female teenager is 16% of her height.

a) Write this rule as an equation that predicts height, y, from forearm length, x. We call this equation the "16% model." Test a few pairs. Does this model make sense?

b) Sketch a graph of the 16% model on the same set of axes as your scatterplot.

c) Compare the 16% model with the data from Ms. Jaczinko's class. Is the model true for all the people in the class? Justify your answer based on your graph.

5. Use the 16% model to make the following predictions.

a) Predict the height of a student whose forearm is 27 cm. Use the data from Ms. Jaczinko's class to check your prediction.

b) Predict the height of a student whose forearm is 33 cm.

c) The forearm lengths of two students differ by 1 cm. Use the 16% model to predict how much their heights differ. What if their forearm lengths differed by 2 cm? Justify your answers.

What is Your Recommendation?

Now that you have a model for the relationship between manatee deaths and powerboat registrations, you are ready to make your recommendation to the Florida Department of Environmental Protection.

1. Present to the class, using your scatterplot and the equation of the least-squares regression line, a convincing case for whether or not to restrict powerboat registrations. Your argument should include the three main steps listed in Section 2.6:

   ❖ Find a model that describes the relationship between manatee deaths and powerboat registrations.

   ❖ Show that the model does a good job of describing the data.

   ❖ Use the model to make your prediction.

2. Say you want to reduce the number of manatee deaths to about 30 per year. How many powerboat registrations should there be? Explain how you can use your model and algebra to answer this question. What number of powerboat registrations do you recommend?

3. Say, instead, that you recommend 700,000 powerboat registrations (slightly below the number of registrations in 1989). Predict the number of manatees that would be killed each year, on average, if this proposal were adopted. Use your scatterplot to see how good your prediction is.

4. Fill in the blank: Every time the number of boat registrations is raised by 50,000, one can predict that, on average, an additional _____ manatees would be killed by powerboats each year. Justify your response.

5. It is possible that the Florida Department of Environmental Protection did not follow your recommendations back in 1991. In 2002, powerboats killed 95 manatees, and there were 923,000 powerboat registrations. Even though the least-squares model should only be used to predict the number of manatee deaths when the number of powerboat

registrations is between 447,000 and 719,000, how well does the least squares model predict the number of manatees killed in 2002?

6. The year 2002 had more manatees killed than the model predicts. **Figure 2.27** shows data for the years 1996 through 2003. Add these data to the previous data.

Year	Powerboat Registrations (in thousands)	Actual Number of Manatees Killed
1996	751	60
1997	797	54
1998	806	66
1999	805	82
2000	841	78
2001	903	81
2002	923	95
2003	978	73

FIGURE 2.27.
Data on powerboats and manatees from 1996–2003.

*These data are from the Florida Fish and Wildlife Research Institute and the United States Coast Guard.

a) What is an equation of a new line of best fit?

b) Graph both the old and new lines of best fit on the calculator scatterplot.

c) What does the new slope tell you?

d) What is the predicted number of manatee deaths with the new model?

e) Is the new prediction closer to the actual number than the old model?

In this section, you used the least-squares model that you calculated in Section 2.8 to make predictions. In the process, you were given additional data to test how well your model predicted future events. You observed that a model that fits one data set may or may not fit an expanded data set. A line that fits a set of data well cannot always be trusted to predict outside the range of the data.

Hence, often when you gather additional data, you may need to refine your mathematical model. The more data points that you can collect, the more accurate your model will be.

Not every data set can be modeled well with a linear regression equation. There are many other types of relationships between variables. We will explore some of these relationships in the following chapters.

Assignment

How would you choose which of two models makes better predictions? Let's explore some data to help you decide. The data in **Figure 2.28** were collected from a ninth grade class. The asterisks (*) indicate missing values.

FIGURE 2.28.
Data on height, stride length, and forearm length.

Height (cm)	Stride Length (cm)	Forearm Length (cm)
166.0	58.2	28.5
164.5	55.9	27.2
175.0	59.1	28.6
184.0	68.9	30.5
161.0	72.5	26.5
164.0	*	28.2
171.0	*	29.0

1. a) Make a scatterplot of height versus stride length. Make a second scatterplot of height versus forearm length. (For both plots, height should be on the vertical axis. Use the same scale on the vertical axis for each plot.) Which of the two scatterplots shows the stronger relationship? How can you tell?

   b) Find the least-squares line that represents each of these relationships.

   c) Use the equations of your two least-squares lines to predict the height of a person whose stride is 73 cm and whose forearm length is 27 cm. Which of the two estimates is more reliable? Explain.

   d) State, in your own words, how you would choose, in general, between two different models that predict the same quantity.

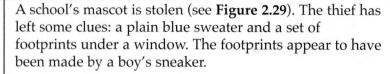

A school's mascot is stolen (see **Figure 2.29**). The thief has left some clues: a plain blue sweater and a set of footprints under a window. The footprints appear to have been made by a boy's sneaker.

The distance between the footprints, from the back of the heel on the first footprint to the back of the heel on the second, reveals that the thief's stride is approximately 58 cm. The length of the thief's forearm can be estimated by measuring the sweater from the center of a worn spot on the elbow to the turn where the cuff meets the sleeve. The thief's forearm is between 26 cm and 27 cm.

FIGURE 2.29.
The missing manatee.

School officials suspect that the thief is a student from a rival high school. **Figure 2.30** contains data from a 1st period class about students' genders, heights, stride lengths, and forearm lengths. Use these data to answer the following questions.

2. Find the mean, median, and mode for the whole class and then find the mean, median, and mode for the boys and the girls.

   a) Is there a difference between the mean stride length for a boy and the mean stride length for a girl?

3. The footprint was made by a boy's sneaker. However, sometimes girls wear boys' sneakers. So you'll need to decide whether or not the thief was male. Use what you know about the thief's stride length and forearm length.

   a) Do the class stride length data tend to confirm that the footprints belonged to a boy and not a girl? Use the data you gathered to support your answer.

   b) Do the forearm data tend to confirm that the sweater belonged to a boy and not a girl? Use the data you gathered to support your answer.

Next you will predict the height of the thief. You have two possible independent variables you can use for your prediction: stride length and forearm length. You will try to decide which is better. Divide the work in Questions 4 and 5 among the members of your group.

Name	Gender	Height (in cm)	Stride Length (in cm)	Forearm Length (in cm)
Scott	Male	166.0	58.25	28.5
John	Male	178.0	68.5	29.0
Matt	Male	171.0	58.5	27.2
Will	Male	165.0	50.125	28.0
Michael	Male	177.5	58.75	31.3
Jeffrey	Male	166.0	62.875	28.3
Even	Male	175.5	59.125	28.6
Brad	Male	171.0	67.75	31.5
Lonnie	Male	184.0	68.875	30.5
William	Male	184.5	66.25	30.8
Robert	Male	183.5	79.5	30.5
Karim	Male	172.0	70.5	30.3
Meredith	Female	164.5	55.875	24.2
Lee	Female	166.0	52.375	27.3
Pilar	Female	168.0	55.375	28.0
Ansley	Female	178.5	59.75	29.1
Julie	Female	166.0	48.375	27.9
Becton	Female	159.0	57.125	28.0
Elizabeth	Female	166.0	64.0	27.4
Shannon	Female	154.5	57.75	25.8
Janie	Female	161.0	63.5	27.0
Jeris	Female	177.0	69.75	30.1
Kat	Female	161.0	72.5	26.5
Blaie	Female	164.0	75.25	28.2
Frances	Female	174.0	58.5	28.4
Eliz	Female	164.0	59.75	26.8
Baily	Female	168.0	55.25	26.4

FIGURE 2.30. Height, stride length, and forearm length for students in 1st period class.

4. a) Use your data to find models for the relationship between height and forearm length. First find a model based on the entire data set. Then use the girls' data to find a second model, and finally use the boys' data to find a third model.

b) Compare the three models. Is there much difference? Explain.

c) Which model gives a better prediction if, in fact, the thief were male? Justify your answer.

d) Which model gives a better prediction if, in fact, the thief were female? Justify your answer.

5. Repeat Question 4 for the relationship between height and stride length.

6. Select the best model for the job of predicting the height of the thief. Support your choice.

7. Use this model to predict the height of the thief. Explain your answer.

8. **Figure 2.31** shows the winners of the Indianapolis 500 Auto Race and their average speed every 4 years from 1912 to 1972. No race was run in 1944 due to World War II.

Year	Number of Years Since 1912	Winner	Winner's Average Race Speed (mph)
1912	0	Joe Dawson	79
1916	4	Dario Resta	84
1920	8	Gaston Cheverolet	89
1924	12	L. L. Corum & J. Boyer	98
1928	16	Louis Meyer	99
1932	20	Fred Frame	104
1936	24	Louis Meyer	109
1940	28	Wilbur Shaw	114
1944	32	—	—
1948	36	Mauri Rose	120
1952	40	Troy Ruttman	129
1956	44	Pat Flaherty	128
1960	48	Jim Rathmann	139
1964	52	A. J. Foyt	147
1968	56	Bobby Unser	153
1972	60	Mark Donohue	163

FIGURE 2.31. Indianapolis 500 Auto Race results.

Finding a linear regression equation is easier if you use the number of years since 1912 as the independent variable. This also gives meaning to the y-intercept.

a) Make a scatterplot of the data.

b) Find the equation of a linear regression model.

c) What do the slope and y-intercept in this model show?

d) The speed of the winner in 1957 was 136 mph. How well does the linear regression model predict this value?

e) Even though you know that 200 mph is outside the range of the data, you would like to use your model to predict when a driver will reach a race speed of 200 mph. Based on the linear regression model, when would you expect a driver to average 200 mph in the Indianapolis 500 race?

f) The actual race speed of the winner, Buddy Rice, in 2004 was 139 mph. Use your line of best fit to predict the speed in 2004. Is the least-squares line a good predictor? Why or why not?

Line Fitting with Domain Restrictions

Let's take what we have learned about building linear models thus far, and apply it to a new situation.

Marietta has been working as a waitress for 7 weeks. Since she earns an hourly wage plus tips, there is no set amount she makes per hour. However, after examining her 7 pay stubs she notices an apparent trend. **Figure 2.32** shows how many hours she worked and how much money she makes per week.

FIGURE 2.32.
Marietta's data table.

Time (*T*) (in hours)	8	15	23	21	16	20	22
Money (*M*) (in dollars)	53	101	155	131	124	138	157

1. Label the axes appropriately and draw a scatterplot of money (*M*) versus time (*T*).

2. Use a straightedge to approximate a trend line for the data. Write an equation for your trend line.

3. Use a graphing calculator to make a scatterplot, and then find the line of best fit using linear regression.

4. How are your trend line and the calculator's line of best fit alike and how are they different?

5. What do the slope and *y*-intercept of the line of best fit mean in this context?

6. Marietta's boss tells her that she must work at least 10 hours every week in order to keep her job. Furthermore, state law says that a restaurant employee cannot work more than 60 hours during any week. Considering these restrictions, sketch a more appropriate trend line. Do not re-plot the data. Number and label your axes appropriately.

7. Use the second graph to predict how much Marietta will make if she works for 30 hours. Explain your answer.

8. Use the second graph to predict how many hours Marietta needs to work to make $75. Explain your answer.

Modeling Project Who Am I?

You are an anthropologist that finds bones of two or more individuals. The table in **Figure 2.33** contains the information about those bones.

Skeleton 1 (possibly female)	Skeleton 2 (taller of the two)	Uncertain
Femur: 413, 414	Femur: 508	Skull: 230
Ulna: 228	Ulna: 290	Humerus: 357
		Radius: 215
		Tibia: 416

FIGURE 2.33. Classification of bones, measurements in millimeters.

Figure 2.34, on the following page, contains actual data from the Forensic Anthropology Data Bank (FDB) at the University of Tennessee. The FDB contains metric, nonmetric, demographic, and other kinds of data on skeletons from all over the United States. These individuals most likely are unidentified bodies that went to forensic anthropologists for analysis and identification.

Use the data in Figure 2.34 to answer the following items. Present your findings in a formal report. All of your conclusions must be supported by statistical analysis of the data in Figure 2.34.

1. Determine two models to predict people's height from the lengths of various long bones in their arms and legs. Explain which of these models you would prefer to use and why.

2. Based on these data, do you agree that Skeleton 1 is female? Do the data provide any information that would help you decide whether Skeleton 2 is male or female?

3. Find models for relationships between pairs of long bones that would help you decide whether the bones in the uncertain column belong to Skeleton 1 or Skeleton 2. (Or is there strong evidence that one of these bones belongs to a third person?)

4. Predict the heights of Skeleton 1 and Skeleton 2. Explain why you chose the model that you used to make your predictions.

sex	height	humerus	radius	ulna	femur	tibia	fibula		sex	height	humerus	radius	ulna	femur	tibia	fibula		sex	height	humerus	radius	ulna	femur	tibia	fibula
1	168	307	240	258	448	384	368		1	165	307	230	248	452	363	355		2	178	337	272	272	475	393	390
1	178	336	247	261	463	404	390		1	163	297	240	260	435	356	356		2	172	344	255	281	470	400	393
1	161	294	213	227	413	335	322		1	143	282	216	233	398	334	318		2	188	360	269	283	510	422	416
1	155	324	262	279	465	395	375		1	154	297	228	248	423	344	334		2	189	347	272	283	547	432	445
1	165	314	243	258	432	364	364		1	171	342	272	290	485	418	407		2	177	330	246	262	462	386	370
1	168	303	223	244	441	355	342		1	162	303	237	262	433	367	364		2	166	322	242	258	442	373	374
1	165	311	231	254	436	362	360		1	150	308	220	247	383	352	341		2	186	332	267	283	478	391	388
1	173	312	248	266	483	405	401		1	157	288	201	215	429	363	350		2	177	322	245	265	457	397	395
1	165	322	229	246	448	368	352		1	158	314	239	263	432	371	358		2	176	332	259	274	458	382	378
1	163	298	221	245	443	355	361		1	162	306	250	268	444	355	352		2	180	323	251	275	448	390	387
1	153	280	218	234	410	345	344		1	159	310	238	255	449	362	352		2	173	335	253	273	497	404	389
1	165	294	220	235	448	354	353		2	169	337	254	273	460	396	385		2	175	330	253	274	470	384	382
1	170	311	235	253	440	360	347		2	153	296	223	243	407	337	338		2	169	313	252	265	472	391	385
1	160	316	214	226	437	356	348		2	175	339	256	271	470	390	381		2	175	336	256	274	464	388	377
1	159	292	223	233	419	346	336		2	179	343	242	263	464	378	371		2	181	390	284	303	521	440	435
1	163	315	228	251	438	356	347		2	179	352	253	269	484	407	397		2	193	356	297	318	522	451	433
1	165	303	237	249	451	356	348		2	198	354	263	292	508	417	412		2	182	362	275	293	499	424	405
1	165	308	234	248	439	348	344		2	173	327	256	276	463	383	387		2	169	322	249	266	426	366	356
1	165	315	227	240	448	363	353		2	180	357	268	278	494	401	390		2	180	337	265	281	482	412	399
1	175	316	244	260	473	390	374		2	178	344	254	269	464	371	366		2	185	363	286	302	520	429	420
1	180	333	256	278	475	391	381		2	175	339	245	272	456	374	366		2	180	355	274	292	490	422	424
1	168	321	230	248	450	365	362		2	177	343	250	266	483	361	365		2	170	378	272	291	512	404	390
1	163	299	219	236	435	357	339		2	180	353	260	281	490	420	415		2	180	370	278	292	523	429	420
1	165	304	246	264	467	392	383		2	170	303	235	249	435	366	361		2	175	333	260	273	484	398	386
1	160	309	236	248	432	364	358		2	191	364	263	278	511	430	417		2	168	342	262	280	484	404	385
1	158	319	246	268	442	371	364		2	188	349	269	288	498	427	423		2	170	347	269	291	476	396	393
1	165	325	242	250	448	378	365		2	179	323	256	276	486	398	400		2	166	315	240	260	456	377	362
1	170	335	248	263	474	400	382		2	180	350	263	280	480	419	418		2	185	363	295	309	524	446	427
1	182	334	254	273	514	420	407		2	181	350	263	282	488	391	381		2	191	382	299	316	537	479	466

FIGURE 2.34. Data from Forensic Anthropology Data Bank (FDB).

Key to Data (in order from left to right): sex (1 = female, 2 = male), height (cm), humerus (mm), radius (mm), ulna (mm), femur (mm), tibia (mm), fibula (mm)

Practice Problems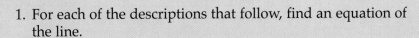

1. For each of the descriptions that follow, find an equation of the line.

   a) The line through the point (2, 3) that has slope $\frac{1}{2}$.

   b) The line that has y-intercept 3 and slope $-\frac{1}{2}$.

   c) The line passing through points (3, 2) and (2, 5).

   d) The line has y-intercept 5 and is parallel to the line $y = 7x - 1$.

2. On December 18, 1994, three amateur spelunkers (cave explorers) found a cave in France, now known as Chauvet Cave, that had ancient paintings. In later explorations, human footprints were found. According to prehistorian Michel-Alain Garcia, the footprints belonged to a boy who was about 4.5 feet tall.

   a) Scientists have used simple models to predict height from footprints since the mid-1800s. One model, still in use today, predicts height by dividing the maximum foot length by 0.15. Write an equation for this model.

   b) The model in a can be expressed as a member of the $y = mx$ family. What is the value of m?

   c) Use this model to predict the maximum length of the footprints found in Chauvet Cave. Give your answer in centimeters. (Remember 2.54 cm = 1 in.)

   d) Suppose that another footprint was found and it measured 1 cm more than the one discussed by Garcia. By how much would you increase your estimate for height? (Be careful of the units you use to report your answer.)

3. Anthropologists have refined early models for estimating a person's height from the length of a footprint. One revision suggests different models depending on whether a footprint is from the right foot or from the left. Here are two models designed for use with adult footprints.

For right foot: $H = 3.641L + 72.92$
For left foot: $H = 4.229L + 56.49$
(where all measurements are in cm)

a) Suppose a footprint measures 22 cm. Predict the person's height, first by assuming the print is from the right foot and then by assuming it is from the left foot.

b) Use your calculator to graph each function. Based on your graphs, what footprint length gives the same height prediction from both models? How can you determine this value using algebra? Try it and check to see if you get the same result.

c) Both the right-foot and left-foot models are from the $y = mx + b$ family. For each of the models, interpret the meaning of m in the height-footprint context. What does b mean in this context?

4. a) Solve the right-foot equation in Question 3 for L in terms of H. Then do the same for the left-foot equation.

b) Suppose a person is 153 cm tall. Use your equations from a to predict the lengths of the person's right and left footprints. How much longer is the larger foot than the shorter foot?

5. Regular Chips Ahoy® chocolate chip cookies boast "1000 chips in every bag." You can also buy reduced fat Chips Ahoy cookies. Do you think both types of Chips Ahoy cookies contain roughly the same number of chips? To find out, a statistics class opened bags of regular Chips Ahoy and reduced fat Chips Ahoy cookies, randomly selected 15 cookies from each bag, and counted the number of chips in each cookie. Their data appear in **Figure 2.35**.

Reduced Fat	13	15	14	12	15	17	13	10	15	18	19	18	20	21	16
Regular	18	20	17	22	20	16	21	18	16	19	27	22	19	16	24

FIGURE 2.35. Chip counts in reduced fat and regular Chips Ahoy cookies.

a) From the data in Figure 2.35, make two dot plots using the same scaling on each. Place one dot plot directly above the other.

b) Suppose you select a cookie from the regular Chips Ahoy bag. Predict the number of chips in the cookie. How far off might your prediction be?

c) Suppose instead, you select a cookie from the reduced fat bag. Predict the number of chips in the cookie. How far off might your prediction be?

d) Does it appear that the number of chips is changed in order to produce a lower fat product? Explain.

e) George wants to make a scatterplot of the row 1 data in Figure 2.35 versus the row 2 data. Do you think his scatterplot would reveal useful information about these two types of cookies? Explain.

6. A newspaper article in the *Worcester Sunday Telegram* reported that scores on intelligence tests are going up at a rate of 3 IQ points per decade.

a) In 1932 the average IQ test score for Americans was 100 points. Use the information given in the article to write a function rule that predicts the average IQ score for years after 1932. Let $x = 0$ represent 1932 to simplify your model.

b) **Figure 2.36** provides data on the average IQ scores of Americans. Make a scatterplot of these data. (Let $x = 0$ represent 1932.)

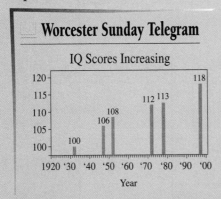

FIGURE 2.36. Average IQ scores from 1932 to 1997.

c) Based on these data, do IQ scores and years have a positive or negative relationship? Explain.

d) Fit a least-squares line to these data. What is the equation for this line? Add this line to your scatterplot in b.

e) Based on the least-squares equation from d, was the newspaper correct in reporting that IQ scores are rising at a rate of 3 points per decade? Explain.

f) A genius is a person with an IQ over 140. If the rate of increase continues, in what year will the average IQ be at the level of genius? Do you believe this? Explain.

The Boston Marathon is the world's oldest and best-known marathon. The first Boston Marathon was held in 1897. Fifteen men participated. Over 17,000 people entered the 104th Boston Marathon held April 17, 2000. Questions 7–8 are based on the first-place times in **Figure 2.37**.

Men's		Women's	
Marathon Number	Time (hr: min: sec)	Marathon Number	Time (hr: min: sec)
31	2:40:22	82	2:34:28
36	2:33:36	83	2:26:46
41	2:33:20	84	2:29:33
46	2:26:51	85	2:22:42
51	2:25:39	86	2:29:28
56	2:31:53	87	2:34:6
61	2:20:5	88	2:24:55
66	2:23:48	89	2:25:21
71	2:15:45	90	2:24:30
76	2:15:30	91	2:24:33
81	2:14:46	92	2:25:24
86	2:8:51	93	2:24:18
91	2:11:50	94	2:23:43
96	2:8:14	95	2:25:27
		96	2:21:45
		97	2:25:11
		98	2:27:13
		99	2:58:0
		100	2:40:10

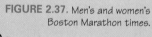

FIGURE 2.37. Men's and women's Boston Marathon times.

7. The men's data in Figure 2.37 contain first-place times for selected marathons from 1927, marathon 31, to 1992, marathon 96.

a) Make a scatterplot of the men's times versus the marathon number. (What unit did you use for time?)

b) Fit a least-squares line to the data in your scatterplot. What is the equation? Add a graph of the least-squares line to your scatterplot.

c) Interpret the slope of your linear model in the context of the Boston Marathon.

d) Use your model to predict the time of the first-place finisher for the 104th Boston Marathon. The actual winning time was 2:09:47. How close is your prediction?

8. Figure 2.37 shows women's times for marathons 82–100, held in years 1980–1996.

a) Fit a least-square's line to these data. Interpret the slope in the context of the marathon. Does this seem reasonable? Explain.

b) Make a scatterplot of the women's data. Add the least-squares line to your plot. Does the least-squares equation do a good job of describing the women's data?

c) Remove the times corresponding to the 99th and 100th marathons. Refit the least-squares line to the remaining data. What affect did the removal of these points have on the slope?

Men's		Women's	
Marathon Number	Time (hr: min: sec)	Marathon Number	Time (hr: min: sec)
82	2:40:10	82	3:48:51
83	2:26:57	83	3:52:53
84	2:38:59	84	3:27:56
85	1:55:0	85	2:49:4
86	2:0:41	86	2:38:41
87	1:51:31	87	2:12:43
88	1:47:10	88	2:27:7
89	2:5:20	89	2:26:51
90	1:45:34	90	2:5:26
91	1:43:25	91	2:9:28
92	1:55:42	92	2:19:55
93	1:43:19	93	2:10:44
94	1:36:4	94	1:50:6
95	1:29:53	95	1:43:17
96	1:30:44	96	1:42:42
97	1:26:28	97	1:36:52
98	1:22:17	98	1:34:50
99	1:25:59	99	1:40:41
100	1:30:11	100	1:52:54

FIGURE 2.38. Men's and women's wheelchair times in the Boston Marathon.

9. a) Make scatterplots for the men's and women's wheelchair times in **Figure 2.38**. So that you can compare the men's and women's data, use the same scale on the axes of both plots.

   b) Is there a positive or negative relationship between marathon number and times for the men's data? What about for the women's data? Interpret what this means in the context of the marathon.

   c) Comment on the form of the data. Does the pattern of the data appear linear or nonlinear? Explain.

The data in **Figure 2.39** provide information on the heights of people as children and again as adults. Use the data in this figure to answer Questions 10–13.

Girl's Height at 1.5	Girl's Adult Height	Boy's Height at 2	Boy's Adult Height
78.0	157.0	89.0	178.0
79.4	158.4	89.9	177.1
80.4	161.4	90.3	179.6
81.3	164.7	90.8	181.8
81.3	160.4	90.9	184.0
82.1	163.7	91.0	180.5
83.2	164.4	91.1	182.0
83.2	170.2	91.2	183.1
83.9	170.5	91.4	180.1
84.9	166.5	91.5	185.1
86.2	171.3	92.9	182.0
87.9	170.7	93.3	186.3
88.2	179.7	94.7	187.4
89.4	176.9	95.4	187.9
90.1	176.9	96.1	189.4

FIGURE 2.39.
Height data on children and adults.

10. Suppose one of the people from this study plans to visit your school.

   a) If you find out the visitor is a woman, predict her height. How did you decide on your prediction?

   b) What if the visitor is a man?

11. Create a display that compares the men's heights with the women's heights. Write a description of the information your display conveys.

12. Suppose you want to predict how tall a 1 1/2-year-old girl will be when she reaches adulthood.

   a) Which is the independent variable and which is the dependent variable?

   b) Make a scatterplot of women's adult heights versus their heights when they were 1 1/2 years old.

c) Would you describe the relationship between women's heights and girls' heights as linear or nonlinear? Does this relationship appear to be positive or negative?

d) Fit a least-squares line to the data. Write its equation, and sketch its graph on your scatterplot.

e) Use your equation to predict the adult height of a 1 1/2-year-old girl who is 82.5 cm tall.

13. a) Determine the least-squares line for predicting men's heights from their heights when they were 2 years old.

b) If two 2-year-old boys differ in height by 1 cm, predict how much their heights will differ when they are adults.

c) What if their heights as 2-year-olds differ by 2 cm?

d) Does the y-intercept of the least-squares line have any meaning in this context? Explain.

e) Does the slope of the least-squares line have any meaning in this context? Explain.

14. The grades in **Figure 2.40** are from a high school statistics class.

a) What is the average grade on the midterm exam? On the final exam?

b) Make a scatterplot of these data.

c) Does the pattern in your scatterplot appear to have linear form? Does there appear to be a positive or negative relationship between students' midterm grades and students' final grades?

d) Fit a least-squares line to the data.

Name	Midterm exam grade	Final exam grade
	50	54
	53	52
	57	59
	62	70
	62	68
	68	70
	69	68
	73	79
	73	82
	74	83
	74	80
	75	82
	78	88
	79	89
	79	90
	81	88
	81	89
	81	92
	82	91
	85	94
	87	98
	100	99
	100	100

FIGURE 2.40.
Midterm and final exam grades.

e) Add a graph of your model from d to your scatterplot. Do you think your line does a good job in describing these data? Explain.

f) One student missed the final exam and had to take it at a later date. She got 77 on the midterm. Predict her grade on the final exam.

Glossary

KEY CONCEPTS

Average: To find the average of a data set, sum the data and divide by the number of data in the set.

Dot plot: Display in which dots are placed above a number line to represent the values of a single variable. (Also called line plot.)

Least-squares line: The line with the smallest sum of squared residual errors.

Linear equation: An equation relating two variables, x and y, that can be put in the form $y = mx + b$.

Linear form: The form of a scatterplot for which it is possible to draw a line that describes the general trend of the data.

Linear regression: The process of fitting a line to data using the least-squares criterion.

Mean: To compute the mean, sum the data and divide by the number of data. Sometimes mean and average are used interchangeably.

Median: The middle value in a set of data.

Mode: The value in a set of data that appears most often.

Nonlinear form: The form of a scatterplot for which the general pattern of the data is not well described by a straight line.

Outlier: In a collection of data, an individual data point that falls outside the general pattern of the other data.

Residual error: Actual value of the dependent variable minus the predicted value.

Strong linear relationship: A scatterplot of the data that lies in a narrow band.

3

CHAPTER

The World of Business: Systems of Equations

Running a Business

Four years ago, Warren Brown left his job as a lawyer to bake cakes. It all started when he decided to become an expert baker. After work, he tried his hand cake making. Then he had dessert parties for his friends. His friends loved his cakes. This made him want to see if he could support himself as a baker. He rented a small kitchen and found customers mostly by word of mouth. His business went so well that he moved into a storefront that he called Cake Love. At first he used his credit cards to get the business started. Then he got a $125,000 loan from the Small Business Administration. Now, Cake Love sells around 40 cakes per day at about $55 each. Cake Love was often very crowded, so Warren opened the Love Cafe across the street. The café serves sandwiches and soups as well as cake.

Warren plans to open more stores this year. Even though managing takes most of his time, he still bakes most mornings. He also works on new types of cakes. His business is now so successful that he has the same salary he used to make as a lawyer.

People start their own businesses for lots of reasons. They know that if they succeed the rewards will be great. Also this gives them a way to make more money than if they worked for someone else, and they can be the person in charge.

Of course there is no guarantee that all businesses will be successful. In fact, for every successful business, there are many that fail. Often a business fails not because it makes a poor product but because of poor financial decisions.

Founder/Owner: Warren Brown
Company: Cake Love and Love Café
Location: Washington, D.C.
Industry: Retail baked goods

In this chapter you will examine a start-up business that makes and sells athletic wear. You will have the opportunity to create mathematical models that can be used to make good financial decisions.

Do You Have Issues?

In this section you will assume the role of a businessperson like Warren Brown. Imagine that last year you started making baseball caps in your house. You gave them to a few close friends as birthday gifts. Your friends liked the caps and offered to pay you to make caps for a few of their friends. Soon you were getting hundreds of requests from people wanting to buy caps. Seeing this response as a great opportunity, you decide to start a business. There is not enough room in your house or time for you to make all these caps on your own. What issues, both practical and financial, do you need to consider before you start your business and agree to produce the caps?

Talk with your group members and make a list of issues that need to be considered before starting the business.

Comparing Pay Structures

One important issue a businessperson must consider is employee pay. After doing some research you find that there are two companies in your city that make caps. Hats, Inc., pays its employees $1000.00 a month plus $5.00 for every hat they complete in the month. Caps, Inc., pays its employees $500.00 a month plus $10.00 for every hat they complete in the month.

In the next activity you will compare these pay structures and see if one will fit your company.

1. Use the description of the pay structure from Hats, Inc., to complete a table like **Figure 3.1.** Then use the process column to develop an algebraic rule for this pay structure.

Number of Caps Completed During the Month (c)	Process Column	Pay (p)

FIGURE 3.1.
Hats, Inc., pay structure.

2. What algebraic rule models the pay structure for Hats, Inc.?

3. Enter your rule into your graphing calculator. Set the table to begin at 0 and increase by 10. Compare the table generated by the calculator to the table you created for Hats, Inc. Do the tables have the same values? What does this tell you about your algebraic rule?

   If the tables do not have the same values, adjust them to match.

4. Use the description of the pay structure from Caps, Inc., to complete a table like **Figure 3.2.** Then use the process column to develop an algebraic rule for this pay structure.

Number of Caps Completed During the Month (c)	Process Column	Pay (p)

FIGURE 3.2.
Caps, Inc., pay structure.

5. What algebraic rule models the pay structure for Caps, Inc.?

6. Enter your rule into your graphing calculator. Set the table to begin at 0 and increase by 10. Compare the table generated by the calculator to the table you created for Caps, Inc. Do the tables have the same values? What does this tell you about your algebraic rule?

7. Look at the table values for both companies. What patterns do you observe?

8. If an employee makes 97 hats in one month, how much will she earn under each pay structure?

9. If an employee makes 103 hats in one month, how much will she earn under each pay structure?

SOLVE THE SYSTEM OF EQUATIONS USING A TABLE

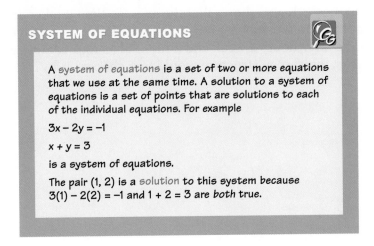

SYSTEM OF EQUATIONS

A system of equations is a set of two or more equations that we use at the same time. A solution to a system of equations is a set of points that are solutions to each of the individual equations. For example

$3x - 2y = -1$

$x + y = 3$

is a system of equations.

The pair (1, 2) is a solution to this system because $3(1) - 2(2) = -1$ and $1 + 2 = 3$ are both true.

10. Use the table feature on your calculator to solve the system of equations. In other words, for what number of hats will two employees, who are on different pay structures, make the same amount of money?

11. As a businessperson, you want to keep the cost of doing business as low as possible. So, which pay structure is best for you? Why?

12. If you were an employee, which pay structure would you prefer? Why?

13. If it takes somewhere between one and three hours for an employee to complete one hat, what is a reasonable domain for this situation? Why?

14. What is a reasonable range for this situation? Why?

15. Use your calculator. Find an appropriate viewing window then graph both functions on the same grid. Sketch your graph and record your window.

SOLVE THE SYSTEM OF EQUATIONS GRAPHICALLY

16. What are the coordinates of the point where the lines intersect? What does this point tell us?

17. If you sell hats for $20 each, which pay structure would you select? Why?

18. If you sell hats for $10 each, which pay structure would you choose? Why?

19. Your research also shows that the average time it takes an employee to finish one hat is two hours. If every employee works forty hours a week and a month has four complete weeks, which pay structure would you choose? Why?

Dolly Wagner plans to open a T-shirt shop in Galveston, Texas. She plans to sell each of her T-shirts for the same low price of $15.00. Use a table like **Figure 3.3** to find her income from T-shirt sales. Then use the process column to write an algebraic rule to represent her total income (y) in terms of the number of T-shirts sold (x).

Number of T-Shirts Sold (x)	Process Column	Total Income (y)
10		
20		
30		
50		
100		
x		

FIGURE 3.3.
Income from T-shirt sales.

1. Write the algebraic rule that models the amount of income earned as a function of the number of T-shirts sold.

2. How much income will Dolly earn for selling 200 T-shirts?

3. How many T-shirts would she have to sell to earn $1050?

As a businesswoman, Dolly knows that it takes money to make money. She needs $525 to get her business started. She also knows that each T-shirt costs $8 to produce. Use a table like **Figure 3.4** to find the cost of producing the T-shirts. Then use the process column to write an algebraic rule to represent her total cost (y) in terms of the number of T-shirts produced (x).

Number of T-Shirts Produced	Process Column	Total Cost
10		
20		
30		
50		
100		
x		

FIGURE 3.4.
Cost of producing T-shirts.

4. Write the algebraic rule that models the cost of producing T-shirts as a function of the number of T-shirts produced.

5. What is the total cost of producing 60 T-shirts?

6. Has Dolly made a profit yet if she has only produced 60 T-shirts and sold them all? Why or why not?

7. Use your graphing calculator to determine an appropriate viewing window, then graph both the income and the cost functions on the same grid. Sketch your graph and record your window.

CHECK THIS!

Remember here an ordered pair is made up of the variables x and y, or the input and output variables. We write the ordered pair as (x, y).

8. What are the coordinates of the point of intersection of the two lines? Write this point as an ordered pair.

9. Use words to describe the meaning of this point in terms of Dolly's business.

SECTION 3.3

Do the Combo!

In the last section you made decisions about employee pay structures by solving systems of equations using tables and graphs. Solving systems of equations is important because it helps make decisions when there are several things to think about.

Let's take another look at the pay structure equations for Hats, Inc., and Caps, Inc.

1. In the algebraic rule for Hats, Inc., $1000 + 5c = p$, what does 1000 represent? What does 5 represent?

2. In the algebraic rule for Caps, Inc., $500 + 10c = p$, what does 500 represent? What does 10 represent?

3. How did you use your table to solve (find the number of hats and the pay where both pay structures are the same) the system of equations $1000 + 5c = p$ *and* $500 + 10c = p$?

4. How did you decide on a reasonable domain and range for this situation?

5. How did you use your graph to solve (determine the number of hats and the pay where both pay structures are the same) the system of equations $1000 + 5c = p$ and $500 + 10c = p$?

There are two other methods to consider when solving systems of equations: linear combination and substitution.

LINEAR COMBINATION METHOD

A system of equations can be solved by the **linear combination** method. This method uses the following steps to eliminate or *cancel out* one of the variables by adding the equations together.

Write the equations one above the other.

$$500 + 10h = p$$
$$1000 + 5h = p$$

| Decide what to multiply by, so that when the equations are added one variable is eliminated. | $500 + 10h = p$
$-2[1000 + 5h = p]$ |

| Multiply every term in the equation by that number. | $500 + 10h = p$
$-2000 + (-10h) = -2p$ |

| Add the equations and solve for the remaining variable. | $-1500 = -p$
$1500 = p$ |

| Substitute that value into one of the original equations and solve. | $500 + 10h = p$
$500 + 10h = 1500$
$-500 + 500 + 10h = 1500 - 500$
$10h = 1000$
$\dfrac{10h}{10} = \dfrac{1000}{10}$
$h = 100$ |

| Use words to describe the meaning of your answer. | The solution to this system of equations is (100, 1500). This means if an employee completes 100 hats in a month, he or she will make $1500 on either pay structure. |

SUBSTITUTION METHOD

A system of equations can be solved by the substitution method. This method uses the following steps to isolate one of the variables in one equation and then substitute into the other equation.

| Isolate one variable. (Note: In the pay structure system of equations, p is already by itself on one side of the equal sign.) | $500 + 10h = p$ |

| Substitute the expression that is equivalent to that variable into the other equation. | $1000 + 5h = 500 + 10h$ |

Solve the equation.

$$1000 + 5h = 500 + 10h$$
$$1000 + 5h - 5h = 500 + 10h - 5h$$
$$1000 = 500 + 5h$$
$$-500 + 1000 = 500 + 5h - 500$$
$$500 = 5h$$
$$\frac{500}{5} = \frac{5h}{5}$$
$$100 = h$$

Substitute that value into one of the original equations and solve.

$$1000 + 5h = p$$
$$1000 + 5(100) = p$$
$$1500 = p$$

Use words to describe the meaning of your answer.

The solution to this system of equations is (100, 1500). This means if an employee completes 100 hats in a month he or she will make $1500 on either pay structure.

After you have made decisions and hired workers, one of your workers comes to you with a question. She has a box containing 144 silver stars. Her job is to complete each hat by attaching 2 stars to every red hat and 6 stars to every green hat. She must complete 50 hats and use all of the stars. She wants to know how many of each color hat she will need to complete.

6. Write a system of equations to determine the solution to this problem.

7. Solve the system using a table.

8. Solve the system graphically.

9. Solve the system using linear combination.

10. Solve the system using substitution.

Assignment

For each of these problems, define the variables and write a system of two equations. Solve the system using a table, a graph, and an algebraic method (linear combination or substitution). Then use words to explain the meaning of your solution in the context of the problem.

1. Jed's market has a total of 26 employees. Each full-time employee makes $275/week, and each part-time employee makes $140/week. Jed pays a total of $5935 in wages each week. How many full-time and part-time employees does he have?

2. Tony has $100 in the bank and is spending it at a rate of $3 each day. Kai has $20 in the bank and is adding to it by saving $5 each day. How many days will it take for the two boys to have the same amount of money?

3. A wildlife manager has rabbits and pheasants in a cage. All together there are 35 heads and 94 feet in the cage. Every rabbit has 4 feet and every pheasant has 2 feet. How many rabbits and how many pheasants are in the cage?

4. A carnival booth has small stuffed bears and large stuffed bears that it uses for prizes. Each small bear is worth $2.50 and each large bear is worth $5. If the booth contains a total of 200 bears with a total value of $625, how many bears are there of each size?

5. An order of MP3 players and CD players totals $4003 without any taxes or other charges. The cost of each MP3 player is $194.50 and the cost of each CD player is $159.50. The shipment contains a total of 24 players, how many of each is in the order?

Buyer Beware!

Now that you've explored how to pay your employees, you have to decide where to buy the materials for your hats.

There are several options. First, you can buy the materials from Cloth City, a retail store in your town where you have been buying them up to this point. Since it is local, Cloth City must charge 8.25% sales tax. Another option is ordering your materials from Discount Fabrics, a wholesaler in another state. By doing so, you avoid paying sales tax. However, the wholesaler charges a processing fee of $13 per order, and 5% of the cost of materials for shipping and handling.

You must decide which option is the most cost-effective way to purchase your materials.

1. Define your variables.

2. Write a system of equations in terms of x and y.

3. Solve the system using a table, a graph, and an algebraic method of your choice.

4. What is the intersection point, or solution to the system? What does it mean in this context?

5. Which option is the most cost-effective way to buy your materials?

6. While shopping on the Internet, you discover Textile City, a textile factory in another state that charges a $37 processing fee, but only 1% of the cost of materials for shipping and handling. In light of this new option, where should you buy your materials?

   To answer this question, write an equation in terms of x and y for the Textile City offer.

7. Use a graph, a table, and the algebraic method of your choice to compare Cloth City with Textile City. Between these two, who has the better offer?

8. Use a graph, table, and the algebraic method of your choice to compare Discount Fabrics with Textile City. Who has the better offer and why?

9. If you plan to order $300 worth of materials, who has the best offer?

10. If you plan to order $450 worth of materials, who has the best offer?

11. If you plan to order $550 worth of materials, who has the best offer?

12. If you plan to order $700 worth of materials, who has the best offer?

13. What are the intersection points where the companies with the best offer change?

14. Which of the three options is the most cost-effective way to purchase your materials?

Assignment

For each of these situations, define variables and write a system of two equations. Solve the system using a table, a graph, and an algebraic method (linear combination or substitution). Then use words to explain the meaning of your solution in the context of the problem. Also state how the solution could be used to help decide which option to select in each situation.

1. The Zippy Photo Store charges $1.60 to develop a roll of 35mm film plus $0.10 for each print, while SuperPIX charges $1.20 for development plus $0.15 for each print. For how many prints is the cost the same?

2. Carlo is trying to decide between two used cars. The sports car costs $5000, but the insurance is $2300 per year. The SUV costs $8000, and the insurance is $800 per year. After how many years would the cost of owning either car be the same?

3. Rita is moving to a new apartment. She can rent a truck to move for $59.88 a day plus $0.49 a mile. Strong Men Movers would charge $81 a day plus $0.38 a mile. It will take one day to move. How many miles will make the costs of the two options (renting a truck or hiring Strong Men Movers) the same?

4. Flash Rent-a-Car rents a compact car for $24.95 per day plus $0.23 per mile. King Car rents the same car for $22.95 per day plus $0.31 per mile. For a one-day rental, how many miles would make the cost of renting the car from either company the same?

What About Those Bills?

Let's apply what you have learned thus far about solving systems of linear equations to a new situation.

1. Sometimes a businessperson needs to find creative ways to sell merchandise quickly.

   You just realized that in two days, you must pay a bill for supplies in the amount of $12,750. Your bank account is empty but you have 1000 hats in stock. You decide to have a two-day tent sale. You sell plain hats for $10 and logo hats for $12. At the end of the first day you sell a total of 500 hats and collect $5750.

   You realize that if you sell the remaining 500 hats at the same prices you will not make enough money. Thus, for the second day, you decide to sell plain hats for $11 and logo hats for $15. Assuming that you sell exactly the same number of each hat as the day before, will you make enough money to pay your bill? Justify your answer.

SUMMARY

In these sections we used the solutions to systems of linear equations to make a variety of business decisions. These decisions involve understanding costs and pay scales, choosing suppliers, and developing pricing.

Cracking the Code

Guarding plans and ideas is important in business. If another company discovers what you plan to do, they can steal your ideas. People try to create codes to hide information. Baseball coaches have elaborate methods for giving signs to keep the opposing team from knowing their plans.

Military leaders also need to keep their plans secret. During World War II, the Germans developed a machine called Enigma that coded messages. The ability of the Allies to break this code contributed to their winning the war. Navajo code talkers contributed to the success of the armed forces in the Pacific by translating messages into the Navajo language, which was impossible for the Japanese to crack. One of the first military leaders to use coded messages was Julius Caesar. His code involved shifting the letters of the Roman alphabet. He shifted the letters three spaces forward as shown in **Figure 3.5** ($x + 3$). Each letter is assigned a number based on its position in the alphabet. So, if you add 3 to the position number of a letter you will get the position number of the coded letter. X, Y and Z (the 24th, 25th and 26th) loop around to the beginning of the alphabet, so these letters are coded A, B, and C (1, 2, and 3).

CHECK THIS!

A shift transformation moves every point the same distance in the same direction.

x	1	2	3	4	5	6	7	8	9	10	11	12	13	14	15	16	17	18	19	20	21	22	23	24	25	26
Original Letter	A	B	C	D	E	F	G	H	I	J	K	L	M	N	O	P	Q	R	S	T	U	V	W	X	Y	Z
Coded Letter	D	E	F	G	H	I	J	K	L	M	N	O	P	Q	R	S	T	U	V	W	X	Y	Z	A	B	C
x + 3	4	5	6	7	8	9	10	11	12	13	14	15	16	17	18	19	20	21	22	23	24	25	26	1	2	3

FIGURE 3.5. An x + 3 shift code.

The following quotation from Julius Caesar,

"As a rule men worry more about what they can't see than about what they can,"

would be coded as,

"Dv d uxoh phq zruub pruh derxw zkdw wkhb fdq'w vhh wkdq derxw zkdw wkhb fdq."

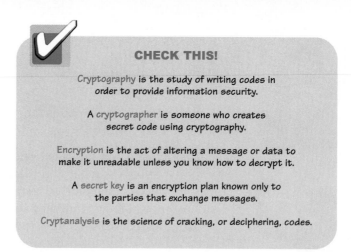

Since it is difficult to keep a key secret, you can make a shift transformation more secure by changing the number of spaces or the direction in which the letters are shifted, giving the secret key only to the person to whom you are sending the message.

In Dan Brown's book *Digital Fortress*, the two main characters send notes to each other using a variation of the Caesar code. They use a shift transformation, moving the letters one space forward, $(x + 1)$. Since both of them know the secret key, they decode each other's messages without anyone else knowing what they are saying.

I LOVE YOU is written as J MPWF ZPV.

You can make your own shift cipher device (like the one shown in **Figure 3.6**) by inserting a sliding slip of paper into a "mask" upon which the letters of the alphabet have been written. You can slide the slip of paper back and forth for different shift transformations.

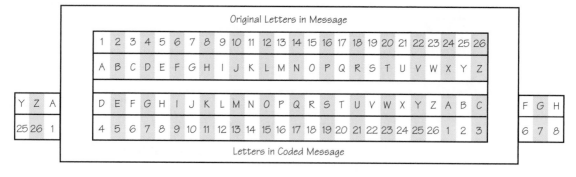

FIGURE 3.6.
A shift cipher device like this one is sometimes called a St. Cyr Strip.

Assignment

1. Caesar might have used the shift three spaces forward ($x + 3$) code to send this secret message to one of his generals. Translate the message back into its original form.

   Dwwdfn Yhuflqjhwrula dw gdzq.

2. Use your St. Cyr Cipher Strip to decode this coded message. The message was encrypted with a shift transformation of $x - 10$, ten places backward.

   Dusuiiyjo yi jxu cejxuh ev ydludjyed.

3. Use the St. Cyr Cipher Strip to encode this message with an $x + 5$, five places forward, shift transformation.

   HAVE INVENTED SELF CLEANING
   FOOTBALL UNIFORM.

   MEET ME AT NOON TO DISCUSS SALES TO TEXANS.

4. Write a school-appropriate note to someone in your group using the shift one forward ($x + 1$) code. Trade notes with someone in your group and decode their message.

5. Interview at least one adult and some of your friends; ask them why encryption techniques are important in today's world. Write down their answers.

The Matrix Code

Shift codes have a big problem: They are easy to crack. Players on the Wheel of Fortune game show guess letters based on how common they are. Code breakers work in much the same way. For example, if they notice that *h* has a high **frequency** (the number of times the letter is used) in a coded message, they might guess that it represents a letter such as *e* or *a* that is common in English. Once code breakers know the shift of one letter, they can decode a message quickly.

So it makes sense to find a way to code that is harder to crack. A matrix can be used to make code cracking more difficult.

CHECK THIS!

A matrix is a rectangular arrangement of numbers. The numbers are arranged in rows and columns and appear to make a rectangular shape. A row is a horizontal group of numbers. A column is a vertical group of numbers.

1. Write the message LOOK BEFORE YOU LEAP as a matrix using the pattern A = 1, B = 2, C = 3, and so on. Use 0 for the blanks between words.

   Number form of message:

   Matrix A =

You are scrambling your message by multiplying by the coding matrix $C = \begin{bmatrix} 2 & 5 \\ 1 & 3 \end{bmatrix}$. The coding matrix has 2 rows and 2 columns.

You are going to use the matrix feature of a graphing calculator to multiply these two matrices. Your teacher will explain to you how to find the product of matrices A and C by entering the matrices into your graphing calculator.

2. Find the product of the two matrices and write it in the space below.

Scrambled message:

Matrix B =

Note that the first time O is coded it is a 30 and the next time it is 155. This means that code breaker needs more than the frequency of a letter to break the code.

3. Does the order in which you multiply the message and coding matrix to get matrix B matter and why?

The message in matrix B was coded with matrix C. That is, if A is a matrix containing the original message, the coding was done with this equation:

$$[C] \cdot [A] = [B]$$

Scrambled message:

$$\text{Matrix B} = \begin{bmatrix} 24 & 49 & 15 & 40 & 69 & 4 & 43 & 9 & 70 & 8 \\ 43 & 76 & 25 & 62 & 113 & 6 & 71 & 14 & 115 & 16 \end{bmatrix}$$

4. How do you think you would find matrix A, the original message, when you know matrices B and C?

5. Matrix equations are solved in much the same way as algebraic equations. How do you use multiplication to solve the equation $2x = 7$?

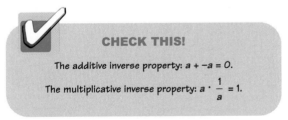

CHECK THIS!

The additive inverse property: $a + -a = 0$.

The multiplicative inverse property: $a \cdot \dfrac{1}{a} = 1$.

You can use multiplicative inverses of matrices to solve matrix equations on a graphing calculator. Your teacher will show you how to do this.

6. Will it matter in which order we multiply the inverse matrix by the coded matrix?

7. Write your product matrix.

Number form of the message:

Matrix A =

8. Write your decoded message below.

Assignment

Blank	0
A	1
B	2
C	3
D	4
E	5
F	6
G	7
H	8
I	9
J	10
K	11
L	12
M	13
N	14
O	15
P	16
Q	17
R	18
S	19
T	20
U	21
V	22
W	23
X	24
Y	25
Z	26

Create and code a secret message. Record your work on Handout 3.2. Follow the instructions below.

1. Your message must be 20 characters (letters or blanks) or less. Write your message in a table like the one below. Place one letter or a blank in each square.

2. Assign each letter in your message a number from the chart at the right. Write the number message in a table like the one below.

3. Create a 2 × 2 coding matrix. Write your coding matrix on the paper titled Secret Message.

Coding matrix: $\begin{bmatrix} a & b \\ c & d \end{bmatrix}$

To avoid problems with fractions, be sure that $ad - bc$ is equal to 1.

4. Multiply your coding matrix times your message matrix. Record the coded message on the Secret Message paper.

5. Exchange your secret message with another student. They will decode your message while you decode theirs. Write the matrix equation you used to decode the message.

6. Translate the message matrix into letters and write it on the Secret Message paper.

What's in a Matrix?

Tables are a common way to show data. **Figure 3.7** shows a table of the average times in minutes for the national physical fitness test endurance run in 1989 (U.S. Department of Education, 2001). The runs were 3/4 mile for ages 10 and 11 and 1 mile for ages 12 through 17:

	Boys	Girls
10–11-year-olds	7.3	8.0
12–13-year-olds	9.1	10.5
14–17-year-olds	8.6	10.7

FIGURE 3.7.
1989 times (in minutes).

When you remove the labels, in this case the boys and the girls and their ages, you have a group of numbers called a **matrix**. The plural of this word is **matrices**. Matrices are often represented by a capital letter. Here is the matrix for the table above:

$$T = \begin{bmatrix} 7.3 & 8.0 \\ 9.1 & 10.5 \\ 8.6 & 10.7 \end{bmatrix}$$

Here T is the label for the matrix, *times*. Matrices are often described by their size or their **matrix dimensions**, the number of rows and columns. This is a 3 × 2 (read "3 by 2") matrix because it has 3 rows and 2 columns. Each entry in a matrix is called an **element or a component**. The symbol $t_{2,1}$ points to the element in the second row, first column. In this table that is 9.1.

Corresponding elements are in the same position (row and column) in each matrix.

	Boys	Girls
10–11-year-olds	7.5	8.2
12–13-year-olds	9.4	10.4
14–17-year-olds	8.8	10.9

FIGURE 3.8.
2000 times (in minutes).

If you take the matrices from the 1989 table and the 2000 table
you get:

$$T = \begin{bmatrix} 7.3 & 8.0 \\ 9.1 & 10.5 \\ 8.6 & 10.7 \end{bmatrix} \quad U = \begin{bmatrix} 7.5 & 8.2 \\ 9.4 & 10.4 \\ 8.8 & 10.9 \end{bmatrix}$$

Matrix U is the student times in 2000.

Matrices can be labor saving tools when you have a graphing
calculator or a spreadsheet program. Suppose you want to
compare how fast students ran in 1989 to how fast they ran in
2000 (Figure 3.7 to Figure 3.8). This involves performing a **matrix
operation.** A matrix operation allows you to perform a process
such as multiplication, addition, or subtraction with one or more
matrices.

If you want to see how students' times have changed you could
do **matrix subtraction.** Matrix subtraction allows us to subtract
elements of one matrix from those of another. To subtract two
matrices, subtract the corresponding elements.

$$T = \begin{bmatrix} 7.3 & 8.0 \\ 9.1 & 10.5 \\ 8.6 & 10.7 \end{bmatrix} \quad U = \begin{bmatrix} 7.5 & 8.2 \\ 9.4 & 10.4 \\ 8.8 & 10.9 \end{bmatrix} \quad D = T - U = \begin{bmatrix} -0.2 & -0.2 \\ -0.3 & 0.1 \\ -0.2 & -0.2 \end{bmatrix}$$

Values from matrix D, or *difference,* can then be put in a table:

	Boys	Girls
10–11-year-olds	–0.2	–0.2
12–13-year-olds	–0.3	0.1
14–17-year-olds	–0.2	–0.2

FIGURE 3.9.
Times 1989 vs. 2000 (in minutes).

This table allows you to see that student times, except for 12- and
13-year-old girls, are slower than in 1989.

In this case it made sense to use matrix subtraction. Sometimes it
makes sense to add elements of matrices. This is called **matrix
addition.**

For example:

$$\begin{bmatrix} 2 & 3 \\ 1 & 0 \end{bmatrix} + \begin{bmatrix} -3 & 5 \\ 7 & -3 \end{bmatrix} = \begin{bmatrix} -1 & 8 \\ 8 & -3 \end{bmatrix}$$

For matrix addition and subtraction, both matrices must be the same size. This means that they have same number of rows and columns.

There are other operations you can perform on matrices. You can multiply a matrix by a number called a **scalar**. To perform **scalar multiplication**, multiply every element in the matrix by the scalar.

EXAMPLE:

$$-5\begin{bmatrix} 3 & 0 \\ -10 & 8 \end{bmatrix} = \begin{bmatrix} -15 & 0 \\ 50 & -40 \end{bmatrix}$$

You can use the definition of equal matrices to solve **matrix equations**. **Equal matrices** have the same dimensions and corresponding elements are equal. To solve for x and y, write the following equations and solve each equation.

EXAMPLE:

Solve for x and y.

$$\begin{bmatrix} 2x - 5 & 4 \\ 3 & 3y + 12 \end{bmatrix} = \begin{bmatrix} 25 & 4 \\ 3 & y + 18 \end{bmatrix}$$

$$2x - 5 = 25 \qquad\qquad\qquad 3y + 12 = y + 18$$
$$2x - 5 + 5 = 25 + 5 \qquad\qquad 3y - y + 12 = y - y + 18$$
$$2x = 30 \qquad\qquad\qquad 2y + 12 = 18$$
$$\frac{2x}{2} = \frac{30}{2} \qquad\qquad\qquad 2y + 12 - 12 = 18 - 12$$
$$\boxed{x = 15} \qquad\qquad\qquad 2y = 6$$
$$\boxed{y = 3}$$

To multiply two matrices, the number of columns in the first matrix must be the same as the number of rows in the second matrix.

Each element in row 1 of the first matrix is multiplied by an element of column 1 in the second matrix, and the products are added together. The result becomes the row 1, column 1 element in the answer matrix. The process is continued until every row of the first matrix is paired with a column of the second matrix.

EXAMPLE:

Multiply.

$$\begin{bmatrix} 2 & -1 & 5 \\ 7 & 3 & -2 \end{bmatrix} \cdot \begin{bmatrix} 0 & -3 \\ -4 & 2 \\ 5 & 9 \end{bmatrix}$$

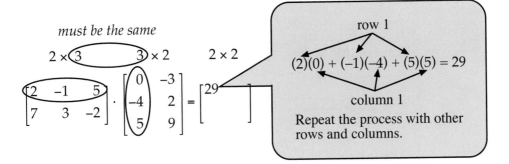

In general, when you pair row m of the first matrix with column n of the second matrix in this way, you get the row m, column n element of the product matrix.

Let's see if you get the same answer when you multiply matrix P by matrix Q as you do when you reverse the order of multiplication.

$$P = \begin{bmatrix} 2 & 3 \\ 0 & -1 \end{bmatrix} \qquad Q = \begin{bmatrix} -3 & 1 \\ 2 & 0 \end{bmatrix}$$

$$P \cdot Q = \begin{bmatrix} 2 & 3 \\ 0 & -1 \end{bmatrix} \cdot \begin{bmatrix} -3 & 1 \\ 2 & 0 \end{bmatrix} = \begin{bmatrix} 2(-3)+3(2) & 2(1)+3(0) \\ 0(-3)+(-1)(2) & 0(1)+(-1)(0) \end{bmatrix} = \begin{bmatrix} 0 & 2 \\ -2 & 0 \end{bmatrix}$$

$$Q \cdot P = \begin{bmatrix} -3 & 1 \\ 2 & 0 \end{bmatrix} \cdot \begin{bmatrix} 2 & 3 \\ 0 & -1 \end{bmatrix} = \begin{bmatrix} -3(2)+1(0) & -3(3)+1(-1) \\ 2(2)+0(0) & 2(3)+0(-1) \end{bmatrix} = \begin{bmatrix} -6 & -10 \\ 4 & 6 \end{bmatrix}$$

Multiply $P \cdot Q$ and $Q \cdot P$ with your calculator to check your answers.

Does order matter when multiplying matrices?

Yes. $P \cdot Q = \begin{bmatrix} 0 & 2 \\ -2 & 0 \end{bmatrix}$, but $Q \cdot P = \begin{bmatrix} -6 & -10 \\ 4 & 6 \end{bmatrix}$, so $P \cdot Q \neq Q \cdot P$.

Therefore, multiplication of matrices is not commutative.

Assignment

Fill in each blank with an appropriate word or number.

1. A _____ is a rectangular arrangement of numbers. It is arranged in _____ and _____. The plural form of this is _____. They are labeled with a _____ and are written within brackets.

2. The _____ of a matrix define the size of the matrix. The number of rows is always stated _____.

3. The matrix $A = \begin{bmatrix} 27 & -3 & 14 \\ 3.8 & -9 & 45 \end{bmatrix}$ is referred to as a ___ × ___ matrix.

4. Each number in a matrix is called an _____.

5. Two matrices are _____ matrices if and only if they have the same dimensions and their corresponding elements are equal. _____ are the elements in the same position (row and column) of each matrix.

6. You can multiply a matrix by a number called a _____ by multiplying every element in the matrix by it.

7. You can use the definition of equal matrices to solve a _____.

8. If two matrices have the _____ dimensions, you can _____ or _____ the corresponding elements to find the sum or difference of the matrices.

9. To multiply two matrices, the number of _____ in the first matrix must be the same as the number of _____ in the second matrix.

Use matrices A, B, C, D, or E for Questions 10–16.

$$A = \begin{bmatrix} 5 & -2 \\ 1 & 7 \end{bmatrix} \qquad B = \begin{bmatrix} 12 & 4 \\ -3 & -1 \end{bmatrix} \qquad C = \begin{bmatrix} 1 & 0 & 5 \\ 7 & 2 & 4 \end{bmatrix}$$

$$D = \begin{bmatrix} -3 & 9 & -1 \\ 5 & 0 & 12 \end{bmatrix} \qquad E = \begin{bmatrix} 0 & 1 & 2 \\ 3 & 4 & 5 \\ 6 & 7 & 8 \end{bmatrix}$$

10. What are the dimensions of matrix C?

11. Which matrix is a 3 × 3 matrix?

12. If matrix A is equal to $\begin{bmatrix} 5 & 3y - 11 \\ 2x + 9 & 7 \end{bmatrix}$, solve for x and y.

13. Multiply –3B.

14. Add C + D.

15. Simplify 5A – B.

16. Multiply A · D.

Using Matrices to Solve Systems of Equations

A square matrix with ones along the diagonal that goes from the upper left corner to the lower right corner and with zeros in every other position is called an **identity matrix**.

2×2 identity matrix:

$$I = \begin{bmatrix} 1 & 0 \\ 0 & 1 \end{bmatrix}$$

3×3 identity matrix:

$$I = \begin{bmatrix} 1 & 0 & 0 \\ 0 & 1 & 0 \\ 0 & 0 & 1 \end{bmatrix}$$

EXAMPLE:

What is the product of the identity matrix and matrix L?

$$L = \begin{bmatrix} -5 & 4 & 3.2 \\ -1 & 0 & 6 \end{bmatrix}$$

$$\begin{bmatrix} 1 & 0 \\ 0 & 1 \end{bmatrix} \cdot \begin{bmatrix} -5 & 4 & 3.2 \\ -1 & 0 & 6 \end{bmatrix} = \begin{bmatrix} & & \end{bmatrix}$$

If you multiply a matrix by the identity matrix, the product is the original matrix.

If the product of two matrices is the identity matrix then the matrices are __inverses__ of each other.

EXAMPLE:

$$A = \begin{bmatrix} -3 & 5 \\ 5 & -8 \end{bmatrix} \qquad A^{-1} = \begin{bmatrix} 8 & 5 \\ 5 & 3 \end{bmatrix}$$

$$A^{-1} \cdot A = I$$

$$\begin{bmatrix} 8 & 5 \\ 5 & 3 \end{bmatrix} \cdot \begin{bmatrix} -3 & 5 \\ 5 & -8 \end{bmatrix} = \begin{bmatrix} 8(-3)+5(5) & 8(5)+5(-8) \\ 5(-3)+3(5) & 5(5)+3(-8) \end{bmatrix} = \begin{bmatrix} 1 & 0 \\ 0 & 1 \end{bmatrix}$$

$$A \cdot A^{-1} = I$$

$$\begin{bmatrix} -3 & 5 \\ 5 & -8 \end{bmatrix} \cdot \begin{bmatrix} 8 & 5 \\ 5 & 3 \end{bmatrix} = \begin{bmatrix} -3(8)+5(5) & -3(5)+5(3) \\ 5(8)+(-8)(5) & 5(5)+(-8)(3) \end{bmatrix} = \begin{bmatrix} 1 & 0 \\ 0 & 1 \end{bmatrix}$$

If you do not know the inverse of a matrix, you can find it with a graphing calculator. This is especially helpful for matrices that are larger than 2 × 2. Your teacher will give you instructions on how to find the inverse matrix with the graphing calculator.

EXAMPLE:

Use a graphing calculator to find the inverse of matrix W.

$$W = \begin{bmatrix} 3 & 1 \\ 4 & 2 \end{bmatrix}$$

Check to be sure that the two matrices are inverses by multiplying them on the calculator.

EXAMPLE:

Use a graphing calculator to find the inverse of matrix F.

$$F = \begin{bmatrix} 4 & 3 & 0 \\ 5 & 2 & -3 \\ -3 & -1 & 2 \end{bmatrix}$$

Multiply $F \cdot F^{-1}$ (and $F^{-1} \cdot F$) to see if the product is the identity matrix.

You have already learned several methods for solving a system of equations: graphing, tables, substitution, and linear combinations. Inverse matrices are an additional method to help you find the solution.

EXAMPLE:

$$2x + y = 9$$
$$-2x - 3y = -7$$

If we look at the coefficients of x and y in the two equations above, we can represent them with a matrix called the **coefficient matrix**.

$$A = \begin{bmatrix} 2 & 1 \\ -2 & -3 \end{bmatrix}$$

The variables are represented with a different matrix called the **variable matrix**.

$$X = \begin{bmatrix} x \\ y \end{bmatrix}$$

Finally, the constants on the right side of each equation also have their own matrix called the **constant matrix**.

$$B = \begin{bmatrix} 9 \\ -7 \end{bmatrix}$$

You can rewrite the equations as the product of two matrices:

$$\begin{bmatrix} 2 & 1 \\ -2 & -3 \end{bmatrix}\begin{bmatrix} x \\ y \end{bmatrix} = \begin{bmatrix} 2x + y \\ -2x - 3y \end{bmatrix} \text{ and is equal to } \begin{bmatrix} 9 \\ -7 \end{bmatrix}.$$

So, the equations could be written as

$$\begin{bmatrix} 2 & 1 \\ -2 & -3 \end{bmatrix}\begin{bmatrix} x \\ y \end{bmatrix} = \begin{bmatrix} 9 \\ -7 \end{bmatrix}$$

$[A][X] = [B]$.

Multiply each side of the matrix equation by the inverse of matrix A.

$[A]^{-1}[A][X] = [A]^{-1}[B]$

Remember that matrix multiplication is not always commutative, so you must left-multiply on both sides of the matrix equation.

Don't forget that the product of a matrix and its inverse is the identity matrix, commonly labeled matrix I.

$[I][X] = [A]^{-1}[B]$

If you multiply a matrix by the identity matrix the product is the original matrix.

$[X] = [A]^{-1}[B]$

Let's do the same steps with the actual numbers in the matrix.

$$\begin{bmatrix} 2 & 1 \\ -2 & -3 \end{bmatrix}\begin{bmatrix} x \\ y \end{bmatrix} = \begin{bmatrix} 9 \\ -7 \end{bmatrix}$$

$$\begin{bmatrix} 2 & 1 \\ -2 & -3 \end{bmatrix}^{-1}\begin{bmatrix} 2 & 1 \\ -2 & -3 \end{bmatrix}\begin{bmatrix} x \\ y \end{bmatrix} = \begin{bmatrix} 2 & 1 \\ -2 & -3 \end{bmatrix}^{-1}\begin{bmatrix} 9 \\ 7 \end{bmatrix}$$

$$\begin{bmatrix} 1 & 0 \\ 0 & 1 \end{bmatrix}\begin{bmatrix} x \\ y \end{bmatrix} = \begin{bmatrix} 2 & 1 \\ -2 & -3 \end{bmatrix}^{-1}\begin{bmatrix} 9 \\ -7 \end{bmatrix}$$

$$\begin{bmatrix} x \\ y \end{bmatrix} = \begin{bmatrix} 5 \\ -1 \end{bmatrix}$$

So, the solution to the system of equations is $x = 5$ and $y = -1$.

Inverse matrices are especially useful when solving systems of equations with three variables.

EXAMPLE:

Solve the system of equations using inverse matrices on the graphing calculator.

$$2x - 3y + z = -7$$

$$x + 2y + 3z = 0$$

$$-3x - 5y + 9z = 1$$

$$\begin{bmatrix} x \\ y \\ z \end{bmatrix} = \begin{bmatrix} 2 & -3 & 1 \\ 1 & 2 & 3 \\ -3 & -5 & 9 \end{bmatrix}^{-1} \begin{bmatrix} -7 \\ 0 \\ 1 \end{bmatrix}$$

$$\begin{bmatrix} x \\ y \\ z \end{bmatrix} = \begin{bmatrix} -2 \\ 1 \\ 0 \end{bmatrix}$$

So the solution set is $x = -2$, $y = 1$, and $z = 0$.

EXAMPLE:

This month, Franco's checking account had a total of 24 checks written and ATM withdrawals. The bank charges $0.30 for each check and $1.50 for each ATM withdrawal. If his service charge was $12.00, how many of each type of activity did Franco carry out this month? Write a system of equations to represent this situation. Clearly define your variables.

This problem can be represented with the system:

$$x + y = 24$$

$$0.30x + 1.50y = 12$$

where x is the number of checks written and y is the number of ATM withdrawals.

Solve the system using inverse matrices.

$$\begin{bmatrix} x \\ y \end{bmatrix} = \begin{bmatrix} 1 & 1 \\ 0.30 & 1.50 \end{bmatrix}^{-1} \begin{bmatrix} 24 \\ 12 \end{bmatrix}$$

$$\begin{bmatrix} x \\ y \end{bmatrix} = \begin{bmatrix} 20 \\ 4 \end{bmatrix}$$

Franco wrote 20 checks and had 4 ATM withdrawals.

Assignment

1. Are these matrices inverses of each other? Why or why not?

   $$A = \begin{bmatrix} 3 & -5 \\ -3 & 6 \end{bmatrix}$$

   $$B = \begin{bmatrix} 2 & \dfrac{5}{3} \\ 1 & 1 \end{bmatrix}$$

2. Use a graphing calculator to find the inverse of this 2 × 2 matrix.

   $$C = \begin{bmatrix} -4 & -2 \\ 5 & 2 \end{bmatrix}$$

3. Use a graphing calculator to find the inverse of this matrix.

   $$D = \begin{bmatrix} 1 & 1 & 0 \\ -3 & -2 & 2 \\ 3 & 1 & -3 \end{bmatrix}$$

4. A softball team bought 24 jerseys and 15 hats for a total cost of $1713. Later, they bought 5 more jerseys and 12 more hats for a total cost of $490. If the cost per item was the same for each order, what was the cost of each jersey and each hat?

5. A collection of nickels, dimes, and quarters has a value of $3.40. There are twice as many dimes as quarters, and there are 32 coins in all. How many of each kind are there?

6. A lawn and garden factory makes rakes, shovels, and clippers at a monthly cost of $6850 for a total of 2150 items. It costs $2 to make a rake, $3 to make a shovel, and $4 to make a clipper. Wholesale, a rake sells for $3, a shovel sells for $4.50, and a clipper sells for $5.50. If the total income is $9825, how many of each item did the factory make?

7. A college student earned $177.50 last week working three part-time jobs. During the 30-hour workweek, she spent twice as many hours babysitting at $5 an hour as she did tutoring at $12 an hour. If she earned $4.50 an hour working in her grandmother's garden, how many hours did she spend at each job last week?

Opening Day

The promotions manager for the Texas Rangers is planning a special opening day giveaway. Each of the first 5000 fans will receive either a souvenir cap or a Rangers blanket. The manager knows that the caps cost $5 each and the blankets cost $12 each. What combination of caps and blankets can he purchase for a cost of $32,000?

After deciding how many caps and blankets to buy, the manager must also find the shipping costs. If the cost is $0.25 to ship each cap and $0.60 to ship each blanket, what will be the total cost for the opening day giveaway (including purchase price and shipping costs)?

Play Ball!

A LOOK AHEAD

In 1965 the world's first sports stadium with a domed roof was built. The Astrodome in Houston, Texas, was the home of the Houston Astros. It was originally built with glass windows as ceiling tiles. However, during the first game, outfielders discovered that the sunlight shining through the windows made it difficult to see fly balls. Thus, the glass was painted in order to let some light in yet reduce glare. Without direct sunshine, the grass died. To solve this problem, engineers created an artificial grass called Astroturf, and artificial turf was born.

After the construction of the Astrodome, other cities built domed stadiums such as the Metrodome in Minneapolis, Minnesota, or the Superdome in New Orleans, Louisiana. During the 1990s, the people of Houston decided to build a new baseball stadium in downtown Houston. The new stadium, now known as Minute Maid Park, has a retractable roof. When the weather is nice, the Astros can play baseball with an open roof. However, if it is hot or rainy, the roof can be closed, and baseball can be played indoors.

Your sporting goods company plans to sell merchandise at a local baseball stadium. In this section, you will examine information regarding prices and expenses and make decisions about selling your merchandise.

At the stadium souvenir stand, you will sell hats with the home team's logo for $12, and T-shirts with the home team's logo for $20. The stadium charges $500 to rent the space to sell items during a home game, and the team charges $1.50 per item with their logo in royalties. Each hat costs $7 in materials and labor to produce, and each T-shirt costs $9 in materials and labor to produce. There is only enough space in the stand to stock 200 total items. You want to know which combinations of hats and shirts will generate more income than expenses; that is, will create a profit.

1. Use a table like **Figure 3.10** to identify possible combinations of hats and shirts that satisfy the requirement to have no more than 200 items in stock.

Number of Hats	Number of Shirts	Total Number of Items

FIGURE 3.10.
Stadium sales table.

2. Neglecting the stadium stand rent, what is the net profit (difference between income and expenses) generated by the sale of one hat?

3. Neglecting the stadium stand rent, what is the net profit generated by the sale of one shirt?

4. Use your combinations of hats and T-shirts and a table like **Figure 3.11** to identify the total income and the total expenses for each combination.

Income Generated by Hats	Income Generated by Shirts	Total Income

FIGURE 3.11. Income and expenses table.

Expenses Generated by Hats	Expenses Generated by Shirts	Stand Rent	Total Expenses

5. How many possible combinations of hats and shirts can you bring to sell at the stadium that will satisfy the restrictions of the situation?

6. Which combination of hats and shirts is best? Justify your choice.

Not Everything is Created Equal

In the previous section we looked at the following problem.

Your sporting goods company plans to sell merchandise at a local baseball stadium. At the stadium souvenir stand, you will sell hats with the home team's logo for $12 and T-shirts with the home team's logo for $20. The stadium charges $500 to rent the space to sell items during a home game, and the team charges $1.50 per item with their logo in royalties. Each hat costs $7 in materials and labor to produce, and each T-shirt costs $9 in materials and labor to produce. There is only enough space in the stand to stock 200 total items. What possible combinations of hats and shirts will generate more income than expenses; that is, will create a profit?

You created a table of possible combinations of numbers of hats and T-shirts that satisfy this situation. Use your table and the context of the problem to answer these questions.

1. Let h represent the number of hats and s represent the number of T-shirts. Write an expression to represent the amount of income that the sale of hats (h) and T-shirts (s) will generate.

2. Write an expression to represent the amount of expenses that the sale of a number of hats (h) and a number of T-shirts (s) will generate.

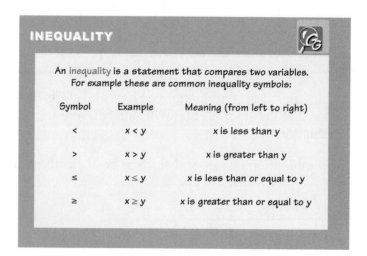

INEQUALITY

An inequality is a statement that compares two variables. For example these are common inequality symbols:

Symbol	Example	Meaning (from left to right)
$<$	$x < y$	x is less than y
$>$	$x > y$	x is greater than y
$\leq$	$x \leq y$	x is less than or equal to y
$\geq$	$x \geq y$	x is greater than or equal to y

3. Use the words **income** and **expenses** to write an inequality that describes the conditions under which a profit will be generated (income will exceed expenses).

4. Substitute your expressions from Questions 1 and 2 into your inequality from Question 3.

5. Solve this inequality for s in terms of h.

6. For this situation and ordered pairs (h, s), which quadrant of the coordinate plane should we consider? Why?

7. Use your graphing calculator to graph this inequality. Describe your window, and sketch your results.

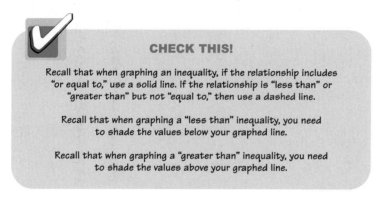

CHECK THIS!

Recall that when graphing an inequality, if the relationship includes "or equal to," use a solid line. If the relationship is "less than" or "greater than" but not "equal to," then use a dashed line.

Recall that when graphing a "less than" inequality, you need to shade the values below your graphed line.

Recall that when graphing a "greater than" inequality, you need to shade the values above your graphed line.

8. Write an inequality representing the possible combinations of hats (h) and T-shirts (s) that you can store in your souvenir stand.

Just like systems of equations, you can work with two inequalities at the same time.

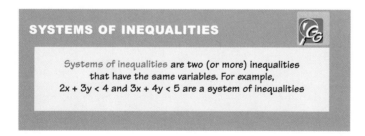

SYSTEMS OF INEQUALITIES

Systems of inequalities are two (or more) inequalities that have the same variables. For example, $2x + 3y < 4$ and $3x + 4y < 5$ are a system of inequalities

9. Use your graphing calculator to graph the number of T-shirts vs. the number of hats for the inequality from Question 8 in the same window as the inequality from Questions 4 and 5. Record your window and sketch your results.

10. Describe the graphs of both inequalities.

11. In this situation, what do the x-coordinate (h-coordinate) and y-coordinate (s-coordinate) represent?

12. Choose a point that lies above and to the right of both lines. Substitute the h-value and s-value for this point into both inequalities. Which inequalities does the point satisfy?

13. Choose a point that lies in the region between both lines. Substitute the h-value and s-value for this point into both inequalities. Which inequalities does the point satisfy?

14. Choose a point that lies below and to the left of both lines. Substitute the h-value and s-value for this point into both inequalities. Which inequalities does the point satisfy?

15. Choose two more points from the region where one point satisfies both inequalities. Do these points also satisfy both inequalities? Justify your answer.

16. Make a conjecture about the region of the coordinate plane where both inequalities are true.

Assignment

1. Graph the following system of inequalities.

$$\begin{cases} y \le 2x - 4 \\ 2x + 3y < 12 \end{cases}$$

   a) Is the point (4, –3) a part of the solution to the system of inequalities? Why or why not?

   b) Is the point (7, 5) a part of the solution to the system of inequalities? Why or why not?

   c) Is the point (3, 2) a part of the solution to the system of inequalities? Why or why not?

   d) Is the origin a part of the solution to the system of inequalities? Why or why not?

2. Graph the following system of inequalities.

$$\begin{cases} y \le 2x - 1 \\ x - 3y \ge 18 \end{cases}$$

   a) At what point do the lines intersect? Is this point a part of the solution to the system of inequalities? Why or why not?

   b) Describe the solution set to the system of inequalities in terms of x and y.

3. To meet shipping requirements at the local U Pack It We Ship It store, the perimeter of the base of a rectangular box must be less than 30 inches. The length must be at least the width of the base of the box.

   a) Write two inequalities that describe this situation.

   b) What are a reasonable domain and range? Assume width is the independent variable and length is the dependent variable.

c) What are some possible values for length and width that satisfy the requirements of the shipping store? Use a table like **Figure 3.12** to organize your data.

Length	Width	Perimeter

FIGURE 3.12.
Box measurement table.

d) Graph the inequalities relating length and width. Use width on the *x*-axis and length on the *y*-axis.

e) Are the lengths and widths from your table in the solution set according to your graph? How do you know?

4. The local petting zoo charges $1 for children's admission and $3 for adults. An elementary school class is going on a field trip to the zoo and has $50 budgeted for admission. The small bus they will take holds up to 30 people. What possible combinations of children and adults could attend the field trip? To find out, answer these questions.

a) Define your variables.

Let $x =$ _____

Let $y =$ _____

b) Write two inequalities representing this situation.

c) What are a reasonable domain and range?

d) Graph the two inequalities.

5. Use your graph to fill in each blank with a value so that the combination can attend.

a) 15 children, _____ adults

b) 3 children, _____ adults

c) ____ children, 2 adults

d) _____ children, 5 adults

Solutions to Systems of Inequalities

In the previous section you solved a system of inequalities graphically by identifying a region of the coordinate plane where the points describe pairs that make both of the inequalities true.

Now let's look at some important ideas about systems of inequalities and compare them to what we know about solving systems of equations.

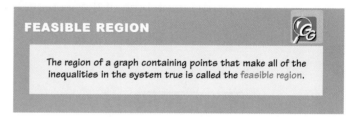

FEASIBLE REGION

The region of a graph containing points that make all of the inequalities in the system true is called the *feasible region*.

1. When you made a table of possible data points that satisfied all of the conditions of your problem, how did you decide which points to use?

2. How did you develop your expressions for income and expenses?

3. When you wrote your inequalities, how did you decide which relation to use: less than, greater than, or equal to?

4. When you graphed your inequalities, how did you decide which region to shade?

5. How did you decide which region satisfied both inequalities?

6. Are the lines in your graph part of the feasible region? Explain.

7. What does the point of intersection of the lines in the graph represent? Is it a part of the feasible region?

8. How does the solution of a system of inequalities compare to the solution of a system of equations?

9. Suppose a customer orders hats and T-shirts. They want no more than 60 items and no more than twice as many T-shirts as hats.

a) What are a reasonable domain and range for this situation? Justify your answer.

b) Write an inequality in terms of h, the number of hats, and s, the number of T-shirts, to describe the condition "they want no more than 60 items."

c) Write an inequality in terms of h and s to describe the condition "no more than twice as many T-shirts as hats."

d) Graph both of these inequalities together on your graphing calculator. Choose an appropriate window and sketch your results. Identify the solution to the system.

SUMMARY

Solving Systems of Inequalities

Systems of inequalities have more than one solution. So a good way to represent the solution is with a graph. To create a graph for the solution to a system of inequalities,

* identify an appropriate domain and range for the situation. Use these to set up your independent and dependent variable axes;

* graph each inequality. Use a solid line if the inequality uses the symbol $\geq$, "greater than or equal to," or the symbol $\leq$, "less than or equal to." Use a broken line if the inequality uses the symbol $>$, "greater than," or $<$, "less than";

* shade the appropriate region for each inequality. Shade the y-values that are greater than (above) the line if the inequality uses "greater than." Shade the y-values that are less than (below) the line if the inequality uses "less than"; and

* identify the feasible region. The feasible region is the part of the graph where the shaded regions of the individual inequalities overlap.

In this section you examined systems of linear inequalities.

1. You looked at tabular and graphical representations of the solution to systems.

2. You compared the solution for a system of equations to the solution for a system of inequalities.

1. A men's medium baseball jersey requires $2\frac{1}{4}$ yards of fabric. A men's large baseball jersey requires $3\frac{3}{4}$ yards of fabric. You have 75 yards of fabric and want to make no more than 30 jerseys. How many of each type of baseball jersey can you make?

   a) Define your variables. Identify the independent and dependent variables.

   b) What are a reasonable domain and range for this situation? Justify your answer.

   c) Write two inequalities to describe the situation.

   d) Solve each inequality for your dependent variable in terms of your independent variable.

   e) Graph both of these inequalities together. Choose an appropriate window and sketch your results. Identify the solution to the system.

2. A loaf of rye bread requires 3 cups of flour and $\frac{1}{2}$ cup of sugar. A loaf of pumpernickel bread requires $2\frac{1}{2}$ cups of flour and 1 cup of sugar. You have 40 cups of flour and 10 cups of sugar. How many of each kind of loaf can you bake?

   a) Define your variables. Identify the independent and dependent variables.

   b) What are a reasonable domain and range for this situation? Justify your answer.

   c) Write two inequalities to describe the situation.

   d) Solve each inequality for your dependent variable in terms of your independent variable.

   e) Graph both of these inequalities together. Choose an appropriate window and sketch your results. Identify the solution to the system.

3. A ballpark offers two packages for birthday party favors. Package A includes 5 tickets for a ball toss game and 6 pennants for the home team. Package B includes 15 tickets for a ball toss game and 3 pennants for the home team. The ballpark management want to have at least 30 packages available. They also want to include at least 300 tickets but no more than 180 pennants. How many of each package can they assemble?

a) Define your variables. Identify the independent and dependent variables.

b) What are a reasonable domain and range for this situation? Justify your answer.

c) Write three inequalities to describe the situation.

d) Solve each inequality for your dependent variable in terms of your independent variable.

e) Graph both of these inequalities together. Choose an appropriate window and sketch your results. Identify the solution to the system.

Make the Most of It

In Sections 3.12 and 3.13, you used a system of inequalities to represent a problem about your sports apparel business. A local baseball stadium has invited you to sell hats and T-shirts at home games. Thus far, you have used linear inequalities to represent constraints on the situation. The system of inequalities yielded a feasible region with possible combinations of hats and T-shirts that satisfy all of the constraints.

Which one of those combinations can you use to generate the most income for your business? In this section you will explore a way to use the graphs of linear inequalities to answer this question.

Earlier in this chapter we have explored the following problem.

Your sporting goods company plans to sell merchandise at a local baseball stadium. At the stadium souvenir stand, you will sell hats with the home team's logo for $12, and T-shirts with the home team's logo for $20. The stadium charges $500 to rent the space to sell items during a home game, and the team charges $1.50 per item with their logo in royalties. Each hat costs $7 in materials and labor to produce and each T-shirt costs $9 in materials and labor to produce. There is only enough space in the stand to stock 200 total items. What possible combinations of hats and shirts will generate more income than expenses; i.e., what combinations will create a profit?

For now, let's ignore the limit of 200 items, since we are thinking about production at the factory. Consider that the cost for hats is $2 in material and $5 in labor. Also, the cost for T-shirts is $4 in material and $5 in labor. There is still a licensing fee of $1.50 per item for the use of the team's logo.

Upon further analysis of your business accounts, you decide that you can spend no more than $1000 in material and $1500 in labor for one week. In order to maximize your income, how many hats and how many T-shirts should you make?

In order to solve this problem, answer these questions.

1. First, create a table like **Figure 3.13** to organize the information:

	Cost of Materials per Item	Cost of Labor per Item	Expected Income per Item
Hats			
T-shirts			
Maximum Allowed			

FIGURE 3.13.
Table of cost data.

2. Let h represent the number of hats sold and s represent the number of T-shirts sold. Write a function rule that describes the amount of income (I) in terms of the number of hats (h) and the number of T-shirts (s) sold.

3. Write a system of inequalities to represent each of these costs in terms of h and s:

a) Cost of Materials

b) Cost of Labor

4. What domain and range restrictions do we need to consider in the context of this problem? How do you know?

5. Solve the inequalities from Question 3 for s in terms of h.

6. Graph the two inequalities from Question 5. Consider your answer to Question 4 to set an appropriate window for your graph. Describe your window and sketch your results.

7. What is the feasible region for this system?

8. What are the vertices of the feasible region? Are these points included in the solution to the system of inequalities?

9. Use your income function to find the income that is generated by these four combinations of hats and T-shirts. Use a table like **Figure 3.14** to organize your data.

Points (h, s)	Income Function $I = 12h + 20s$

FIGURE 3.14.
Income possibilities.

10. According to our table, how many hats and T-shirts should we produce in order to maximize our income?

11. Is it possible, within the constraints of this problem, to earn an income of $7000? Justify your answer.

12. Is it possible, within the constraints of this problem, to earn an income of $3000? Justify your answer.

13. Are there other points from the feasible region that generate more income than one of the vertices? Expand your table to record your findings.

If you want to minimize or maximize a quantity when you have two constraints that can be represented by linear inequalities, you can use a method called **linear programming**. Here are the steps:

❖ *Define a system of inequalities.* Use the information in the problem to write a system of inequalities that describe the constraints. A table, such as the one in Figure 3.14, will help you organize the data and write your inequalities.

❖ *Write an equation describing the quantity you wish to minimize or maximize.* In the example used in this section, you tried to maximize income, so you wrote an income function.

❖ *Graph both inequalities on the same coordinate plane.*

❖ *Identify a feasible region.* The feasible region is the area of overlapped shading between the inequalities. The domain and range of the situation will also help define the feasible region. In the example used in this section, you can only have positive numbers of hats and T-shirts, so the feasible region is restricted to Quadrant I.

❖ *Find the coordinates of the vertices of the feasible region.* Using technology such as a graphing calculator can help you to find the intersection point of two lines.

❖ *Substitute the coordinates of each of the vertices into the function you are trying to minimize or maximize.* You will need to substitute each ordered pair into your function and calculate the desired quantity.

❖ *Identify the coordinates that yield the minimum or maximum value of the function.* One of the vertices of the feasible region will yield a minimum or maximum for the quantity. This ordered pair is the solution.

 Assignment

1. Two machines, Machine A and Machine B, produce two types of chili: beef and vegetarian. One machine does the processing of the raw ingredients and mixes them together. The other machine cooks the chili and packages it. There is a profit of $150 per case of vegetarian chili and $100 per case of beef chili. To make a case of vegetarian chili, Machine A must run 2 hours and Machine B must run 9 hours. To make a case of beef chili, Machine A must run 4 hours and Machine B must run 2 hours. Each machine runs 24 hours a day. What combination of cases of vegetarian chili and beef chili will result in a maximum profit?

   Let x = the number of cases of vegetarian chili and y = the number of cases of beef chili.

   a) What domain and range make sense for this situation?

   b) Write an income function in terms of x and y.

   c) Write four inequalities, in terms of x and y, that describe the constraints.

   d) Graph the system of inequalities. Identify the feasible region. If you use a graphing calculator, also record your viewing window.

   e) Identify the vertices of your feasible region. Use your income function to calculate the income for each combination of cases of vegetarian chili and cases of beef chili.

   f) What combination of cases of vegetarian chili and beef chili will result in a maximum profit?

2. A trucking company can ship the hats and T-shirts produced by your sporting goods company. Their trucks can hold a maximum of 250 cubic feet in volume and 2030 pounds in weight. One box of T-shirts has a volume of 5 cubic feet, weighs 35 pounds, and will generate $114 in profit. One box of hats has a volume of 3 cubic feet, weighs 35 pounds, and will generate $70 in profit. What combination of boxes of hats and T-shirts can you ship to maximize your profit?

Let x = the number of boxes of hats and y = the number of boxes of T-shirts.

a) What domain and range make sense for this situation?

b) Write four inequalities, in terms of x and y, that describe the constraints.

c) Write a profit function in terms of x and y.

d) Graph the system of inequalities. Identify the feasible region. If you use a graphing calculator, also describe your viewing window.

e) Identify the vertices of your feasible region. Use your income function to calculate the profit for each combination of boxes of hats and boxes of T-shirts.

f) What combination of boxes of hats and boxes of T-shirts will result in a maximum profit?

Count the Trees

Let's apply what you have learned thus far about building and using systems of linear inequalities to a new situation.

1. Your friend, Judy, has started a tree farm. She grows oak and pine seedlings and then ships them to customers. Because your business has become so successful, she comes to you for advice.

   A customer wants to order a combination of oak and pine seedlings. The customer wants at least 8 seedlings to include at least 5 pine seedlings. The box your friend usually uses for shipping can hold no more than 18 pounds. One oak seedling weighs 1 pound and costs $3 to ship. One pine seedling weighs 2 pounds but because of favorable shipping regulations only costs $2 to ship. What combination of oak and pine seedlings should Judy advise that the customer order to minimize shipping costs? Justify your answer.

Modeling Project Growing a Business

Titangrow is an amazing new fertilizer developed by Newton Farm Products. Demand is expected to be high, but there are many issues to think about besides sales.

First of all, the company must think about making the product. A key ingredient they use in Titangrow is salmon bone meal. This ingredient is expensive: Each increase of 5% in the quantity of salmon bone meal adds 20% to the manufacturing costs.

In testing the product, the company found that when the product has 25% bone meal it is most effective. At 15% bone meal it is very effective. And at 10% bone meal it is somewhat effective. Below 10% bone meal, the product is not effective. The company has decided to create 3 products:

Titangrow Super (25%)

Titangrow Advanced (15%)

Titangrow Standard (10%)

Each 10-lb bag of Titangrow Standard costs the company $3 to make.

Market research makes the company think that demand for Titangrow Super will be equal to Titangrow Advanced and Standard combined.

The company must also keep track of inventory at three warehouses.

Create a model for manufacturing, pricing (and profit), and keeping track of the inventory for these three products. Then create a summary report that addresses these issues for the entire product line. When you develop your model, remember that systems of equations, matrices, and linear programming are powerful business tools. Use them wisely.

Practice Problems

1. Suppose that one music subscription service, Dozster, charges $11 per month plus $0.85 per song; and a second service, Melody, charges $8 per month plus $1.00 per song. Let c be the monthly cost in dollars and n be the number of songs downloaded in a month.

   a) Write a linear equation that represents the monthly subscription cost for Dozster.

   b) Write a linear equation that represents the monthly subscription cost for Melody.

   c) Compare the monthly subscription cost for each company if you download 10 songs. Which is cheaper?

   d) Compare the monthly subscription cost for each company if you download 30 songs.

   e) How many songs could you download from each subscription service for $50?

   f) Use a system of equations to find the number of downloads and the cost at which the two services are the same.

   g) Graph this system of equations and record your window.

   h) For what number of songs is Melody a better deal? For what number of songs is Dozster a better deal?

2. A coffee shop sells several different kinds of coffee. The shop also uses some of its coffees to make its own custom blends. Coffee A sells for $6.00 a pound. Coffee B sells for $10.00 a pound. The shop's manager wants to create a blend of the two types A and B that sells for $7.00 a pound. The manager wants to make 10 pounds of this blend.

   Use a for the number of pounds of coffee A in the blend. Use b for the number of pounds of coffee B.

   a) Write an equation that describes the total number of pounds of coffees A and B in the blend.

b) Write an equation that describes the total dollar value of the blend.

c) What is the number of pounds of each coffee in 10 pounds of the blend?

3. Joel is saving his money to buy a used car. The price of the car is $5525, but it will be reduced by $150 for each month that the car remains unsold. Joel currently has $3250 in his savings account and will be able to save an additional $175 each month.

a) Write an algebraic expression to represent each of the following:

the price of the car in n months

the amount of money Joel will have saved after n months

b) Determine the number of months until Joel will be able to buy the car. Show your work.

4. Using only 34-cent stamps and 20-cent stamps, Peggy put $3.52 postage on a package. She used twice as many 34-cent stamps as 20-cent stamps. Determine how many of each type of stamp she used.

5. Mr. Parson sells nuts at a corner stand. He sells peanuts for $1.00 per pound and walnuts for $2.00 per pound. He wants to make 50 pounds of peanut-walnut mix. He will maintain the cost per pound of each type of nut and sell the mix for $1.60 per pound. How many pounds of each nut should he put in the mix? Show how you arrive at your answers.

6. A florist is offering two different package deals of roses and carnations. One package contains 20 roses and 34 carnations for $50.40. The other package contains 15 roses and 17 carnations for $32.70. This information can be represented by the system of equations below, where r is the cost of one rose and c is the cost of one carnation.

$20r + 34c = 50.40$

$15r + 17c = 32.70$

Solve the system of equations to find the cost of one rose.

7. Sketch a graph of this system of equations.

$$y = -x - 3$$

$$y = x - 3$$

8. In the following system of equations, what would be the first step in eliminating the variable y?

$$x + y = 7$$

$$2x - y = 5$$

9. This graph was made to compare the costs of renting copy machines from Company A and from Company B. What information is given by the point of intersection of the two lines?

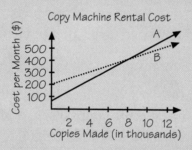

10. A clothing company receives orders from three shops. The first shop orders 25 jackets, 75 shirts, and 75 pairs of pants. The second shop orders 30 jackets, 40 shirts, and 35 pairs of pants. The third shop orders 20 jackets, 40 shirts, and 35 pairs of pants. Display this information in a matrix with rows representing the shops and columns representing the type of clothing ordered. Label the rows and columns of your matrix.

11. Through July 20, 1997, the three baseball players with the highest batting average in the National League had these batting statistics.

	AB	R	H	HR	RBI	Avg.
L. Walker (Colorado)	343	88	138	27	79	.402
Gwynn (San Diego)	372	64	147	15	84	.395
Piazza (Los Angeles)	332	56	118	19	62	.355

These are the statistics for the same players through September 30, 1997.

	AB	R	H	HR	RBI	Avg.
L. Walker (Colorado)	568	143	208	49	130	.366
Gwynn (San Diego)	592	97	220	17	119	.372
Piazza (Los Angeles)	556	104	201	40	124	.355

Find and label a matrix that displays the changes in these statistics. Notice that the batting averages for two of the three players decreased. How is a decrease shown in your matrix?

12. A state university announced a 7% raise in tuition. The current rates per semester hour are shown in the following table.

	Undergraduate	Graduate
Resident	$75.00	$99.25
Nonresident	$204.00	245.25

a) Write and label a matrix that represents this information.

b) Find a new matrix that represents the tuition rates per semester hour after the 7% raise goes into effect. Label your matrix.

c) Find a matrix that represents the dollar increase for each of the categories. Label your matrix.

d) Which matrix operation did you use in b? Which matrix operation did you use in c?

13. Nancy has jewelry shops in Westmarket, Eastmarket, and Oldmarket plazas. Her sales of cultured pearls for July are shown in the following table.

	Old	West	East
Earrings	10	8	12
Pins	6	5	4
Necklaces	3	2	2
Bracelets	4	3	2

Earrings sell for $40 a pair, pins for $35 each, necklaces for $80 each, and bracelets for $45. Use matrix multiplication to find Nancy's total sales at each location.

14. Use matrices to find (x, y) if

$$5x + 3y = 2$$

$$7x - 2y = 3$$

15. A small company makes unfinished tables and chairs. Each table uses 40 board feet of wood, and each chair uses 20 board feet. It takes 6 hours of labor to make a table and 8 hours to make a chair. There are 2000 board feet of wood and 600 labor hours available for the next week.

a) Let T represents the number of tables made in the next week and C the number of chairs. Write inequalities that relate T and C to the total available amount of wood, and to the total available labor.

b) The inequalities $T \geq 0$ and $C \geq 0$ also describe this situation. Why?

c) Draw a graph that represents this situation.

d) Can the company produce 30 tables and 35 chairs in the next week? Defend you answer.

e) Can the company produce 50 tables and 10 chairs in the next week? Defend you answer.

16. Eric wants to buy chips and soda for the guests at his party. Each large bag of chips costs $2.40 and each large bottle of soda costs $2.00. He has $48.00 to spend.

a) Write a system of inequalities that describes the limits on Eric's choices for number of bags of chips C and number of bottles of soda S that he can buy.

b) Draw a graph that represents the possible numbers of bags of chips and bottles of soda that Eric can buy.

c) If Eric buys 10 bags of chips, what is the largest number of bottles of soda that he can buy?

17. Write a system of inequalities that describes the graph in **Figure 3.15.**

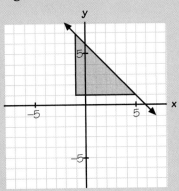

FIGURE 3.15.
Graph for Question 17.

18. A to Z Auto Service claims that their car repair rates are the lowest in town. They charge an initial fee of $50 plus $30 per hour. Assuming their claim is true, draw a graph that describes the rates charged by A to Z's competitors.

19. Julie does not want to spend more than $300 on ice skating. Her skates will cost $42, her lessons will cost a total of $56, and the practice time will cost $7.50 per hour. Write an inequality Julie could use to determine the maximum number of hours, h, she can practice without spending more than $300.

20. Principal Greene has received a $1500 grant to buy new printers at his school. He has a choice between $50 color printers and $130 high-resolution laser printers. He wants at least 20 new printers. If C represents the number of color printers and L represents the number of laser printers, write a set of inequalities that models his choice.

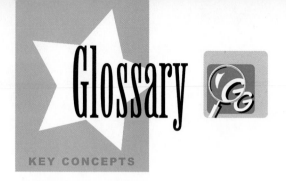

Glossary

KEY CONCEPTS

Additive Inverse: The additive inverse of a number a is the number b, such that $a + b = 0$. So the additive inverse of 2 is –2.

Coefficient Matrix: A matrix that contains the coefficients from a system of equations.

Constant Matrix: A matrix that contains the constant from a system of equations.

Corresponding Elements: Elements in matrices that are in the same position (row and column) of each matrix.

Equal Matrices: Two matrices are equal matrices if and only if they have the same dimensions and their corresponding elements are equal.

Expense: An amount paid for a good or service. In business, expenses must be deducted from income to find profit.

Feasible Region: The area of a graph containing all the points that make all of the inequalities in the system true.

Frequency: In this chapter, frequency is the number of times a value occurs. If it occurs often the frequency is high. If it does not occur often the frequency is low.

Identity Matrix: A square matrix (the same number or rows and colums) that contains ones running diagonally from the upper left corner to the lower right corner and has zeroes in all of the other entries. Multiplying a square matrix by an identity matrix does not change the matrix.

Income: In most cases the amount earned from work. In business, income minus expenses equals profit.

Inequality: A type of statement that compares two values. For example these are common inequality symbols:

Symbol	Example	Meaning (from left to right)
$<$	$x < y$	x is less than y
$>$	$x > y$	x is greater than y
$\leq$	$x \leq y$	x is less than or equal to y
$\geq$	$x \geq y$	x is greater than or equal to y

Intersect or Intersection: In this chapter, the point (x, y) in a graph where two lines cross.

Inverse Matrix: The product of a square matrix and its inverse matrix is an identity matrix. Not every square matrix has an inverse matrix.

Linear Combination Method: A way to solve systems of equations in which one of the variables is eliminated by adding equations.

Linear Programming: The study and processes for maximizing or minimizing a function based on linear constraints. Linear programming is a tool used in business to make financial decisions.

Matrix (Matrices): An arrangement of numbers written between brackets. The numbers are arranged in rows and columns and make a rectangular shape. A row is a horizontal group of numbers. A column is a vertical group of numbers. The plural of matrix is matrices.

Matrix Addition: A process of adding two matrices. Matrix addition can only be performed on matrices with the same dimensions.

Matrix Dimensions: The number of rows and columns in a matrix. A matrix with 3 rows and 2 columns is a 3×2 matrix (read "3 by 2").

Matrix Element or Component: A single entry in a matrix.

Matrix Multiplication: A process of multiplying two matrices.

Matrix Operation: A process, such as addition or scalar multiplication, carried out on one or more matrices.

Matrix Subtraction: A process of subtracting two matrices. Matrix subtraction can only be performed on matrices with the same dimensions.

Multiplicative Inverse: The multiplicative inverse of a number a is the number b, such that $a \times b = 1$. So the multiplicative inverse of 2 is $\frac{1}{2}$.

Profit: The amount left over after expenses have been deducted from income. This can be a negative number (often called "loss").

Scalar Multiplication: A process where all of the elements in matrix are multiplied by a number. The number is called a scalar.

Solution to a System of Equations: A set of values that are solutions to every individual equation in a system.

Substitution Method: A way to solve systems of equations by isolating a variable in one equation then substituting into the other equation.

System of Equations: Two or more equations. A system of linear equations is a group of two or more linear equations.

Systems of Inequalities: Two (or more) inequalities that have the same variables. For example, $2x + 3y < 4$ and $3x + 4y < 5$ are a system of inequalities.

Variable Matrix: A matrix that contains the variables from a system of equations.

4

CHAPTER

Art: Transformations, Symmetry, and Proportions

Picture Perfect

Visual Art takes many forms. A pencil drawing and a graphic design are examples of art. A computer-generated graphic, a quilt, and a painting are also forms of art. Many occupations make use of art as well. For example, architects use drawings to represent building designs.

Artists and architects use flat surfaces to create pictures that appear to be three-dimensional. In using a flat surface, such as a canvas or paper, to represent three-dimensional space, artists and architects use mathematical ideas and geometric principles to make a flat image look like the real object.

What does mathematics have to do with creating art? Have you noticed patterns that repeat in paintings, logos, and even wallpaper? Sometimes the images that make up the pattern are always the same size, sometimes they are made to look 3-D, or three-dimensional, by drawing some objects over others and changing their size.

An architect uses the geometry of one-point or multi-point perspective to help a client visualize what a proposed building will look like. A theater designer uses geometry to create the illusion of depth and distance on a small stage. An animator guides you into and around three-dimensional objects using a two-dimensional screen. A painter uses perspective and shading to add depth to drawings.

Before you begin, take a look around you and notice the lines and shapes that make up your world. Think about how you might sketch the room you are in or the front of your school.

Small children view the world differently than you do. Their sense of depth is not well developed. Just look at a picture drawn by a young child, such as **Figure 4.1.**

FIGURE 4.1.
House by Jackson Barber, age 8.

Notice that the sun is too large and the building appears flat. Now compare Figure 4.1 with an architect's drawing for a new school (**Figure 4.2**).

FIGURE 4.2 Tantasqua Regional High School.

The drawing helped community members imagine what the completed project might look like before they voted on whether or not to build the new school. A realistic drawing is an important tool in their decision.

In this chapter we will explore three questions:

❖ How do artists create patterns by moving or transforming an object?

❖ What geometric principles guide an artist's creation of an accurate drawing?

❖ What principles give a picture depth?

This chapter explores how the answers to these questions help artists, architects, and others develop realistic drawings.

Transformations in Art

Repetition is a technique used often by artists. By repeating a basic element an artist can create a pleasing painting, tapestry, or other work. There are several ways to repeat a basic design element.

Consider the tapestry in **Figure 4.3.** The artist has repeated a basic condor element in two ways.

Figure 4.4 shows how the condor in the upper left corner (condor 1) can slide to its right to produce another condor. A **slide** is one type of repetition used by artists.

FIGURE 4.3.
Condor by Cerapio Vallejo.

FIGURE 4.4.
A slide produces another condor.

FIGURE 4.5. Slide condor 1 to the right, then slide it down, and then flip it to the left.

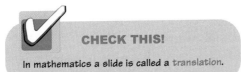

CHECK THIS!

In mathematics a slide is called a translation.

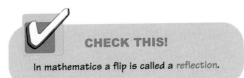

CHECK THIS!

In mathematics a flip is called a reflection.

But not all of the condors can be produced by sliding one of them. For example, the condor directly below condor 1 cannot be obtained by sliding condor 1. It can be obtained from condor 1 by sliding condor 1 to the right, then down, and then flipping it to the left (see **Figure 4.5**). A **flip** is another kind of repetition used by artists.

All of the condors in this tapestry can be obtained from condor 1 by translating or reflecting or a combination of the two.

The mirror in **Figure 4.6** demonstrates a third way artists use repetition.

FIGURE 4.6.
Art Nouveau mirror by Cidinha.

The crescent shape on the right edge of the mirror can be obtained from the one on the left by spinning the left one 180° about a central point. **Figure 4.7** shows how one crescent tip can be rotated to get the corresponding tip of the other crescent. A **spin** is another kind of repetition used by artists.

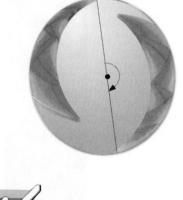

FIGURE 4.7.
Spinning one crescent tip 180°
produces the other.

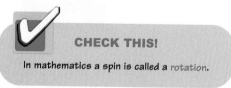

CHECK THIS!

In mathematics a spin is called a rotation.

Translations, reflections, and rotations are types of **transformations**. That is, each is a process that uses a basic image to create new ones. In this lesson you will learn more about transformations and how they are used in art and in mathematics.

In each of the following figures, identify a basic element and describe ways to get the other sections of the figure by translating, reflecting, and/or rotating the basic element. Your basic element should be as small a portion of the entire figure as possible.

1.

FIGURE 4.8. Image for Question 1.

2.

FIGURE 4.9. Image for Question 2.

3.

FIGURE 4.10. Image for Question 3.

4.

FIGURE 4.11. Image for Question 4.

5.

FIGURE 4.12. Image for Question 5.

6.

FIGURE 4.13. Image for Question 6.

7.

FIGURE 4.14. Image for Question 7.

8.

FIGURE 4.15. Image for Question 8.

9.

FIGURE 4.16. Image for Question 9.

10.

FIGURE 4.17. Image for Question 10.

Types of Symmetry

Many works of art have one or more kinds of symmetry. Symmetry is often easy to recognize, but can be hard to define. In Section 4.1 you learned about translations, reflections, and rotations. One way to describe a symmetric image is that a copy of the image can be translated, reflected, or rotated to match the original.

In this section you will create symmetric images. The basic instructions for Questions 1–4 are:

❖ Use two circular coffee filters. Place one on top of the other.

❖ Fold the double-filter as described.

❖ Make a variety of cuts along each straight edge of the folded filters.

❖ Unfold them.

❖ Describe how one filter (the copy) can be translated, rotated, or reflected to obtain the other. If there is more than one way to obtain one from the other, describe all of the ways that you can find.

1. Fold the filters in half once as shown in **Figure 4.18.** Make various cuts along the edge as shown in **Figure 4.19** (but not exactly the same as in the figure).

FIGURE 4.18.
Folded filters.

FIGURE 4.19.
Folded and cut filters.

2. Fold the filters in half and then in half again. (After the two folds, you should have quarter-filters). Make various cuts along the edge.

3. Fold the filters in half, then in half again, and then in half again. (After the three folds, you should have an eighth-filters). Make various cuts along the edge.

4. Fold the filters in half, and then fold into thirds. Make various cuts along the edge.

An image has **reflection symmetry** (also called line symmetry) when a copy of it can be reflected through a line to match the original image. The line is called a **line of symmetry**. For example, **Figure 4.20** shows the line of symmetry for a butterfly image.

FIGURE 4.20.
Reflection symmetry.

An image has **rotation symmetry** when a copy of it can be rotated less than 360° about a central point to match the original image. When you describe rotation symmetry, try to identify the central point and to state the size of the smallest possible rotation for which the copy and original match. For example, the flower in **Figure 4.21** has 72° rotation symmetry.

FIGURE 4.21.
72° rotation symmetry.

72°

5. For each of the objects you created in Questions 1–4, identify all lines of symmetry and the size of any rotation symmetry.

6. Does the flower in Figure 4.21 have any symmetry other than rotation symmetry? Explain.

SECTION 4.3

Transformations, Isometries, and Symmetry

Many computer software programs have drawing features that allow you to create and edit images. For example, you can copy an image and paste it elsewhere. Many programs let you reflect or rotate an image. All of these operations are mathematical transformations. In this section you will take a closer look at how some of these operations work.

When a geometric object is transformed into another one, the original is called the **pre-image**. The new object is called the **image**. The vertices of the pre-image are labeled with capital letters, such as ABC for a triangle. The vertices of the image are often labeled with the same letters, but with an apostrophe after each. The apostrophe is read as **prime**. The transformation is often indicated with an arrow.

CHECK THIS!

The notation △ABC→△A'B'C' says that triangle ABC is transformed into triangle A-prime B-prime C-prime. △ABC is the pre-image. △A'B'C' is the image.

Draw a triangle on a sheet of paper and label its vertices A, B, and C. Be sure the triangle is not special; that is, don't draw an equilateral triangle, an isosceles triangle, or a right triangle. Draw the triangle so that one vertex is pointing upward; use A for the label of this vertex.

1. Place a sheet of patty paper over the triangle and trace it. Slide the patty paper a few inches to the right. On the original paper, mark the locations of the new vertices and connect them. Label the new triangle's vertices A', B', and C'. Discuss the type of transformation that A'B'C' is of ABC. Also discuss the properties of the original that changed and those that are the same.

2. On the paper with the original triangle, draw a vertical line to the right or the left of the triangle. Place the patty paper triangle back on the original triangle. Fold the patty paper along the line. On the original paper, mark the locations of the new vertices and connect them. Label the new triangle's vertices A', B', and C'. Discuss the type of transformation that A'B'C' is of ABC. Also discuss the properties of the original that have changed and those that are the same.

3. On the paper with the original triangle, mark a point somewhere outside of the triangle. Place the patty paper triangle back on the original triangle. Place the point of your pencil on the point that you marked and spin the patty paper about 90°. On the original paper, mark the locations of the new vertices and connect them. Label the new triangle's vertices A′, B′, and C′. Discuss the type of transformation that A′B′C′ is of ABC. Also discuss the properties of the original that have changed and those that are the same.

4. Does your original triangle have any lines of symmetry? That is, does it have reflection symmetry? If not, what type of triangle does?

5. Does your original triangle have rotation symmetry? If not, what type of triangle does?

Translations, reflections, and rotations are all special types of transformations called isometries. An **isometry** is a transformation that does not change size or shape. (Later in this chapter you will examine a transformation that is not an isometry.)

Isometries change the location of an object. Some of them change an object's orientation. You can orient a simple geometric figure by labeling the vertices A, B, C, and so on as you move from vertex to vertex in a clockwise direction. When an isometry changes the orientation, as you move from A′ to B′ to C′, and so on, you move in a counterclockwise direction.

6. Which transformations change orientation: translation, reflection, or rotation?

Each of these three isometries has specific characteristics that are used to identify it.

❖ A translation has a distance and a direction. For example, you might slide ΔABC two inches to the right. Translations change location, but do not change size, shape, or orientation.

❖ A reflection has a line (sometimes called a mirror). For example, you might reflect ΔABC about line $\overleftrightarrow{CD}$. Reflections change location and orientation, but do not change size or shape.

❖ A rotation requires a point and an angle. For example, you might rotate ΔABC 30° counterclockwise about point D. Rotations change location, but do not change size, shape, or orientation.

CHECK THIS!

Mathematicians consider a counterclockwise rotation positive and a clockwise rotation negative. For example, a rotation 30° clockwise is considered a rotation of –30°.

Sometimes two or more isometries are combined. For example, in **Figure 4.22**, △ABC is translated to the right to get △A'B'C'.

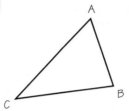

FIGURE 4.22.
A translation.

In **Figure 4.23**, △A'B'C' is reflected across the line shown to obtain △A"B"C".

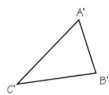

GLIDE REFLECTION

A *glide reflection* is a composite of a translation and a reflection.

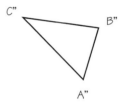

FIGURE 4.23.
A translation followed by a reflection.

△A"B"C" is a called a **glide reflection** of △ABC.

7. What properties of a figure are changed by a glide reflection?

8. Use patty paper to create a composite of two reflections through a pair of parallel lines. That is, draw a line to the right of △ABC and reflect the triangle through it. Then draw a second line parallel to the first and to the right of the image. Reflect the image through this second line. How is the final image related to △ABC?

9. Use patty paper to create a composite of two reflections through a pair of intersecting lines. That is, draw a line to the right of △ABC and reflect the triangle through it. Then draw a second line that intersects the first and that is to the right of the image and below it. Reflect the image through this second line. How is the final image related to △ABC?

Assignment

1. **Figure 4.24** is a design element. Identify the smallest basic portion of it that can be used to obtain the rest and describe isometries that can be used to build up the design.

FIGURE 4.24.
A design element.

FIGURE 4.25.
Fabric design by Debra L. Hayden.

2. **Figure 4.25** shows how artist Debra Hayden used a basic element similar to the one in Figure 4.24 to create a larger work. Describe at least two ways that she could have done this.

3. Describe the symmetries of the basic element in Figure 4.24.

4. Discuss how an artist could use transformations to create **Figure 4.26** from a basic element of the figure.

FIGURE 4.26.
Image for Question 4.

5. **Figure 4.27** is a capital letter A in the Arial font. Some capital letters are symmetric. Identify those that have symmetry and describe the symmetry of each.

FIGURE 4.27.
Image for Question 5.

6. You may hear people talk about point symmetry. The plant in **Figure 4.28** demonstrates this type of symmetry. **Point symmetry** occurs when every point of a figure has an image directly across the figure's center, as shown in **Figure 4.29.**

FIGURE 4.28.
Image for Question 6.

FIGURE 4.29.
Image for Question 6.

a) Do any letters of the alphabet demonstrate point symmetry? If so, which ones?

b) Point symmetry is the same as another type of symmetry that you have studied in this lesson. Explain.

7. **Figures 4.30, 4.31, and 4.32** are oil company logos. Describe the symmetry of each.

FIGURE 4.30.
Shell Oil logo.

FIGURE 4.31.
Gulf Oil logo with letters.

FIGURE 4.32.
Gulf Oil logo without letters.

Algebraic Representations of Symmetries and Isometries

All of the isometries that you have examined change an object's location. People who use isometries in their work often need to describe locations precisely. When precise location is important, people use coordinates. For example, each pixel on a computer screen has coordinates that identify its location and that are used to transform the pixel or an object composed of many pixels in various ways.

In this section you will use coordinates to describe isometries exactly.

PART A:
TRANSLATIONS ON A COORDINATE PLANE

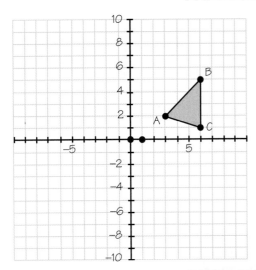

FIGURE 4.33. Triangle ABC.

A **translation** is a transformation that slides an object a given distance and in a given direction. Thus, every translation has a distance and a direction.

1. On a sheet of grid paper, plot ∆ABC as shown in **Figure 4.33.** Record the coordinates of the vertices of ∆ABC, the pre-image, in a table like **Figure 4.34.**

2. Trace ∆ABC and the coordinate axes onto a sheet of patty paper. Label the copy, the image, A′B′C′.

Pre-Image			Image		
A	B	C	A′	B′	C′

FIGURE 4.34. Table for Question 1.

3. Place the copy on top of the original so that the axes are aligned. Slide the copy 3 units left and 2 units up. Record the coordinates of the image in your table.

4. Did the triangle change size or shape?

5. Return the image to its original position and repeat 3 and 4 two more times using different values for the horizontal and the vertical translation.

6. Write equations that describe a translation of (x, y) horizontally h units and vertically k units. Use x' and y' for the coordinates of the image.

7. How does one represent a vertical or horizontal direction and a distance of a translation on a coordinate plane?

A **reflection** is a transformation that flips an object across a line.

PART B: REFLECTION ACROSS THE Y-AXIS

8. Refer back to your original ΔABC. Record the coordinates of the vertices of ΔABC, the pre-image, in a table like **Figure 4.35**.

Pre-Image			Image		
A	**B**	**C**	**A′**	**B′**	**C′**

FIGURE 4.35.
Table for Question 8.

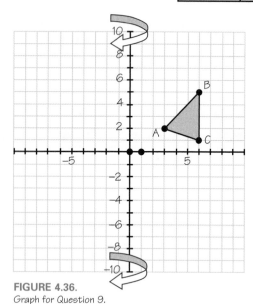

FIGURE 4.36.
Graph for Question 9.

9. Place the copy of ΔABC from Part A on top of the original so that the axes are aligned. Reflect the triangle across the y-axis, the line of reflection, by flipping the paper and realigning the axes (**Figure 4.36**). Record the coordinates of the image in your table.

10. Did the triangle change size or shape?

11. Describe the relationship between the line of reflection and vertices A and A′. Are these relationships true for the other vertices of the triangle?

12. Repeat 8–11 two times drawing the pre-image ΔABC in a different position, the fourth quadrant for example, each time.

13. Write equations that describe a reflection of (x, y) across the y-axis.

14. Do the y-coordinates change? If so, how?

15. Why are x-coordinates of the image opposites of those of the pre-image?

PART C: REFLECTION ACROSS THE X-AXIS

16. Refer back to your original $\triangle ABC$. Record the coordinates of the vertices of $\triangle ABC$, the pre-image, in a table like **Figure 4.37.**

Pre-Image			Image		
A	B	C	A′	B′	C′

FIGURE 4.37.
Table for Question 16.

17. Place the copy of $\triangle ABC$ from Part A on top of the original so that the axes are aligned. Reflect the triangle across the x-axis, the line of reflection, by flipping the copy of $\triangle ABC$ and realigning the axes (**Figure 4.38**). Record the coordinates of the image in your table.

18. Did the triangle change size or shape?

19. Describe the relationship between the line of reflection and vertices A and A′. Are these relationships true for the other vertices of the triangle?

20. Repeat 16–19 two times drawing the pre-image $\triangle ABC$ in a different position (the fourth quadrant for example) each time.

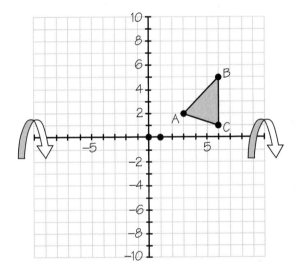

FIGURE 4.38.
Graph for Question 17.

21. Write equations that describe a reflection of (x, y) across the x-axis.

22. Do the x-coordinates change? If so, how?

23. Why are the y-coordinates of the image the opposite of those of the pre-image?

PART D: 90° ROTATIONS ON A COORDINATE PLANE

A **rotation** is a transformation that spins an object about a given center a given number of degrees.

24. Draw triangle ΔABC in the first quadrant of a new sheet of grid paper. Record the coordinates of the vertices of ΔABC, the pre-image, in a table like **Figure 4.39**.

Angle of Rotation	Pre-Image			Image		
	A	B	C	A′	B′	C′

FIGURE 4.39.
Table for Question 24.

25. Trace ΔABC and the coordinate axes onto another sheet of patty paper. Label the image A′B′C′.

26. Place the image on top of the original so that the coordinate axes are aligned. Rotate the copy 90° about the origin by turning the copy counterclockwise (**Figure 4.40**). Record the coordinates of the image in your table.

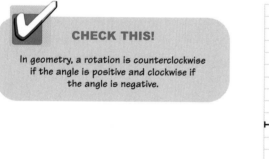

CHECK THIS!

In geometry, a rotation is counterclockwise if the angle is positive and clockwise if the angle is negative.

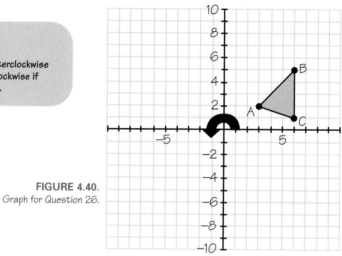

FIGURE 4.40.
Graph for Question 26.

27. Did the triangle change size or shape?

28. Repeat 26 and 27 for rotations of 180°, 270°, and 360°. Record the results in your table.

29. Describe a rotation by multiples of 90° in words.

Coordinate descriptions of rotations that are not multiples of 90° use trigonometry, so you will not do them now. Trigonometry is discussed in Chapter 8.

PART E:
GLIDE REFLECTIONS USING THE X-AXIS

Since glide reflections are a composite of a translation and a reflection they require a distance, a direction, and a line of reflection.

Conduct an investigation into glide reflections using the x-axis as the line of reflection. Proceed as you did in Parts A–D. Begin with a triangle drawn in one of the quadrants. Select a distance to translate it horizontally and a distance to translate it vertically. After you have translated it, reflect the image across the x-axis and record the results. Do the same for more triangles until you think you can answer Question 30.

30. Write equations that describe a glide reflection of the point (x, y) h units horizontally, k units vertically, and across the x-axis.

 Assignment

1. What are four types of isometries?

2. What are the basic attributes of translations?

3. What are the basic attributes of reflections?

4. What are the basic attributes of rotations?

5. Draw △ABC on your paper. If you reflect the triangle across any one of its sides, what shape is formed by combining the original triangle and the new triangle?

6. In **Figure 4.41**, which of these do not produce an image of the rectangle with two of its vertices lying on the *y*-axis?

    A. A reflection across the *x*-axis

    B. A translation right 3 units

    C. A translation up 2 units

    D. A reflection across the *y*-axis

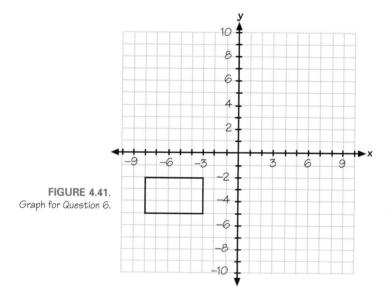

FIGURE 4.41.
Graph for Question 6.

7. In **Figure 4.42**, if triangle ΔABC is reflected across the x-axis, what are the coordinates of A′?

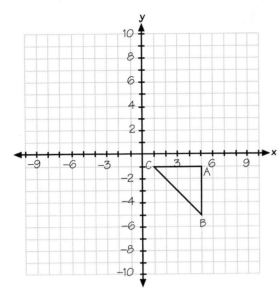

FIGURE 4.42.
Graph for Question 7.

8. In **Figure 4.43**, the vertices of a pre-image are at (–2, –2), (–1, –2), (–1, 0), and (–2, 0).

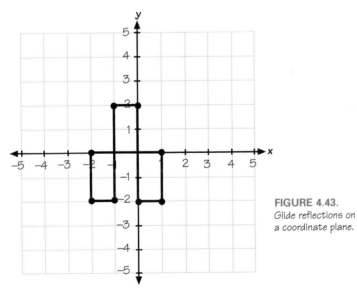

FIGURE 4.43.
Glide reflections on a coordinate plane.

a) Write equations that describe the use of a glide reflection to obtain another part of the figure.

b) Write equations that describe the use of a translation to obtain the remaining part of the figure.

Design a Logo

Create a graphic design for a corporate logo that incorporates translations, reflections, rotations, or glide reflections. The design should include elements of translation, rotation, and reflection symmetry.

Transfer your logo to a coordinate plane. Label points on the pre-image. Use algebraic notation to verify the accuracy of your use of symmetry to create your logo.

Write a short paragraph describing how and why each element of symmetry is incorporated into the design.

SECTION 4.6

It's a Matter of Perspective

Artists, architects, illustrators, and animators are trained to draw objects and scenes. Their drawings need to be accurate and realistic. Small children have not received training to draw objects and scenes. A child's drawing may appear to be less accurate and realistic than the drawing of a trained artist.

Examine the picture assigned to your group by your teacher. Use **Figures 4.44 to 4.47** to answer the following questions:

1. What do you notice about this picture?

2. What do you notice about the sizes of the different objects in the picture?

3. What do you notice about the sizes of the different objects and their placement in the picture?

FIGURE 4.44.
Cave drawings similar to those found at Hamangala.

FIGURE 4.45.
Architectural rendering of a building.

FIGURE 4.46.
Perspective Absurdities by William Hogarth.

FIGURE 4.47. *Ancient Egyptian art.*

4. Which pictures most accurately depict a real-life scene? Why?

5. Which pictures least accurately depict a real-life scene? Why?

The term perspective has several different meanings:

a) A view: "From what perspective are you looking down into the parking lot?"

b) A mental outlook: "What is your perspective on standardized dress codes for high school students?"

c) The relationship of parts to the whole: "What is the perspective on the soccer team's success this year? What role did the players play? What role did their training play? What role did the coach play?"

d) The technique of representing objects in three-dimensional space with a two-dimensional plane: "How does the artist use perspective to make a painting that looks like real life?"

6. Which paintings seem to make use of perspective?

Sometimes an artist may use perspective to get your attention. The painting shown in Figure 4.46 contains several intentional mistakes.

7. Work together with your group to identify the real-world inaccuracies in the painting by William Hogarth (Figure 4.46). Describe each error in detail.

8. Describe what the artist needs to do to correct each item you identified in Question 7.

Question 9 will help you identify elements or principles involved in perspective drawing.

9. Look at *The Bathers* by George Seurat (**Figure 4.48**). The artist has tried to represent a three-dimensional subject on a two-dimensional plane. The two-dimensional plane is the surface of the canvas or paper.

FIGURE 4.48.
*The Bathers at Asnieres
by George Seurat.*

a) Imagine that your group is in a hot-air balloon floating over the subjects in this picture. Draw a bird's-eye view of what you would see. You don't have to sketch every detail. Just draw simple outlines of some of the objects in each picture and label them. (For example, you might use ovals labeled as dog, man lying down, man wearing hat, or sailboat to indicate objects.)

b) Explain what clues in the original picture helped you decide where to place particular objects in your drawing.

c) Discuss how the artist tried to convey depth.

Location and Size

PART 1: WHAT'S IN THE FRONT?

In Section 4.6 you identified several factors that are important to the proper use of perspective in art. One of these is the location of objects with respect to each other: objects close to the viewer should overlap those that are behind them. Another factor is the relative size of objects. In this part of this section you will take a closer look at overlapping. In the second part you will take a closer look at size.

Within your group, use centimeter grid paper to create 4 cubes as described in the table in **Figure 4.49.**

Cube	Side Length (cm)
A	2
B	4
C	6
D	8

FIGURE 4.49.
Measurements of cubes.

Use masking tape or string to mark a 35 centimeter by 35 centimeter square region on a flat surface. Label the sides of the square north, south, east, and west.

Position the cubes according to the diagram shown in **Figure 4.50.**

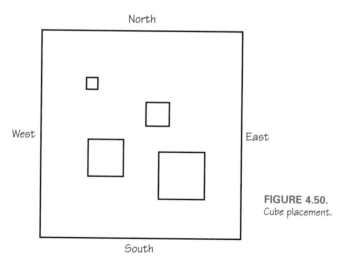

FIGURE 4.50.
Cube placement.

1. Lower yourself so that you are eye level with the cubes. Draw a two-dimensional sketch of what you see from the north, the south, the east, and the west.

2. How are the sketches alike?

3. How are the sketches different?

4. How does overlapping reveal which cube is in front of another cube?

5. **Figure 4.51** shows the top view of a table on which three cubes are sitting. Draw four eye level views of what you would see if you were sitting at positions A, B, C, and D.

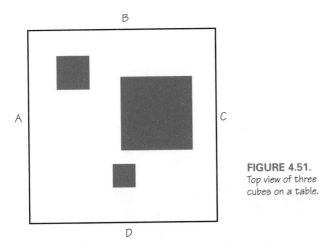

FIGURE 4.51.
Top view of three cubes on a table.

6. What statements can you make about how to draw images of objects so that the person looking at the image knows which objects are closer than other objects?

PART 2: THE RULER METHOD

In Part 1 you investigated the principle of overlapping: objects close to the viewer should overlap those that are behind them. You have also seen that objects far away from the viewer should be drawn smaller than objects close to the viewer. In this part you will take measurements of the size an object appears to be to a viewer.

❖ Hold a cube directly in front of you so all you see is a square.

❖ Move the cube toward you. What happens?

❖ Move the cube away from you. What happens?

❖ What general statement might we make about objects as they move farther away?

You will work with your group to take some measurements that illustrate numerically why objects appear to grow smaller as they move farther away.

Decide which member of your group will do each of the following:

❖ hold the ruler (holder)

❖ stand at a distance (stander)

❖ take measurements (measurer)

❖ record measurements (recorder)

The instructions that follow are written from the holder's point of view.

7. Hold a ruler in your hand, and extend your arm in front of you at eye level so that the ruler is vertical and your arm is extended fully (see **Figure 4.52**).

FIGURE 4.52.
Photograph showing how to hold the ruler.

8. The measurer should carefully measure the distance from your eye to the ruler.

9. Instruct a member of your group (the stander) to stand several meters away from you. The measurer should use a meter stick or tape measure to measure the distance between your eye and the other person.

10. Use the ruler in your extended hand to measure the apparent height of the other person (the height you would draw the person if the paper or canvas coincided with the ruler). See **Figure 4.53**.

FIGURE 4.53.
Measuring the apparent height of a person.

11. Record the measurements in a table like **Figure 4.54.**

Measure	Distance (eye to ruler)	Distance (eye to person)	Apparent Height of Person	Actual Height of Person
1				
2				
3				
4				

FIGURE 4.54.
Table for Question 11.

12. Complete the ratios in the table in **Figure 4.55.**

Measure	Distance Eye to Ruler / Apparent Height	Distance Eye to Person / Actual Height
1		
2		
3		
4		

FIGURE 4.55.
Table for Question 12.

13. What do you notice about the ratios of the measures?

14. **Figure 4.56** shows a viewer looking at a person. The picture plane represents the canvas or paper on which an artist draws or paints. Use the figure to explain your conclusion about ratios in Question 13.

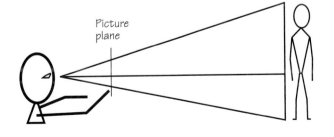

Picture plane

FIGURE 4.56. Image for Question 14.

1. Look at Figure 4.45 on p. 215. Describe at least one object that overlaps another. Which of the two objects is closer to the viewer?

2. Refer to Hogarth's painting (see Figure 4.46 on p. 216). Describe where Hogarth violates the principle of overlapping.

3. **Figures 4.57 and 4.58** show the top view of two different tables on which three cubes are sitting.

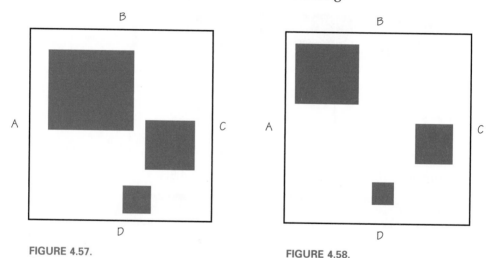

FIGURE 4.57.
Top view of three cubes on Table 1.

FIGURE 4.58.
Top view of three cubes on Table 2.

a) Look at the top view presented in Figure 4.57. Draw what you would see if you were sitting at position D.

b) Look at the top view presented in Figure 4.58. Draw what you would see if you were sitting at position D.

4. Martin built a cityscape using rectangular prisms. He made the following statements about his cityscape as viewed from eye level.

   a) From the southern view, I see a model building that is 2 inches tall. Behind it is a model building that is 6 inches tall. To the right of the 2-inch model, I see a 4-inch tall model building.

   b) From the eastern view, I see a model building that is 4 inches tall. To its right, I see a model building that is 3 inches tall. A 6-inch tall building stands behind this building.

   c) From the northern view, I see a model building that is 4 inches tall that stands behind a 3-inch tall building. I also see a 6-inch tall building.

   d) From the western view, I see a 6-inch tall building. I also see a 2-inch tall building with a 4-inch tall building behind it.

   Sketch the four side views to scale and the top view that match Martin's description. Explain your thinking.

5. Tommy used cubes to build five models of buildings that are each 1 cube wide. The models are described in **Figures 4.59 and 4.60.**

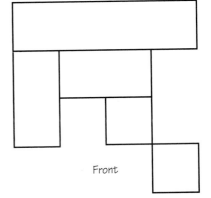

Building	Height (cubes)	Length (cubes)
A	1	4
B	2	2
C	4	1
D	3	1
E	1	2

FIGURE 4.59. Table for Question 5.

Front

FIGURE 4.60. Top view.

How many possible arrangements of these five models match the top view? Sketch the front, right, and back views of each arrangement.

6. **Figure 4.61** is a sketch of your hand held up 2 feet in front of your eye.

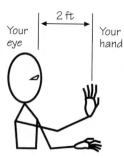

2 ft

Your eye

Your hand

FIGURE 4.61. Figure for Question 6.

Suppose your hand is 7 inches long and just blocks out a friend who is 5′ 6″ tall.

a) Make a sketch showing your eye, your hand, and your friend. Show all of the given measurements in your figure.

b) How far from your eye is your friend standing? Explain how you got your answer.

Principles of Perspective: Diminution and Scale

In the last section you probably noticed that the farther an object is away from you the smaller it appears. This basic principle is called **diminution**.

Diminution

Look at the Seurat painting (Figure 4.48) and think about how he used diminution. For example, Seurat used the principle of diminution when he made people in the boat smaller than the people on the land. But he had to decide how much smaller to make them. The key to this decision lies in a technique called **scaling**. To help you see how scaling works you first need to know what is meant by **lines of sight**. You also need to recall a few facts about **similar triangles**.

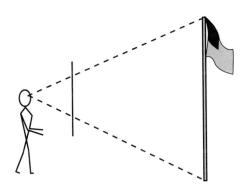

FIGURE 4.62.
Lines of sight to top and bottom of a flagpole.

Suppose a person is viewing a flagpole as shown in **Figure 4.62.** The dashed lines represent the lines of sight from the viewer's eye to the top and bottom of the flagpole. The vertical line in front of the viewer represents a canvas (the picture plane). The lines of sight intersect the picture plane to indicate where the top and bottom of the flagpole should be in a sketch made on the canvas.

Figure 4.62 shows two triangles. One of them is the small triangle with portions of the two lines of sight and the picture plane as sides. Another is the large triangle with the lines of sight and the flagpole as sides. These two triangles appear to be similar. That is, they appear to be the same shape, but of different sizes.

Mathematicians use the following conventions when discussing triangles:

* Vertices of triangles are labeled with capital letters.

* A side of a triangle is labeled with the same letter (but in lower case) as its opposite angle. For example, in **Figure 4.63**, side a is opposite angle A. A side can be named with the letters of its endpoints. Thus, side a is also referred to as $\overline{BC}$.

* An angle is named with the letter of its vertex. If two or more angles shared a vertex, three letters are used to name an angle. For example, ∠A can also be called ∠BAC.

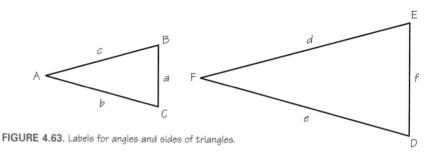

FIGURE 4.63. Labels for angles and sides of triangles.

In Figure 4.63, triangle ABC appears to be similar to triangle FED. Mathematicians write this statement with symbols: ΔABC~ΔFED. When two triangles are similar, corresponding angles are congruent and ratios of corresponding sides are equal. For example, in Figure 4.63, if ΔABC~ΔFED, then

* ∠A ≅ ∠F; ∠B ≅ ∠E; ∠C ≅ ∠D

* $\dfrac{AB}{FE} = \dfrac{BC}{ED} = \dfrac{AC}{FD}$.

Mathematicians have proved a theorem that is helpful in deciding whether two triangles are similar. It says that if two triangles have two pairs of congruent angles, then the triangles must be similar. The proof is based on the fact that the sum of the measure of any triangle's angles is always 180°.

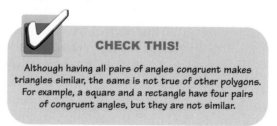

CHECK THIS!

Although having all pairs of angles congruent makes triangles similar, the same is not true of other polygons. For example, a square and a rectangle have four pairs of congruent angles, but they are not similar.

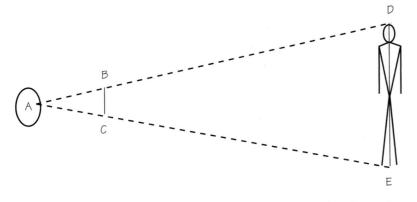

FIGURE 4.64.
Image for Question 1.

1. **Figure 4.64** shows a viewer's eye at A, lines of sight to the top (D) and bottom (E) of a person, and the points where the lines of sight cross the picture plane (B and C). What triangles are similar? Why?

2. In Question 1, what ratios must be equal?

The **scale** of an object in a picture is the ratio of the image height in the picture (the apparent height) to the actual height of the object. That is, the scale is the fraction $\frac{\text{apparent height}}{\text{actual height}}$.

3. In Figure 4.64, what ratio represents the scale?

In **Figure 4.65**, a line representing the distance from the viewer to the person being viewed has been added to Figure 4.64. This line creates an altitude in each triangle.

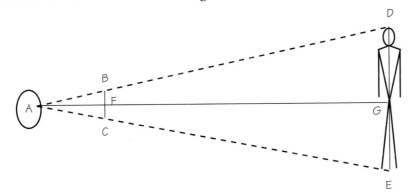

FIGURE 4.65.
Figure for Question 4.

 CHECK THIS!

An altitude of a triangle connects a vertex to the opposite side and is perpendicular to the opposite side.

Mathematicians have proved that corresponding altitudes of similar triangles have the same ratio as corresponding sides.

4. In Figure 4.65, what ratio of altitudes must equal the ratio $\frac{BC}{DE}$?

5. Suppose you are drawing on a canvas that is 2 feet in front of you. You use proper perspective to sketch an image 6 inches tall of an object that is really 3 feet tall.

   a) What is your drawing's scale?

   b) How far away are you from the object?

6. Suppose an object is 6 feet tall. You are 18 feet away from it, and you are sketching it on a canvas that is 2 feet in front of you. How tall should the object be in your picture?

In scaling an object to its proper size, an artist uses a mathematical transformation called **dilation**. A dilation uses a fixed point to project every point of a geometric figure. To an artist, the fixed point is the artist's eye. To a mathematician, the fixed point is the dilation's center.

Figure 4.66 shows a dilation of ∆ABC from point E that doubles the distance between E and each point of ∆ABC. ∆A'B'C' is the image that results. For example, the distance from E to A' is twice the distance from E to A. Since the distance from E to ∆ABC is doubled, the sides of ∆ABC are scaled by a factor of 2.

In a dilation, a segment's **scale factor** is the ratio $\frac{\text{length in image}}{\text{length in pre-image}}$.

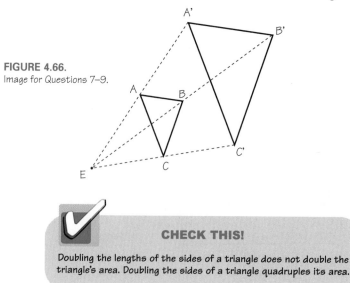

FIGURE 4.66.
Image for Questions 7–9.

✔ **CHECK THIS!**

Doubling the lengths of the sides of a triangle does not double the triangle's area. Doubling the sides of a triangle quadruples its area.

7. Use rulers and protractors to confirm that △ABC and △A′B′C′ are similar and that dilation doubles the lengths of △ABC's sides.

8. Describe a dilation using point E that triples the lengths of △ABC's sides. Trace △ABC and point E onto a sheet of paper and perform the dilation.

9. Describe a dilation using point E that halves the lengths of △ABC's sides. Trace △ABC and point E onto a sheet of paper and perform the dilation.

Assignment

1. The Washington Monument is 169 meters high. How far should you stand from it to block it out with your hand? Assume your hand is 18 cm in length and the picture plane is 50 cm from your eye. Sketch the situation first. (Be careful with the units.)

2. Suppose you are sketching a tree on a piece of paper taped to an easel (see **Figure 4.67**). The paper is 2 feet from your eye. The tree is 10 feet high and 40 feet from your eye. How tall should you make the image of the tree on the paper?

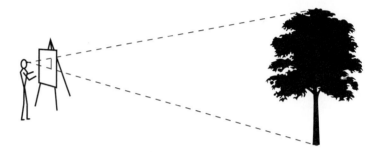

FIGURE 4.67.
Lines of sight for drawing a tree.

3. At many amusement parks and beaches, there are stores that take a photo, enlarge it, and give you a larger-than-life portrait suitable for framing. They also give you a small wallet-sized copy. **Figure 4.68** shows examples using stick pictures (not to the same scales).

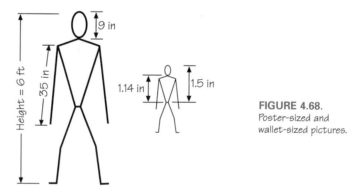

FIGURE 4.68.
Poster-sized and wallet-sized pictures.

a) Is enough information given to find the height of the smaller figure? If not, make up a reasonable value for the information you need.

b) Assume the leg length of the large portrait is 35 inches (so the torso length is 28 inches). Find the leg length and head height for the smaller figure.

c) What fraction of the large figure's height is in its legs? What fraction of the small figure's height is in its legs?

4. Trace **Figure 4.69** onto your paper. Sketch a side-doubling dilation of ΔABC using E as the center.

FIGURE 4.69.
Image for Question 4.

5. In Question 4, does the dilation produce a larger similar triangle if E is inside ΔABC? Does it matter where point E is located? Explain.

6. a) Since two triangles are similar if they have two pairs of congruent angles, two right triangles are similar if they have just one pair of congruent acute angles. Explain.

b) **Figure 4.70** shows a right triangle ABC with right angle at A and an altitude drawn from vertex A to the hypotenuse $\overline{BC}$. List all of the similar triangles. Be sure you have the vertices in proper correspondence.

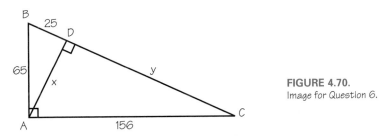

FIGURE 4.70.
Image for Question 6.

c) Use the fact that corresponding sides of similar triangles are proportional to find lengths x and y in Figure 4.70.

d) In Figure 4.70, ∠B measures approximately 67.4°. Find the measures of all the angles of the triangles.

A Parallel Universe (One-Point and Two-Point Perspective)

When you hold a cube at eye level so one face is parallel to your plane of vision, the cube looks like a square. If the cube is transparent, then you see the otherwise invisible sides and edges of the cube (see **Figure 4.71**).

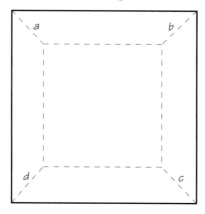

FIGURE 4.71.
Transparent cube.

1. The dotted lines represent the edges of the hidden faces. The edges labeled *a*, *b*, *c*, and *d* are perpendicular to the plane of vision.

   a) What do you notice about the vertical edges?

   b) How do the front and back faces compare?

   c) Why does the back face appear smaller than the front face of the cube?

   d) What do you notice about the top, bottom, and side faces?

   e) What do you notice about the edges labeled *a*, *b*, *c*, and *d*, which are perpendicular to the plane of vision?

FIGURE 4.72.
The inside of a hallway.

2. Refer to the picture of a hallway in **Figure 4.72.**

   a) There are several lines that run the length of the hallway. Select two of these lines. Place a ruler or edge of a paper along each of them. Describe where the two lines intersect.

   b) Repeat a using two lines that run the length of the ceiling. How does their intersection point compare to the intersection point of the floor lines?

In reality, the walls of the hallway in Figure 4.72 are parallel. Yet, in the picture the distance between the walls appears to narrow. If you imagine the walls were extended, they would converge to a single point. This is an example of the artistic principle of **convergence**: The apparent distance between parallel horizontal lines narrows. The single point to which the lines converge is called a **vanishing point**.

The vanishing point for horizontal lines is at the eye level of the viewer (see **Figure 4.73**). This eye level is called the **horizon line**—the line on which the earth and the sky appear to meet. The vanishing point for the railroad ties in Figure 4.73 is a point on the horizon.

FIGURE 4.73. The vanishing point is on the horizon at the eye level of the viewer.

A similar effect is visible when you look down the parallel rails of a railroad track. They also appear to converge in the distance (see **Figure 4.74**).

FIGURE 4.74.
Railroad tracks illustrate convergence.

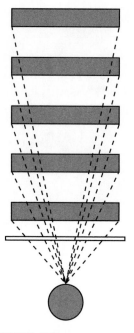

FIGURE 4.75.
Top view of person viewing a series of railroad ties.

3. **Figure 4.75** is a top view of a person observing a railroad track and a series of equally spaced railroad ties. The lines of sight are drawn from the viewer to the ends of the first five ties.

a) To simplify the picture, the rails connecting the ties are not shown. Describe the relationship between the two rails in real life.

b) The **projection** of the first tie in the picture plane is almost as wide as the picture (see Figure 4.74). Compare the length of the projection of the fourth tie to the length of the projection of the first tie. How do they differ?

c) Describe the projection of the railroad tie that is as far as the eye can see from the viewer. (What does it look like? Where is it located in the picture plane?)

d) Each tie connects the parallel rails. What does your answer to c tell you about the images of the rails in the picture?

CHECK THIS!

In this chapter the projection of a real object is its image on the picture plane. The image is formed by lines of sight from the viewer to points on the object.

4. **Figure 4.76** is a side view of the same person viewing the first four railroad ties.

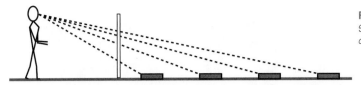

FIGURE 4.76.
Side view observing a series of equally spaced railroad ties.

a) The projection of the first tie appears near the bottom of the picture plane (see Figure 4.74). How is the location of the projection of the fourth tie different from the location of the projection of the first tie?

b) Describe the location in the picture plane of the tie (not shown) that is farthest from the viewer. Assume the ground is level.

Lines of sight may be drawn from a viewer to one or more objects in a scene. A line of sight that is parallel to the ground (assuming the ground is level) is said to be at the eye level of the viewer (see **Figure 4.77**).

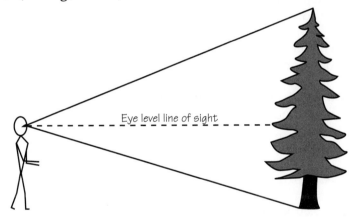

Eye level line of sight

FIGURE 4.77.
Sketch illustrating line of sight at the eye level of the viewer.

5. **Figure 4.78** shows an incomplete picture. The artist has drawn in the **horizon**, the horizontal line at the viewer's eye level where the earth and sky appear to meet. She has added two dots at the bottom of her picture to mark the edges of a roadway that she plans to sketch.

a) Where in the picture do you expect to find the portion of roadway that is as far as the eye can see: above the horizon, below the horizon, or on the horizon? How large will it appear?

Horizon

Roadway

FIGURE 4.78.
Incomplete picture of roadway.

b) Trace a copy of Figure 4.78. Sketch the roadway for the artist. What assumptions did you make about the viewer's location in relation to your roadway?

FIGURE 4.79.
Print Shop Interior by Jan van der Straet.

6. a) On Handout 4.1, which is the same as **Figure 4.79**, draw lines along all edges that represent lines parallel to the center aisle. Extend the lines to identify the locations of any corresponding vanishing points.

b) What is true of the vanishing points for sets of parallel lines from various objects in Figure 4.79? Explain.

Next, you will apply what you have learned about the principles of diminution and convergence to drawings with two vanishing points.

THE CORNER STONE

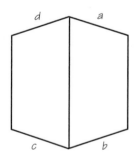

FIGURE 4.80.
View of cube looking directly at a vertical edge.

7. Hold a cube so that the top and bottom are parallel to the ground and one face is parallel to your plane of vision. Keeping the bottom of the cube parallel to the ground, rotate the cube so that no face is parallel to your plane of vision.

a) Trace a copy of **Figure 4.80** onto paper.

b) Which pairs of edges in Figure 4.80 are, in reality, parallel? How is the principle of convergence apparent in the sketch of the rotated cube?

c) For the cube in Figure 4.80, there are two vanishing points. Draw them on your copy of Figure 4.80.

d) The vanishing point is on the horizon for horizontal lines that increase in distance from the viewer. You have drawn two vanishing points in c. Are there two horizons? Explain.

8. a) Locate the vanishing points and horizon in **Figures 4.81 and 4.82.** Draw the vanishing points and horizon on Handout 4.2 provided by your teacher.

FIGURE 4.81.
Photograph for Question 8.

FIGURE 4.82.
Painting for Question 8.

b) How many vanishing points are there? How do you know?

9. Practice drawing a rectangular box using two vanishing points. Use pencil and paper or a geometric drawing utility. Begin with a drawing that looks similar to **Figure 4.83.** Notice the box is drawn entirely below the horizon.

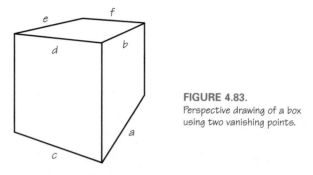

FIGURE 4.83.
Perspective drawing of a box using two vanishing points.

a) Why does edge *e* appear to converge to the same vanishing point as edges *a* and *b*?

b) Why does edge *f* appear to converge to the same vanishing point as edges *c* and *d*?

c) In Figure 4.83, you see two sides and the top of the box. Why is the artist able to see the top of the box in Figure 4.83 but not in Figure 4.80?

d) Describe what happens to the drawing of the box when you bring the two vanishing points closer together.

e) Describe what happens to the drawing of the box when you move the two vanishing points farther apart.

10. Revisit the painting by William Hogarth (Figure 4.46 on p. 216). How many vanishing points do you observe? Describe where they are located and explain how you found them.

1. Draw lines of convergence and identify the vanishing points for **Figure 4.84.**

FIGURE 4.84.
Le Pont de l'Europe 1876 by Gustave Caillebotte.

2. a) Find the vanishing point for **Figure 4.85.** Use Handout 4.3 provided by your teacher.

   b) Is the vanishing point in the center of the picture, to the right of center, or to the left of center? Is the vanishing point inside or outside of the bounds of the picture?

3. Revisit Figure 4.79 on p. 236. Draw the horizon line using a different color pen from the one used to draw the lines of convergence.

4. Revisit Figure 4.85. Draw the horizon line using a different color pen from the one you used in Question 2.

FIGURE 4.85.
The Flatiron Building.

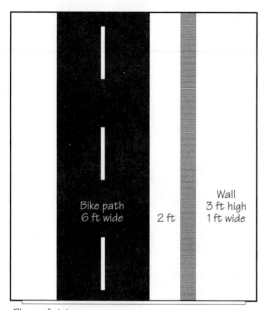

Bike path
6 ft wide   2 ft

Wall
3 ft high
1 ft wide

Plane of vision   You

FIGURE 4.86. Aerial view.

5. **Figure 4.86** shows an aerial view of a scene.

a) Draw a perspective view of the same scene with one vanishing point.

b) Suppose power poles are in line with the viewer's line of sight to the vanishing point on the horizon. Describe what the viewer sees. What principle(s) of perspective is (are) responsible for this phenomenon?

6. Optical illusions can be created when the principle of diminution or the principle of convergence is violated.

a) Describe the effect of not reducing the height of a series of objects and not reducing the spacing between the objects as other objects converge to the vanishing point or points (see **Figure 4.87**).

FIGURE 4.87.
Drawing for 6a.

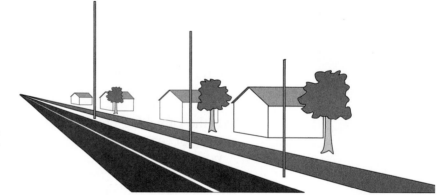

b) Describe the effect of not reducing the height but allowing the spacing between the objects to agree with principles of perspective (see **Figure 4.88**).

FIGURE 4.88.
Drawing for 6b.

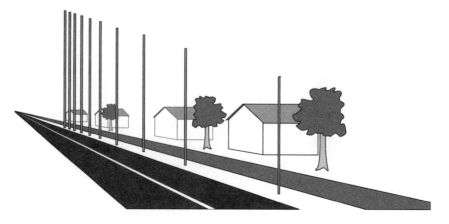

c) Explain how the B.C. cartoon in **Figure 4.89** violates a principle of perspective drawing.

FIGURE 4.89.
B.C. cartoon.

7. **Figure 4.90** shows a sketch of a room containing three rectangles representing a chest, a large wardrobe, and an air purification unit. The drawings of the objects are incomplete. Only the sides facing the viewer are sketched.

a) Trace Figure 4.90 onto paper. Complete the perspective drawings of the chest, wardrobe, and air purification unit so that they appear to have depth. What principles of perspective did you use to guide your drawing?

b) In your completed drawing, you should be able to see the top of one of the rectangles and the bottom of another. For the third object, you should be able to see neither its top nor its bottom. What determines whether the viewer can see the top or the bottom of an object in a perspective drawing? Under what conditions are both the top and the bottom of an object not visible to the viewer?

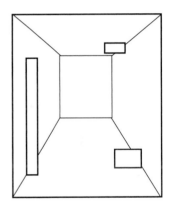

FIGURE 4.90. Incomplete drawing of room containing three objects.

8. **Figure 4.91** shows the front view of a cottage. Draw the cottage in perspective with two vanishing points.

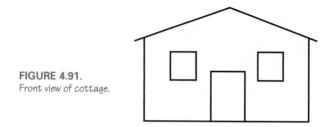

FIGURE 4.91.
Front view of cottage.

9. Eight-year old Jackson Barber drew the picture in **Figure 4.92.** It does not adhere to the principles of perspective. Draw the same scene on your own paper using principles of perspective and two vanishing points.

FIGURE 4.92.
Child's drawing that does not conform to principles of perspective.

Setting Your Sights

FIGURE 4.93.
Trees lining a roadway.

1. In **Figure 4.93** you see a photograph of a roadway.

   a) What do you notice about the lines that form the sides of the road in the picture?

   b) Next, focus on the trees on the right side of the road. Suppose you draw a line connecting the treetops (assume the trees are approximately the same height) and another line connecting the tree bottoms. What would you notice about these lines?

   c) Use the rulers that surround the picture to identify the location of the vanishing point for the lines joining the tops of the trees and the bottoms of the trees. Then imagine drawing a horizontal line through that vanishing point. What does this horizontal line represent?

   d) The shoulders of the road extend beyond the white lines on both edges of the road. Pretend a ladder is placed from the edge of the shoulder to the top of each tree, forming a triangle. Describe the triangles that result from doing this with each tree.

e) Sketch a top view of the road. (Use circles to indicate the trees.) What should be true of the lines that form the sides of the road? What about the lines connecting the treetops and tree bottoms?

FIGURE 4.94.
Chemin de la Machine, Louveciennes by Alfred Sisley.

2. a) On Handout 4.4, which is the same as **Figure 4.94**, draw lines marking the sides of the road and along the treetops on either side of the road. Draw lines along the roofs of buildings on the right side of the road.

b) What is true of these lines? Explain.

3. **Figure 4.95** shows two roads meeting at an intersection. You are an architect, and your client wants to build an office at the corner opposite where the viewer is standing. Draw Figure 4.95 on your own paper, or use Handout 4.5 provided by your teacher.

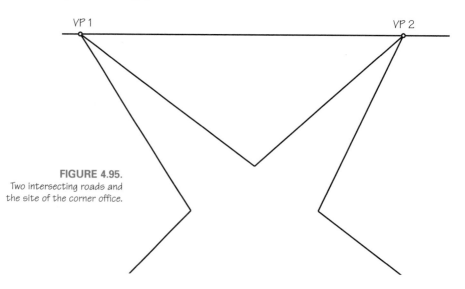

FIGURE 4.95.
Two intersecting roads and the site of the corner office.

a) Make a detailed drawing showing what the office will look like when it is constructed. Include windows, doors, and a sign on the building. (For an extra challenge, try drawing a sloped roof rather than a flat roof.)

b) What is the effect if the vanishing points are brought closer together or moved farther apart? Draw sketches to illustrate your answer.

c) What would have to change in the context to increase the apparent distance between the vanishing points?

Modeling Project It's a Masterpiece!

In this chapter you have seen a number of ways artists use mathematics to create art. Now it is *your* turn. Use tools you have developed in this chapter to create an original piece of art. This drawing or painting should use the principles of symmetry, overlapping, and diminution. While the subjects of the art can be anything you would like to draw, including an architectural drawing of a building or objects around your home and school, your drawing should include notes that describe how you used mathematics to transform your objects in order to create patterns and give the drawing depth. These notes should include the types of transformations you used to create your work and details like the scale factors you chose and why.

For a twist on this project, revisit the Hogarth painting (Figure 4.46 on p. 216) and create your own impossible drawing. Instead of your notes explaining how you used mathematics to create the image, your notes should explain how you could use mathematics to "fix" your drawing.

Finally, be creative and have fun. Be ready to share your drawing with the class and answer questions about your work.

Practice Problems

1. In each of these figures, identify a basic element and describe ways to get the other sections of the figure by applying one or more of the transformations you studied in this chapter. Your basic element should be as small a portion of the entire figure as possible.

a)

b)

c)

d)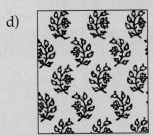

e)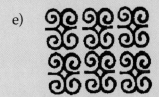

f)

2. The natural world has many examples of symmetry (although it is usually not perfect). Describe the lines of symmetry and the types of symmetry in each of these photos of living things.

3. Three spheres are arranged on a table.

   a) **Figure 4.97** shows the top view. Draw the views from A and B. What principle of perspective did you use in your drawing?

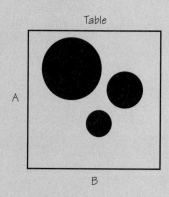

Table

A

B

FIGURE 4.97.
Three spheres
arranged on a table.

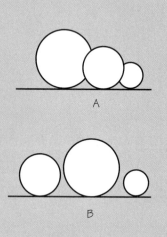

A

B

FIGURE 4.98.
Side views of the
rearranged spheres.

   b) Suppose the spheres are rearranged. Views from sides A and B are shown in **Figure 4.98**. Draw an overhead view of the table. Explain how you arrived at your answer.

4. In each of these questions, suppose a ruler is held at arm's length so the plane of vision is 27 inches from the eye of the viewer.

a) You are 100 yards away from the goalpost on a football field. The image height of the goalpost is 1.6 inches. How high are the goalposts?

b) Close to shore on the opposite side of a lake is a tower that is 50 feet tall. The image height of the tower is 1 inch. How wide is the lake?

c) A certain basketball player stands 86 inches tall. You are standing under the basket and he is attempting a free throw, which means he is about 18 feet from you. How tall is his image?

5. **Figure 4.99** shows the top view of a person looking at five railroad ties.

a) Measure the length of each projection in the picture plane. Then measure the distance between each tie and the viewer's eye. Record your measurements in a table like **Figure 4.100.**

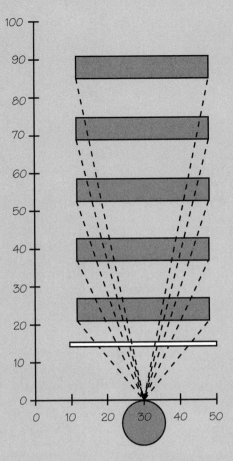

Tie Number	1	2	3	4	5
Distance to Viewer's Eye					
Projection Length					

FIGURE 4.100. Table for recording railroad tie lengths and distances.

b) Based on your measurements, describe the relationship between the length of the projection of each tie and its distance from the viewer's eye.

c) Graph your data. Label the horizontal axis distance to eye and label the vertical axis projection length. Describe the shape of your plot.

6. a) Copy **Figure 4.101** on your own paper and label the unlabeled sides and vertices according to mathematical convention.

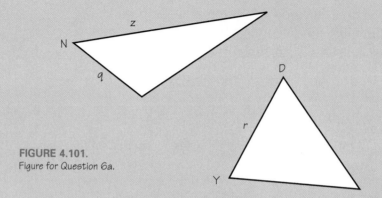

FIGURE 4.101.
Figure for Question 6a.

b) For each of the triangles in **Figure 4.102**, how long is the side opposite angle V?

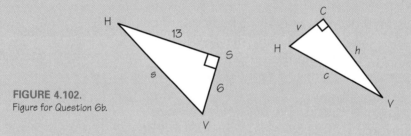

FIGURE 4.102.
Figure for Question 6b.

7. Three equal-height poles are to be placed 100 feet apart, increasing in distance from the viewer. Which drawing (see **Figure 4.103**) seems to be the best representation? Explain your answer.

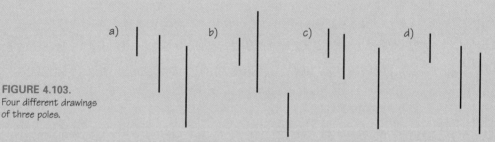

FIGURE 4.103.
Four different drawings of three poles.

8. **Figure 4.104** is a perspective view of a hallway.

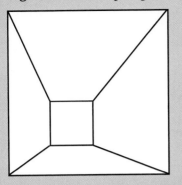

FIGURE 4.104.
Perspective view of hallway.

a) Trace a copy of Figure 4.104 on your own paper. Determine the vanishing point and the horizon line.

b) Raise the vanishing point and move it to the right. Redraw the hallway using the new vanishing point.

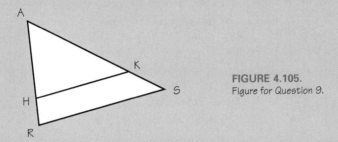

FIGURE 4.105.
Figure for Question 9.

9. In **Figure 4.105**, RS is parallel to HK.

a) Which angles have equal measure?

b) What is the relationship between triangles ARS and AHK? Explain.

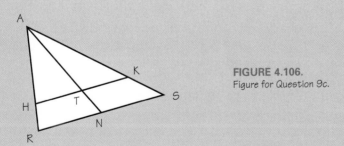

FIGURE 4.106.
Figure for Question 9c.

c) **Figure 4.106** was created by adding the line segment AN to Figure 4.105. Segment AN bisects segment HK. Explain why it must also bisect RS.

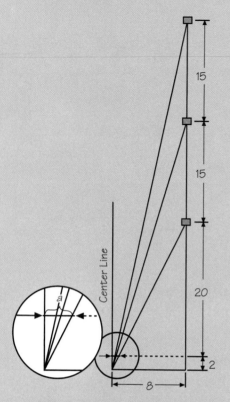

FIGURE 4.107.
Top view of hallway support poles, including artist's viewing position (not to scale).

10. Suppose you want to draw accurately a series of poles that support a cover above a walkway. **Figure 4.107** shows a top view of the poles in relation to the artist and the plane of vision. The artist chooses a viewing position 8 feet to the left of the line of poles and 22 feet from the first pole that is to be drawn in the picture. The 9-foot poles are 15 feet apart. The distance between the artist's eye and the picture plane is 2 feet.

a) The center line is an imaginary line from the artist's eye to the vanishing point. In a top view, this line is parallel to the line of poles. Describe the relationship in a perspective drawing between (1) the center line and (2) a line drawn through the poles at the artist's eye height. If drawn, how would the center line appear in the picture?

b) In the drawing, what is the scale at the first pole? At the second pole? At the third pole?

c) What is the image size of Pole 1? What are the image sizes of Pole 2 and Pole 3?

d) Determine the horizontal distance *a*, the distance between the image of Pole 1 and the center line. How can you use the scale at Pole 1 to determine this distance?

e) Determine the horizontal distance *b*, the distance between the image of Pole 2 and the imaginary center line.

f) Determine the horizontal distance *c*, the distance between the image of Pole 3 and the imaginary center line.

Glossary

KEY CONCEPTS

Convergence: Artistic principle asserting that lines or edges of objects that in reality are parallel appear to come together as they recede from the observer.

Corresponding angles: Angles in corresponding positions within similar figures. Corresponding angles are equal in measure.

Dilation: A transformation in which the distance to every point's image is exactly k times its original distance from some specified, fixed point. The fixed point is called the center of the dilation; k is the scale factor for the dilation.

Diminution: The phenomenon by which an object appears smaller as its distance from the observer increases.

Glide reflection: A composite of translation and reflection. A glide reflection reverses a figure's orientation.

Horizon Line: The imaginary line on the plane of vision (or picture plane) containing the vanishing points of all horizontal (in reality, parallel to the ground) converging lines. It is always on the same level as the viewer's eyes (the eye level).

Hypotenuse: The side opposite the right angle of a right triangle. It is the longest side in the right triangle.

Image: In transformations, a new object created by transforming a pre-image.

Isometry: A transformation that preserves size and shape.

Line of Sight: An imaginary line between the viewer's eye and a point on a three-dimensional object being observed.

Line of Symmetry: A line that divides a figure into two identical mirror images.

One-Point Perspective: Occurs with only one family of horizontal parallel lines not parallel to the plane of vision; all the lines converge to a single vanishing point located on the horizon line.

Overlapping: A technique to achieve a sense of depth and space in drawings by showing which objects are in front and which are in back.

Perspective: A technique of representing objects from three-dimensional space in a two-dimensional plane, like a picture. Through the use of perspective, an artist's two-dimensional drawing or painting imitates the appearance of a three-dimensional object or scene.

Plane of Vision (or Picture Plane): An imaginary plane between the observer and the three-dimensional object(s) being viewed. The image of the object is projected onto the plane. In this chapter the picture plane is always vertical.

Point Symmetry: A figure has point symmetry when it fits exactly on itself after a rotation of 180°.

Pre-image: When a geometric object is transformed into another one, the original is called a pre-image.

Projection: The projection of a real object is its image on the picture plane. The image is formed by lines of sight from the viewer to points on the object.

Pythagorean Formula: A formula relating the lengths of three sides of a right triangle: $a^2 + b^2 = c^2$, where c is the length of the hypotenuse and a and b are the lengths of the other two sides (legs).

Reflection: A transformation that flips a figure across a line. A reflection reverses a figure's orientation.

Reflection (line) Symmetry: When a figure fits exactly on itself after reflection through a line. The line divides the figure into two identical halves.

Rotation: A transformation that spins a figure about a given center a given number of degrees.

Rotation Symmetry: When a figure fits exactly on itself after it spins less than 360° about a center.

Scale: The ratio of an image size (height or width) in a picture to the actual size of the object.

Scale Factor: In a dilation, the ratio of the new lengths to the old lengths.

Similar Triangles: Two triangles are similar when they have equal corresponding angles and equal ratios of corresponding sides.

Transformation: A process that creates a new figure from an original. The original is called the pre-image; the new figure is called the image.

Translation: A transformation that slides a figure a given distance in a given direction.

Two-Point Perspective: Occurs with exactly two families of horizontal parallel lines not parallel to the plane of vision; all the lines converge to one of two vanishing points located on the horizon line.

Vanishing Point: The imaginary point to which parallel lines not in the plane of vision appear to converge.

5

CHAPTER

Motion: Quadratic Functions

It's Show Time

You may have watched scenes like these in the movies:

The star jumps off a roof and lands in the back of a passing truck.

During a chase the star speeds down the street on a motorcycle, through a cross street and just misses a truck.

Auto thrill shows also do stunts using cars and trucks. Jeff Lattimore's stunt in Chittwood's Thrill Show is "The Leap for Life." Jeff has performed it for years. In his stunt Jeff climbs a ladder and stands on an eight-foot stool. Then the ladder is removed leaving Jeff in the air just as a car comes speeding at him. Jeff jumps right before the car hits the stool and knocks it out from under him. Jeff lands safely on the ground.

Since this stunt is complicated, trial-and-error would be a poor method. A mistake could result in loss of life. Using mathematics and understanding the laws of physics are key to planning stunts.

In this chapter you'll see functions that model motion. You'll use mathematical models to predict where an object will be at a specific time. Then you'll use the models to plan and test a small-scale stunt in your class.

Walking the Walk

In this section you will look at distance-versus-time graphs. For these graphs, you will consider time as the independent variable and distance as the dependent variable.

A motion detector can track a student walking toward or away from it. The motion detector along with a graphing calculator can be used to create a distance-versus-time graph of a student walk.

Imagine a student walking away from a motion detector.

1. As time increases, what happens to the student's distance from the motion detector?

2. Imagine a distance-versus-time graph of the walk. Would you expect the graph to be a straight line or a curved line? Why?

3. Would you expect the graph to increase or decrease from left to right? Why?

Use a motion detector and a graphing calculator to collect data and create a distance-versus-time graph for a classmate who walks away from the motion detector.

Sketch your graph.

4. How can you tell from the graph that the student was walking away from the motion detector?

5. Was the student walking at a constant rate? How can you tell from the graph?

Again use a motion detector and a graphing calculator to collect data and create a distance-versus-time graph for a second classmate who walks away from the motion detector.

Sketch your graph.

A sample graph is shown in **Figure 5.1.**

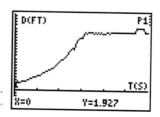

FIGURE 5.1.
A sample distance-versus-time graph.

6. How can you tell from the graph that the second student was walking away from the motion detector?

7. Was the second student walking at a constant rate? How can you tell from the graph?

Now you will analyze data taken from student walks in front of a motion detector. When you track the walk of another student, you can see how you would have to walk to produce the same distance-versus-time graph.

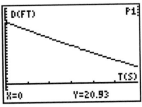
FIGURE 5.2.
Cara's distance-versus-time graph.

Cara walked in a straight line in front of a motion detector. The motion detector recorded her distance from it every 0.1 second for 6 seconds. Cara's distance-versus-time graph appears in **Figure 5.2.**

8. Based on Figure 5.2, did Cara walk toward or away from the motion detector? How can you tell from her graph?

9. Did she begin moving as soon as the motion detector began collecting data or did she pause first? How can you tell from her graph?

10. After she began moving, did she walk at a constant rate, or did she speed up or slow down? Explain how you decided.

11. What plan could you give to a walker so that he or she could walk a graph that is the same as Cara's?

12. Using your group's plans, have one group member walk in front of a motion detector. Is the graph of the walk close to the graph of Cara's walk?

13. How can you modify your plans so that the walk will more closely model Cara's walk?

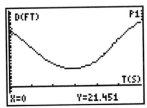
FIGURE 5.3.
Paul's distance-versus-time graph.

Paul walked in a straight line in front of a motion detector. The motion detector recorded his distance from it every 0.1 second for 6 seconds. Paul's distance-versus-time graph appears in **Figure 5.3.**

14. Based on Figure 5.3, did Paul walk toward or away from the motion detector? How can you tell from his graph?

15. After he began moving, did he walk at a constant rate, or did he speed up or slow down? Explain how you decided.

16. What rules could you give to a walker so he or she could walk a graph like Paul's?

17. Using your group's rules, have one group member walk in front of a motion detector. Is the graph of the walk like the graph of Paul's walk?

18. How can you modify your rules so that the walk will more closely model Paul's walk?

19. What differences in Cara's and Paul's walks caused the differences in the two graphs?

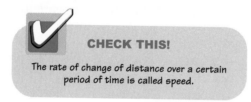

CHECK THIS!

The rate of change of distance over a certain period of time is called speed.

SECTION
5.2

Linear versus Non-Linear

In the last section you saw graphs created by students walking in a straight line in front of a motion detector. Cara's graph was close to a straight line while Paul's graph was curved.

In this section you will compare data that create a straight line (a linear relationship) with data that create a curved graph (a quadratic relationship).

CHECK THIS!

A linear relationship has a constant rate of change and is represented by a function of the form $f(x) = mx + b$.

CHECK THIS!

A quadratic relationship has a variable rate of change and is represented by a function of the form $f(x) = ax^2 + bx + c$.

Victoria walked in front of a motion detector. Some of the data are shown in **Figure 5.4.**

Time, t (seconds)	Distance, d (feet)
0	2
1	5
2	8
3	11
4	14
5	17

FIGURE 5.4.
Data from Victoria's walk.

Ricardo also walked in front of a motion detector. Some of the data recorded by the motion detector for his walk are shown in **Figure 5.5.**

Time, *t* (seconds)	Distance, *d* (feet)
0	25
1	15
2	9
3	7
4	9
5	15

FIGURE 5.5.
Data from Ricardo's walk.

1. What do you see in the data for Victoria's and Ricardo's walks?

2. Use your graphing calculator to create a distance-versus-time scatterplot of Victoria's data. Record your window and sketch your graph.

3. Use your graphing calculator to create a distance-versus-time scatterplot of Ricardo's data. Record your window and sketch your graph.

4. What do you see in the graphs for Victoria's and Ricardo's walks?

5. What do you think might be a cause of the differences in the graphs?

One tool to help develop representation of a function is differences in consecutive table values.

Let's look at the first and second differences in the two tables.

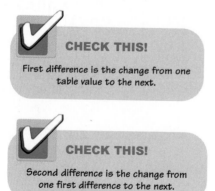

CHECK THIS!

First difference is the change from one table value to the next.

CHECK THIS!

Second difference is the change from one first difference to the next.

Chapter 5 Motion: Quadratic Functions

6. Find the first and second differences in the table values for Victoria's walk. Record them in a table like the one in **Figure 5.6.**

Time (seconds)	Distance (feet)	First Difference	Second Difference
0	2		
1	5	—	—
2	8	—	—
3	11	—	—
4	14	—	—
5	17	—	

(+1 between each time value)

FIGURE 5.6.
Differences for Victoria's walk.

7. What patterns do you see?

The function that models Victoria's walk where d is distance and t is time is $d = 3t + 2$. Recall that an equation in this form ($y = mx + b$) has a linear graph where m is the slope of the line and b is the y-intercept.

8. Look at the table with differences for Victoria's walk. Where do you see the number 3? What does 3 represent here?

9. Look at the table with differences for Victoria's walk. Where do you see the number 2? What does 2 represent here?

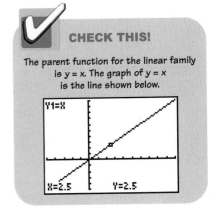

CHECK THIS!

The parent function for the linear family is $y = x$. The graph of $y = x$ is the line shown below.

Y1=X

X=2.5   Y=2.5

10. How do you think the first and second differences in a table of data can be represented by a linear function?

11. Explain how you think first differences relate to the function rule.

12. Find the first and second differences in the table values for Ricardo's walk. Record them in a table like the one in **Figure 5.7.**

13. What patterns do you see?

Time (seconds)	Distance (feet)	First Difference	Second Difference
0	25		
1	15	—	—
2	9	—	—
3	7	—	—
4	9	—	—
5	15	—	

(+1 between each time value)

FIGURE 5.7.
Differences for Ricardo's walk.

The equation for the function that models Ricardo's walk where d is distance and t is time is $d = 2t^2 - 12t + 25$. Recall that $y = ax^2 + bx + c$ where $a \neq 0$ is a **quadratic function**. Quadratic functions are often used to model relationships in the real world such as distance versus time and area versus length. The graph of a quadratic function is U-shaped and is called a **parabola**. The point at which the graph changes between increasing and decreasing is called the **vertex**.

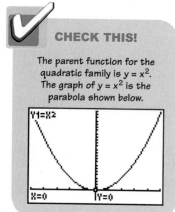

CHECK THIS!

The parent function for the quadratic family is $y = x^2$. The graph of $y = x^2$ is the parabola shown below.

14. Look at the table for Ricardo's walk. What is the relationship between the second differences and the number 2, the coefficient of t^2, in $d = 2t^2 - 12t + 25$?

15. Look at the table for Ricardo's walk. Where do you see the number 25? What does 25 represent here?

16. Look at the table for Ricardo's walk. How does the first, first difference compare to the coefficients of t^2 and t in $d = 2t^2 - 12t + 25$?

17. How can the first and second differences in a table of data be represented by a quadratic function?

18. How do the first and second differences relate to the function rule that models the data in a quadratic relationship?

19. In **Figure 5.8,** one of the tables represents a linear function while the other one shows a quadratic function. Without graphing, find which one is linear and which one is quadratic. How did you decide?

Table 1

x	y
0	0
1	2
2	4
3	6
4	8

Table 2

x	y
0	0
1	2
2	8
3	18
4	32

FIGURE 5.8.
Two tables of x- and y-values.

20. Use the first and second differences in Table 1 to write a function rule.

21. Use the first and second differences in Table 2 to write a function rule.

Chapter 5 Motion: Quadratic Functions

Assignment

INTERPRETING MOTION GRAPHS

The first graph in **Figure 5.9** was recorded when Jasmyne walked in front of a motion detector. The program recorded time in seconds and distance in feet. The TRACE feature of her calculator was used to find the distance for 0.9, 2.9, and 4.9 seconds (the other three graphs).

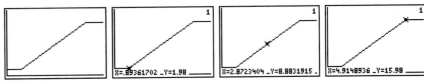

FIGURE 5.9. Graph of Jasmyne's walk.

1. How far was Jasmyne from the motion detector when the program began running?

2. How many seconds did Jasmyne stand still before moving?

3. How much time did Jasmyne spend walking?

4. How far did she walk?

5. During the time that Jasmyne was walking, did she walk at a constant rate, increase her speed, or decrease her speed?

6. During the time that Jasmyne was walking, at what rate did she walk?

7. The two graphs in **Figure 5.10** were recorded by a motion detector when a student walked in front of it. Assuming the scales are the same for both graphs, in which case was the student walking faster? Explain your answer.

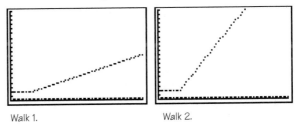

Walk 1.   Walk 2.

FIGURE 5.10.
Two walks recorded
by a motion detector.

A motion detector can be used to create a distance-versus-time graph for a toy car.

8. Imagine a toy car moving along a straight line in front of a motion detector. The motion produces a distance-versus-time graph that is a straight line. What does this tell you about the toy car's rate?

9. How could you determine the toy car's rate?

10. Suppose that the distance-versus-time graph has a negative slope. What does that tell you about the car's motion?

11. Suppose that a toy car is traveling along a straight line in front of a motion detector. The motion produces a distance-versus-time graph that is curved. What does this tell you about the car's rate? Explain.

Suppose a movie stunt calls for a person to step off a roof as soon as a truck reaches a white line painted across a road. To design such a stunt, you need to calculate details about the person's fall. For example, you need to know how long it would take for the person to land on the ground.

12. Imagine that a stunt person steps off a roof 25 feet above the ground and falls vertically to the ground. Sketch a possible height-versus-time graph for the falling body. Add scales and labels to your axes.

13. Explain why the stunt person will fall as you have shown in your graph.

14. How might you collect data that would help you decide whether your graph does a good job of describing the stunt person's fall?

Jackie and Jermaine drew the graphs in **Figure 5.11** as part of their answers to Question 12.

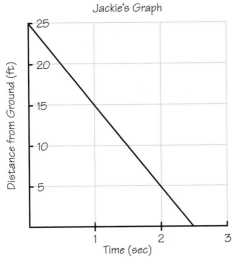

Jackie's Graph

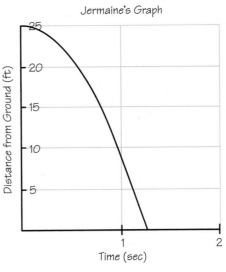

Jermaine's Graph

FIGURE 5.11.
Jackie's and Jermaine's graphs.

15. Suppose these graphs show the motion of a walker in front of a motion detector. Write a plan for a walker so that the graph would look like Jackie's.

16. Write rules for a walker so that the graph would look like Jermaine's.

A motion detector tracked Rashida as she jogged. Some of the data are shown in **Figure 5.12.** Create a scatterplot of the distance-versus-time data. Be sure to add labels and scales to your axes.

17. Sketch and describe the shape of Rashida's graph.

18. Did Rashida jog at a steady pace, speed up, or slow down? How can you tell from the graph?

19. Use first and second differences to find a rule. Test your rule over your scatterplot. Write your rule and sketch the graph.

Time (seconds)	Distance from Sensor (feet)
0	2
1	3.15
2	6.8
3	12.95
4	21.6
5	32.75
6	46.4
7	62.55
8	81.2

FIGURE 5.12.
Distance-versus-time data for Rashida's walk.

Characteristics of Quadratic Functions

In the last section you explored linear and quadratic relationships. In this section you will formalize how and why those relationships work the way they do.

Recall the student walks in front of a motion detector.

1. Describe a walk that creates a linear relationship.

2. Describe a walk that creates a non-linear relationship.

3. Describe a walk that creates a linear relationship with a positive slope.

4. Describe a walk that creates a linear relationship with a negative slope.

5. Describe a walk that has a graph similar in shape to a parabola.

As you discovered in the last section, the use of first and second differences is a good tool to develop a function rule.

Let's take a closer look at linear and quadratic relationships.

CHECK THIS!

Recall that in a linear relationship the first differences are constant.

CHECK THIS!

Recall that in a quadratic relationship the second differences are constant.

6. In a linear relationship, how are the first differences related to the function rule?

7. In a linear relationship, how are the first differences related to the graph of the function rule?

8. In a quadratic relationship, how are the first differences related to the graph of the function rule?

9. In a quadratic relationship, how are the second differences related to the function rule?

x	$y = ax^2 + bx + c$	First Differences	Second Differences
0	$a \cdot 0^2 + b \cdot 0 + c = c$		
		$a + b + c - c =$ $a + b$	
1	$a \cdot 1^2 + b \cdot 1 + c = a + b + c$		$2a$
		$4a + 2b + c - (a + b + c) =$ $3a + b$	
2	$a \cdot 2^2 + b \cdot 2 + c = 4a + 2b + c$		$2a$
		$9a + 3b + c - (4a + 2b + c) =$ $5a + b$	
3	$a \cdot 3^2 + b \cdot 3 + c = 9a + 3b + c$		$2a$
		$16a + 4b + c - (9a + 3b + c) =$ $7a + b$	
4	$a \cdot 4^2 + b \cdot 4 + c = 16a + 4b + c$		

FIGURE 5.13.
Quadratic differences table.

10. Notice that the first of the first differences in **Figure 5.13** is equal to $a + b$ and the second difference is equal to $2a$. How can you use these facts to develop a quadratic model?

11. Use these facts to justify the quadratic model, $d = 2t^2 - 12t + 25$, for Ricardo's walk in Section 5.2.

12. Use first and second differences to find a function rule for each data set.

a)

x	y
0	2
1	10
2	24
3	44
4	70

b)

x	y
0	4
1	6
2	8
3	10
4	12

SUMMARY

Using differences can help write a function rule for linear or quadratic data. Assume that the difference between consecutive x-values in the table is constant. If the relationship is **linear**, then the first differences in the y-values will be constant, indicating a constant rate of change. Linear relationships can be expressed in the form $y = mx + b$ where m and b are real numbers.

If the first differences in the y-values are not constant, then there is not a constant rate of change and the relationship is non-linear. If the second differences, or the differences between the first differences, are constant then the relationship is **quadratic**. Quadratic relationships can be expressed in the form $y = ax^2 + bx + c$ where a, b, and c are real numbers.

Using Finite Differences to Write Linear Functions:

Time (seconds)	Distance (feet)
0	2
1	5
2	8
3	11
4	14
5	17

+1 between each Time value (Δx); +3 between each Distance value (Δy)

FIGURE 5.14.
Finite differences for linear functions.

$y = mx + b$ where

❖ b is the y-value when $x = 0$

❖ m is the ratio of $\frac{\Delta y}{\Delta x}$.

Using Finite Differences to Write Quadratic Functions:

❖ Use the polynomial form $y = ax^2 + bx + c$.

❖ If the x-interval is equal to 1, the value for a is half of the constant second difference ($2a$ = second difference).

❖ If the first x-value is 0, the first, first difference is equal to $a + b$.

❖ The y-value when $x = 0$ is the value of c.

CHECK THIS!

A quadratic is a type of polynomial, a mathematical expression of the form $a_nx^n + a_{n-1} + a_{n-2}x^{n-2} + \ldots + a_1x^1 + a_0$. Polynomials of one, two, or three terms are called monomial, binomial, or trinomial, respectively. The polynomial form of a quadratic function is the same as the general form, $y = ax^2 + bx + c$.

x	y
0	6
1	5
2	10
3	21
4	38

FIGURE 5.15.
Table of x- and y-values.

x	y
0	
1	
2	
3	
4	

FIGURE 5.16.
Table for Question 2.

1. Use first and second differences to find a quadratic function for **Figure 5.15.**

2. Make a quadratic function and create a table.

   a) First choose values for *a*, *b*, and *c*.

   b) Next substitute them into $y = ax^2 + bx + c$. Write the results in a table like the one shown in **Figure 5.16.**

   c) Use first and second differences to check your work.

The table in **Figure 5.17** shows the distance a car travels from the time the driver decides to stop to the time the car comes to a complete stop. These data are for an alert driver, a well-maintained car, and dry road conditions.

3. Find a function rule to model the relationship between each distance and speed.

   a) Reaction distance: _____

   b) Braking distance: _____

   c) Stopping distance: _____

4. Complete Figure 5.17.

Speed	Reaction Distance	Braking Distance	Stopping Distance
0	0	0	0
10	10	5	15
20	20	20	40
30	30	45	75
40	40	80	120
50			
60			
70			

FIGURE 5.17.
Car stopping distance data.

5. According to Figure 5.17, a driver should allow 120 feet to stop safely when driving 40 mph. About how many car lengths is this? In your experience, how much space do most drivers allow?

6. How might an advocate for reducing highway speeds from 70 mph to 55 mph use stopping distances to support his or her case?

7. The Figure 5.17 data are for an alert driver. How might tuning the radio or talking on a cell phone or to other passengers affect these data?

8. Use your knowledge of quadratic models to complete the table in **Figure 5.18.**

FIGURE 5.18.
Stopping distance data
for large trucks and trains.

Speed	Stopping Distance for Large Trucks	Stopping Distance for Trains
0	0	0
10	21	900
20	54	2020
30	99	3360
40	156	4920
50		
60		
70		

9. Find a function rule to model the relationship between each distance and speed.

   a) Stopping distance for large trucks: _____

   b) Stopping distance for trains: _____

10. The average speed of trains traveling through town is about 40 mph. What is the stopping distance for that speed in feet? What is the stopping distance in miles?

11. If the train comes around a bend and the conductor sees a car stalled on the tracks about half a mile ahead, what is the fastest speed the train can travel and be able to stop to avoid a collision?

<table>
<tr><th>SECTION
5.4</th><th colspan="2">Quadratics on the Move </th></tr>
</table>

In Chapter 4 you explored translations, reflections, and dilations. These transformations can be applied to the graphs of functions. Since graphs of functions are done on a coordinate plane, you can use algebraic rules for these transformations.

Knowing about function transformations will help you later in this chapter. If you know how to translate, reflect, and dilate a parabola then you can quickly fit a quadratic function to a set of data.

TRANSLATIONS OF QUADRATIC GRAPHS

1. Use grid paper to sketch the graph of $y = x^2$. Then complete a table like **Figure 5.19**.

The effect of changing $y = x^2$ to $y = x^2 + k$.

2. Complete the table in **Figure 5.20** for $y = x^2$ and $y = x^2 + 3$, then graph both functions on your graphing calculator. Make a sketch of both graphs.

x	$y = x^2$
−3	
−2	
−1	
0	
1	
2	
3	

FIGURE 5.19.
Table for Question 1.

x	Pre-Image $y = x^2$	Image $y = x^2 + 3$
−3		
−2		
−1		
0		
1		
2		
3		

FIGURE 5.20.
Table for Question 2.

3. Write a sentence that describes the transformation.

4. Complete the table in **Figure 5.21** for $y = x^2$ and $y = x^2 - 2$, then graph both functions on your graphing calculator. Make a sketch of both graphs.

x	Pre-Image $y = x^2$	Image $y = x^2 - 2$
−3		
−2		
−1		
0		
1		
2		
3		

FIGURE 5.21.
Table for Question 4.

5. Write a sentence that describes the transformation.

6. Describe what you think the effect will be on the graph of a quadratic function when a positive or negative number is added to the function after squaring the x-value.

The effect of changing $y = x^2$ to $y = (x + h)^2$.

7. Complete the tables in **Figure 5.22** for $y = x^2$ and $y = (x - 3)^2$, then graph both functions on your graphing calculator. Make a sketch of both graphs.

x	Pre-Image $y = x^2$
−3	
−2	
−1	
0	
1	
2	
3	

x	Image $y = (x - 3)^2$
0	
1	
2	
3	
4	
5	
6	

FIGURE 5.22.
Tables for Question 7.

8. Write a sentence that describes the transformation.

9. Describe the effect on the graph of a quadratic function when a positive number is subtracted from the x-value before squaring.

10. Describe the effect on the graph of a quadratic function when a negative number is subtracted from the x-value before squaring.

The effect of changing $y = x^2$ to $y = -x^2$.

11. Complete the table in **Figure 5.23** for $y = x^2$ and $y = -x^2$, then graph both functions on your graphing calculator. Make a sketch of both graphs.

12. Write a sentence that describes the transformation.

The effect of changing $y = (x + 2)^2$ to $y = -(x + 2)^2$.

13. Complete the table in **Figure 5.24** for $y = (x + 2)^2$ and $y = -(x + 2)^2$, then graph both functions on your graphing calculator. Make a sketch of both graphs.

14. Write a sentence that describes the transformation.

15. Describe the effect on the graph of a quadratic function when the opposite sign is applied to the function values.

The effect of changing $y = x^2$ to $y = ax^2$.

16. Complete the table in **Figure 5.25** for $y = x^2$ and $y = 2x^2$, then graph both functions on your graphing calculator. Make a sketch of both graphs.

17. Write a sentence that describes the transformation.

x	Pre-Image $y = x^2$	Image $y = -x^2$
−3		
−2		
−1		
0		
1		
2		
3		

FIGURE 5.23. Table for Question 11.

x	Pre-Image $y = (x + 2)^2$	Image $y = -(x + 2)^2$
−5		
−4		
−3		
−2		
−1		
0		
1		

FIGURE 5.24. Table for Question 13.

x	Pre-Image $y = x^2$	Image $y = 2x^2$
−3		
−2		
−1		
0		
1		
2		
3		

FIGURE 5.25. Table for Question 16.

x	Pre-Image $y = x^2$	Image $y = \frac{1}{2}x^2$
-3		
-2		
-1		
0		
1		
2		
3		

FIGURE 5.26.
Table for Question 18.

18. Complete the table in **Figure 5.26** for $y = x^2$ and $y = \frac{1}{2}x^2$, then graph both functions on your graphing calculator. Make a sketch of both graphs.

19. Write a sentence that describes the transformation.

20. Describe the effect on the graph of a quadratic function when a scale factor greater than 1 is applied to the function values.

21. Describe the effect on the graph of a quadratic function when a scale factor less than 1, but greater than 0, is applied to the function values.

$y = a(x - h)^2 + k$		
	Verbal Description	Visual
$|a| > 1$	Vertical stretch of the graph of the function	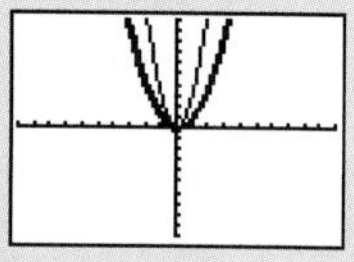
$|a| < 1$	Vertical compression of the graph of the function	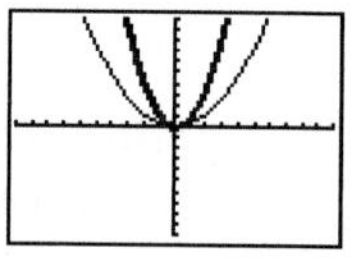
$-a$	Reflection of the graph of the function across the x-axis	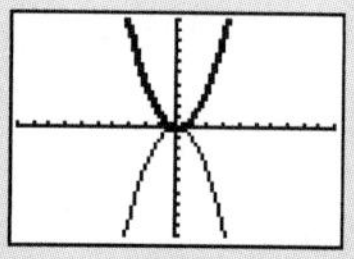
h	Horizontal translation of the graph of the function	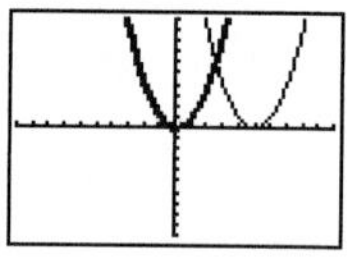
k	Vertical translation of the graph of the function	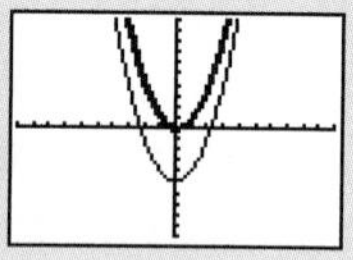

FIGURE 5.27.
Transformation summary.

Transforming functions is a useful tool when you want to fit a function to data. For example, recall Paul's parabola-shaped walk from Section 5.1 shown in **Figure 5.28.**

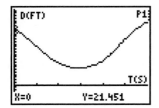

FIGURE 5.28.
Paul's walk.

To fit a function to these data, first compare the graph of the parent function for quadratics, $y = x^2$, to the scatterplot of Paul's walk shown in **Figure 5.29.**

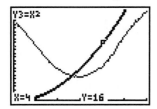

FIGURE 5.29.
Parent function and Paul's walk.

22. The graph of $y = x^2$ must be translated to the right to resemble the walk more closely. Write a function rule that translates the parabola to the right.

23. Now the graph of $y = (x - 3)^2$ must be shifted vertically to more closely resemble the walk. Write a function rule that has the necessary vertical shift.

24. Next the graph of $y = (x - 3)^2 + 7$ must be stretched vertically to resemble the walk more closely. Write a function rule that has the necessary vertical stretch.

Assignment

The graph of the parent function $y = x^2$ is graphed in the window shown in **Figure 5.30**.

FIGURE 5.30.
The parent function $y = x^2$.

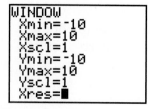

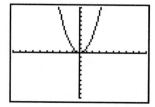

Each graph below was graphed using the window above, and is a transformation on $y = x^2$. The vertex of the parabola is labeled in each case. For each graph, write a function rule **and** a verbal description of the transformation.

1.
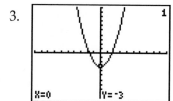

2.
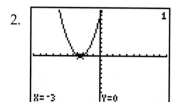

3.
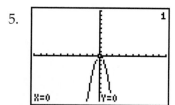

4.
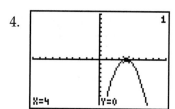

5.
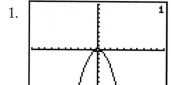

6.
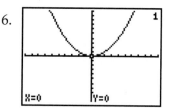

Questions 7–11 each represent a transformation of $y = x^2$. For each, sketch a graph **and** a write a verbal description of the transformation.

7. $y = x^2 - 4$

8. $y = (x - 4)^2$

9. $y = 2(x - 3)^2$

10. $y = \frac{1}{3}x^2$

11. $y = (x - 5)^2 + 3$

12. The scatterplot in **Figure 5.31** was made with the window shown.

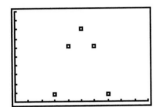 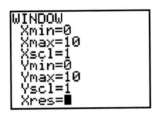

FIGURE 5.31.
Figure for Question 12.

13. Use transformations to develop a function that fits the data in the scatterplot. Explain your process.

Oh, That Traffic!

Let's apply to a new situation what you have learned about quadratic functions.

Sometimes motion isn't as speedy as at other times.

A mass transit organization in a large city is measuring the speed at which vehicles travel on the freeways. By measuring the speed at specific times of day over a long period of time, they can calculate average speeds during the day. This information can be put into a graph (see **Figure 5.32**) that helps commuters decide how much time they should allow to get to work.

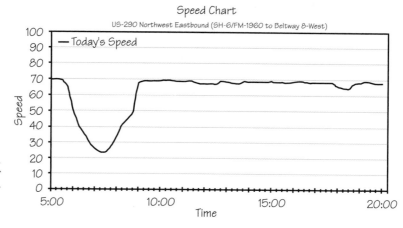

FIGURE 5.32.
Travel times for
a local freeway.

For one part of a freeway in this city, the speeds at half-hour intervals during the morning rush hour have been recorded (see **Figure 5.33**). The start of rush hour, 5:00 a.m., is time interval 0. The time interval increases by 1 for each successive half hour; i.e., 5:30 a.m. is time interval 1, 6:00 a.m. is time interval 2, etc.

Time	Time Interval	Average Speed (mph)
5:00 a.m.	0	70
5:30 a.m.	1	50
6:00 a.m.	2	35
6:30 a.m.	3	25
7:00 a.m.	4	20
7:30 a.m.	5	20
8:00 a.m.	6	25
8:30 a.m.	7	35
9:00 a.m.	8	50
9:30 a.m.	9	70

FIGURE 5.33.
Morning rush hour data.

1. Use Figure 5.33 to make a scatterplot of speed versus time interval. Then build a model to predict the speed in terms of the time interval and graph your model over the scatterplot.

2. Ms. Sanchez and Mr. Williams travel this part of the freeway on their way to work in the morning. Ms. Sanchez reaches this part of the freeway at 6:10 a.m., and Mr. Williams reaches this part of the freeway at 8:10 a.m. If the traffic moves at average speed, who will travel faster? Justify your answer.

3...2...1... Blast Off!

On Earth, gravity pulls objects toward the ground. This force, as you may have learned in your science class, changes an object's speed. If the object is moving away from Earth, gravity slows it down. If the object is moving toward Earth, gravity speeds it up. This change in speed is called acceleration.

CHECK THIS!

Acceleration is the change in an object's speed per unit of time.

When astronauts try to escape Earth's gravity and travel into space, they must overcome this force. One way they do that is to use a rocket. In this section you will observe a model rocket launch. While your rocket won't fly into space, you will observe changes in its speed as a result of the force of gravity.

MODEL ROCKET LAUNCH SIMULATION

Begin by setting the MODE and the WINDOW on your graphing calculator as shown in **Figure 5.34**.

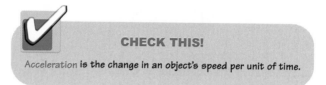

FIGURE 5.34.
Settings for
graphing calculator.

You've now changed your calculator setting from FUNCTION mode (what we normally use to graph, in terms of x and y) to PARAMETRIC mode. In parametric mode, we graph two functions, x_T and y_T, both in terms of *time*.

1. Press the Y= key on your graphing calculator. Enter each of these into your calculator.

$$X_{1T} = 2.5$$

$$Y_{1T} = 1.6T^2 + 80T$$

2. Press the GRAPH key on your calculator. Sketch the graph produced on the axes.

3. Is this graph what you expected? Why or why not?

4. In Section 5.1 you examined distance-versus-time walking graphs. One of the graphs had a parabolic shape. How does the flight of the rocket compare to the walk that produced a parabolic graph?

1. Based on the information provided, graph a parabola.

   ❖ Line of symmetry is $x = 2$.

   ❖ The vertex has a y-coordinate of -1.

   ❖ $(3, 0)$ is an x-intercept.

   ❖ $(0, 3)$ is the y-intercept.

CHECK THIS!

A line of symmetry in a parabola is a line that passes through the vertex as shown in Figure 5.35.

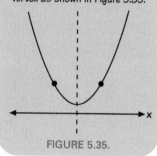

FIGURE 5.35.

Each point on the parabola can be reflected across the line of symmetry to match a point on the other side of the parabola.

2. What are the coordinates of the vertex?

3. What are the coordinates of the other x-intercept?

4. Name one other point that is on the parabola and explain how you found this point.

5. Write an equation of this parabola.

In Chapter 1 equations of the form $y - y_1 = m(x - x_1)$ are introduced as a way of describing linear models. This form is called point-slope form.

The **vertex form** of a quadratic function is $y = a(x - h)^2 + k$. This form can be modified by subtracting k from both sides to get $y - k = a(x - h)^2$.

6. Compare the point-slope form for linear functions with the modified vertex form for quadratics. How are they similar? How are they different?

There Could Be Several Factors

Quadratic functions can be used to model many things that occur in the real world. In Section 5.6 you used a graphing calculator to simulate a rocket launch.

Aerospace engineers use quadratic functions to model the motion of a rocket, much like you did in Section 5.6. Then they use these quadratic functions to get quadratic equations to help solve problems.

Recall that we used $Y_{1T} = -16T^2 + 80T$, where T represents time and Y represents the height of a rocket, to describe the motion of the rocket (see **Figure 5.36**).

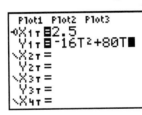
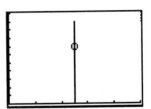

FIGURE 5.36.
Modeling a rocket's height.

$Y_{1T} = -16T^2 + 80T$ can be rewritten as $y = -16x^2 + 80x$ if we let x be time and y be the height of the rocket.

1. Write an equation that can be solved to find the times when the rocket was 64 feet above the ground.

2. Rewrite this equation in $ax^2 + bx + c = 0$ form, and completely factor the polynomial.

CHECK THIS!

Factoring Polynomials

Step 1: Find the largest factor common to all the terms of the polynomial.

Step 2: Use the distributive law to write the polynomial as a product where one factor is the largest common factor.

EXAMPLE:

Factor $18x^5 - 9x^3 + 27x^2$.

Although there are other factors common to the three terms (18, 9, and 27), such as $3x$, the largest common factor is $9x^2$.

So you factor the polynomial using $9x^2$.

$18x^5 - 9x^3 + 27x^2$

$= (9x^2)(2x^3) + (9x^2)(3)$

$= 9x^2(2x^3 - x + 3)$

3. Use your graphing calculator to graph $y = ax^2(2x^3 - x + 3)$. Sketch the graph and record your viewing window.

4. At which x-values does the graph cross the x-axis?

5. Use your graphing calculator to make a table of values for $y = -16x^2 + 80x - 64$. At what x-values is $y = 0$? How do these values relate to the x-values where the graph crosses the x-axis?

6. How do the x-values from Questions 4 and 5 relate to the factors in the factored form from Question 2?

7. What happens when you substitute each x-value into the factored form?

8. What is the meaning of $y = 0$ in this situation?

9. Write an equation that can be solved to find the times when the rocket was on the ground.

10. Completely factor the polynomial you wrote in Question 9.

11. Use your graphing calculator to graph the function in factored form. Sketch the graph and record your viewing window.

12. At which x-values does the graph cross the x-axis?

13. Use your graphing calculator to make a table of values for $y = -16x^2 + 80x$. At what x-values is $y = 0$? How do these values relate to the x-values where the graph crosses the x-axis?

14. How do the x-values from Questions 11 and 12 relate to the factors in the factored form from Question 9?

15. Substitute the x-values from Question 13 into the factored form and simplify. Describe your results.

16. Write a sentence that describes the relationship between the solutions to a quadratic equation of the form $ax^2 = bx + c = 0$ and the factors of $ax^2 + bx + c$.

Solving Quadratic Equations by Factoring

$y = ax^2 + bx + c$ is called the **general form** of a quadratic function. This form does not provide a lot of information about the graph of the function. Other forms are more useful because they display some important features of the graph. One form of a quadratic function is called the **factored form** and is written $y = a(x - r_1)(x - r_2)$, where r_1 and r_2 are integers.

1. Consider $y = -0.5(x - 1)(x - 5)$, which is written in factored form.

    a) Use a graphing calculator to graph the function. Sketch the graph and record your viewing window.

    b) Where does the graph cross the x-axis?

    c) Substitute the x-values where the graph crosses the x-axis into the function. What y-values do they yield?

Factored form $y = a(x - r_1)(x - r_2)$ is a useful way to write a quadratic function. The numbers r_1 and r_2 are the x-coordinates of the points where the graph crosses the x-axis. These x-coordinates are called the **roots** of the quadratic equation $-0.5x(x - 1)(x - 5) = 0$. They are also known as **x-intercepts** of the graph or **zeros** of the function $y = -0.5x(x - 1)(x - 5)$.

    d) Why is an x-intercept also called a zero?

    e) How can you tell if a point is an x-intercept from just the coordinates of the point?

In the factored form of a quadratic, the only things that change are the numbers. In general, it can be written like this:

$y = \underline{\quad} (x - \underline{\quad})(x - \underline{\quad})$

The challenge is to fill in the blanks with the correct numbers.

2. Write a quadratic function in factored form that has x-intercepts at -4 and $+2$. Use your graphing calculator to graph the function, sketch the graph, and record your window.

3. Adjust your answer so that the vertex is located at the point (−1, −3.6). Explain how you determined this answer.

4. a) Change your answer into polynomial form $(y = ax^2 + bx + c)$.

   b) Use your graphing calculator to display the graph of both the factored form and the polynomial form. Compare the two graphs and explain any similarities or differences.

   c) If you are graphing two functions, but only one graph appears, what does this mean?

You can use factored form to solve quadratic equations. You may have noticed that substituting a zero of a function causes one of the factors to be 0. Multiplying any number or factor by 0 yields a value of 0. So, if you have a quadratic in polynomial form that equals 0, such as $x^2 − 6x + 5 = 0$, you can use factoring to solve the equation.

Here are the steps to use factored form to solve quadratic equations.

❖ Factor the polynomial completely.

$$x^2 − 6x + 5 = 0$$

$$(x − 5)(x − 1) = 0$$

❖ Set each factor equal to 0.

$$x − 5 = 0 \qquad x − 1 = 0$$

❖ Solve each equation for x.

$$x − 5 = 0 \qquad x − 1 = 0$$

$$x = 5 \qquad x = 1$$

This method of solving a quadratic equation uses the **Zero-Product Property**. This property states that if the product of two real numbers is zero, then one or both of the numbers must be equal to zero. In other words, if $ab = 0$, then either $a = 0$ or $b = 0$.

If a polynomial is factored and the polynomial is equal to zero, then at least one of its factors must also be equal to zero. In other words, if $(x - 5)(x - 1) = 0$, then $x - 5 = 0$ or $x - 1 = 0$. As long as one of these two factors is equal to 0, then that value of x will solve the equation.

Let's take another look at our rocketry example from Section 5.7.

Recall that the function describing our rocket launch simulation from Section 5.6 was $y = -16x^2 + 80x$. We wrote an equation that could be solved to find the times when the rocket was 64 feet above the ground, $-16x^2 + 80x = 64$.

5. Rewrite this equation in $ax^2 + bx + c = 0$ form, and completely factor the polynomial.

6. Use the Zero-Product Property to solve the equation.

7. The height (in feet) of a rocket, y, after x seconds of flight can be found using $y = -16x^2 + 80x$. At what time(s) will the rocket be 84 feet above the ground?

8. What types of problems are best solved with the Zero-Product Property?

SUMMARY

If a quadratic that can be factored equals 0, then the equation can be solved using the Zero-Product Property.

❖ The Zero-Product Property states that for all real numbers a and b, if $ab = 0$, then $a = 0$ or $b = 0$.

To solve an equation by factoring:

❖ First, write the quadratic equation in polynomial form, $ax^2 + bx + c = 0$.

❖ Second, completely factor the polynomial.

❖ Third, set each factor equal to zero.

❖ Fourth, solve each of these equations for x.

❖ Since each of these x-values makes at least one factor equal to zero, by the Zero-Product Property, each of these x-values solves the equation.

Assignment

1. One way to see factored form is with an area model that represents the various parts of a quadratic expression. The pieces needed to represent $2x^2 + 7x + 3$ are shown in **Figure 5.37**.

FIGURE 5.37.
Area model for a quadratic.

  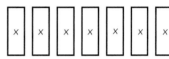

a) What are the dimensions for each type of piece?

b) $2x^2 + 7x + 3$ can be in factored form by arranging the pieces to form a rectangle. On a piece of paper, trace the pieces in Figure 5.37, and then arrange them to form a rectangle. Sketch the results.

c) The length and width of the rectangle represent the factors. What are the factors of $2x^2 + 7x + 3$?

d) Write $y = 2x^2 + 7x + 3$ in factored form.

e) Solve $2x^2 + 7x + 3 = 0$ using the Zero-Product Property.

f) Write $y = 4x^2 + 12x + 9$ in factored form.

g) Solve $4x^2 + 12x + 9 = 0$ using the Zero-Product Property.

2. A flare is launched from a life raft. The relationship between height and time is modeled by $h = 192t - 16t^2$ where h is height in feet and t is time in seconds. When is the flare 512 feet above the raft?

3. A toy rocket is fired in the middle of a field. The relationship between height and time is modeled by $h = -16t^2 + 256t$ where h is height in feet and t is time in seconds. When does the rocket's height equal 540 feet?

4. A rectangle has a length, l, and a width, w. The area of a rectangle can be found using $A = lw$. The perimeter of a rectangle can be found using $P = 2l + 2w$. The perimeter of a certain rectangle is 48 centimeters.

a) Use the perimeter formula to write a function rule for the length of the rectangle in terms of its width.

b) Write a function for the area of the rectangle in terms of its width.

c) If the rectangle has an area of 143 square centimeters, what are its length and width?

So What's Your Point?

Quadratic functions can be used to model many things that occur in the real world. In Section 5.6 you used a graphing calculator to simulate a rocket launch.

Suppose that a rocket launch can be simulated with $y = -16x^2 + 128x$, where x represents the time since launch in seconds and y represents the height of the rocket in feet.

1. Use your graphing calculator to graph the function. Sketch your graph and record your window.

2. Use the TRACE feature to estimate the coordinates of the vertex.

3. Use your graphing calculator to find the location of the vertex. How did you find your answer?

4. Recall that the vertex form of a quadratic function is $y = a(x - h)^2 + k$. Use the vertex you found in Question 3 to write a vertex form for this function.

5. Use transformations to fit an appropriate function, in vertex form, to the parabola. Use your graphing calculator to confirm your answer.

6. Use algebra to show that your function from Question 5 is equivalent to $y = -16x^2 + 128x$.

7. Here is a quadratic expression: $2(x + 3)^2 + 2$.

   a) List the operations you would use, in order, to evaluate this expression. Then evaluate the expression for $x = 4$.

   b) Knowing the order of operations can help in solving equations. Use your answer from a to list the opposite operations (i.e., opposite of multiplication is division).

c) If you want to solve $2(x + 3)^2 + 2 = 20$ you can start with 20 and apply the operations that you listed in b in reverse order. What numbers do you get? (Hint: there are two answers!)

d) To formalize your work, apply the same steps to both sides of $2(x + 3)^2 + 2 = 20$, one at a time, in the same order you used in c. Write down what the equation becomes after each of the steps.

8. a) A ball is shot straight up from a cannon. After 6 seconds, it reaches a height of 576 feet. Using $a = -16$, write a function in vertex form that can be used to model this rocket's flight.

b) Use this function to find the height of the ball 4 seconds after launch.

c) At what time will the ball be 320 feet above the ground?

Solving Quadratic Equations Using Inverse Operations

There are many ways to solve a quadratic equation. The best method depends on the form of the equation. Section 5.9 uses vertex form.

The Tower of the Americas in San Antonio, Texas, was built for the 1968 World's Fair. At 750 feet it is taller than both the Washington Monument in Washington, DC and the Space Needle in Seattle, Washington.

1. A penny is dropped from the top of the Tower of the Americas. The height of the penny, h, after t seconds can be described by $h = -16t^2 + 750$. At about what time will the penny be 100 feet from the ground?

One way to look at the vertex form, $y = a(x - h)^2 + k$, is by transformations from Section 5.4. Vertex form is the result of starting with the parent function, $y = x^2$, stretching or compressing it by a factor of a, reflecting it if $a < 0$, and then translating it horizontally h units and vertically k units. In each translation the only point that needs to be tracked is the vertex. The dilation (compression or stretch) affects how quickly or slowly the graph grows as the values for x increase or decrease.

Another way to look at the vertex form is through order of operations. They are shown as an arrow diagram in **Figure 5.38.**

FIGURE 5.38.
Arrow diagram.

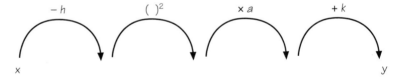

Because the operations can be done in sequence, one at a time, vertex form can be useful for solving equations using inverse operations.

2. $y = a(x - h)^2 + k$, if you know the values of y, a, h, and k, describe how to solve the equation for x. (Hint: Look back at work in Question 1.)

Every year for New Year's Eve, a fireworks display is launched from the Tower of the Americas. Suppose that a firework is launched from the top of the tower. Its height, h, above the ground can be found from the time, t, since launch using $h(t) = -16t^2 + 64t + 750$.

Planners need to know how high the firework will go before it begins to descend. Also, they must know how long the firework will take to reach this height. One way that we can answer these questions is to rewrite the function in vertex form using a method called **completing the square**.

Before using completing the square to answer these questions, let's consider a simpler problem.

Algebra	Tiles
a) Represent the function using tiles. $f(x) = x^2 + 4x + 1$	
b) Group the x-terms using parentheses. $f(x) = (x^2 + 4x) + 1$	
c) Find $\frac{1}{2}$ of the coefficient of x. Square this value, then add it to your x-terms inside the parentheses. $f(x) = (x^2 + 4x + \_\_\_) + 1$ $\frac{4}{2} = 2$ $2^2 = 4$	
d) You want to manipulate the equation, not change its value. So, take the value from step c) that you added and subtract it outside the parentheses. Factor your trinomial, and combine like terms to write the vertex form. To add 4, then, we also need to subtract 4. $f(x) = (x^2 + 2x + 4) + 1$ $f(x) = (x + 2)^2 - 4 + 1$ Combine like terms. $f(x) = (x + 2)^2 - 3$	

FIGURE 5.39.
Table for Question 3.

3. Write $f(x) = x^2 + 4x + 1$ in vertex form. To do this, use algebra tiles to represent the polynomial. In your notebook, make a table like **Figure 5.39** to record the steps and sketch the tiles.

4. Write $g(x) = 2x^2 + 4x + 3$ in vertex form. To do this, use algebra tiles to represent the polynomial. In your notebook, make a table like **Figure 5.40** to record the steps and sketch the tiles.

Algebra	Tiles
a) Group the x-terms: $g(x) = (2x^2 + 4x) + 3$	
b) We just followed a procedure for completing the square when $a = 1$, so let's factor 2 out of both of the x-terms. $g(x) = 2(x^2 + 2x) + 3$	
c) Find $\frac{1}{2}$ of the coefficient of x. Square this value, then add it to your x-terms inside the parentheses. $g(x) = 2(x^2 + 2x + \underline{\quad}) + 3$ $\frac{2}{2} = 1$ $1^2 = 1$	
d) You want to manipulate the equation, not change its value. So, take the value from step c) that you added and subtract it outside the parentheses. Factor your trinomial, and combine like terms to write the vertex form. *To add 1, we need to subtract 1. Don't forget, however. We are adding TWO groups of 1, or 2, and need to subtract 2 to compensate.* $g(x) = 2(x + 1)^2 - 2 + 3$ $g(x) = 2(x + 1)^2 + 1$	

FIGURE 5.40.
Table for Question 4.

Completing the square can be thought of in two ways.

Geometrically, you are physically completing a square. The tiles reveal this process. In Question 3 we were completing the square for $x^2 + 4x$. We built what we could out of algebra tiles then asked ourselves, "What number of unit tiles do I need to complete the square?" The answer is 4 tiles (see **Figure 5.41**). The factors of the resulting polynomial are $(x + 2)$ and $(x + 2)$.

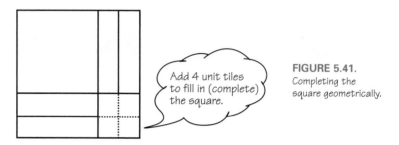

Add 4 unit tiles to fill in (complete) the square.

FIGURE 5.41.
Completing the square geometrically.

Algebraically, you are completing a binomial perfect square. In Question 3, we were completing the square for $x^2 + 4x$. We want to write this expression as a binomial squared; that is, a polynomial of the form $(x + \#)^2$.

$x^2 + 4x + \underline{\hspace{1cm}}$

$x^2 + 4x + \left(\dfrac{4}{2}\right)^2$

$x^2 + 4x + 4$

$(x + 2)(x + 2)$

$(x + 2)^2$

Solving a quadratic equation by completing the square has two parts.

Part 1. Rewrite the polynomial in vertex form using completing the square.

Part 2. Use inverse operations to isolate the variable.

EXAMPLE:

Solve $x^2 - 3x = 10$ by completing the square.

Step 1: The left-hand side needs to be written in vertex form. Since $-3 \div 2 = -1.5$, and $(-1.5)^2 = 2.25$, add 2.25 to both sides of the equation.

$x^2 - 3x + 2.25 = 10 + 2.25$

Step 2: Rewrite the left-hand side as a binomial squared.

$(x^2 - 3x + 2.25) = 12.25$
$(x - 1.5)^2 = 12.25$

Step 3: Use inverse operations to solve for x.

Take the square root of each side:

$\sqrt{(x-1.5)^2} = \sqrt{12.25}$
$x - 1.5 = \pm 3.5$

Add 1.5 to each side of the equation.

$x - 1.5 = \pm 3.5$
$x - 1.5 + 1.5 = \pm 3.5 + 1.5$

Simplify.

$x = \pm 3.5 + 1.5$
$x = 3.5 + 1.5 \qquad x = -3.5 + 1.5$
$x = 5 \qquad\qquad x = 2$

SUMMARY

If a quadratic equation is in vertex form, $y = a(x - h)^2 + k$, then it can be solved by using inverse operations.

❖ Start with y.

❖ Subtract k.

❖ Divide by a.

❖ Take the square root.

❖ Add h.

Sometimes a quadratic equation can be solved by a method called **completing the square**. To solve $x^2 + bx + c = 0$ using this method:

❖ Write the equation in the form $x^2 + bx = -c$.

❖ Add $\left(\frac{b}{2}\right)^2$ to both sides:

$$x^2 + bx + \left(\frac{b}{2}\right)^2 = -c + \left(\frac{b}{2}\right)^2$$

❖ Rewrite the left side as a perfect square:

$$\left(x + \frac{b}{2}\right)^2 = -c + \left(\frac{b}{2}\right)^2$$

❖ Take the square root of each side and solve for x.

Assignment

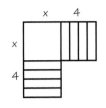

FIGURE 5.42.
An incomplete area model.

1. **Figure 5.42** is an incomplete area model. Complete the square.

   a) What algebraic expression does Figure 5.42 represent?

   b) How many small squares (area of each = 1) do you need to add to complete the square?

   c) What algebraic expression does the model then represent?

   d) What algebraic expression represents the sum of the pieces being used to complete the square?

2. Complete each square, then write the polynomial as a binomial squared.

   a) $x^2 - 16x +$ _____

   b) $x^2 + 7x +$ _____

3. Write $h(x) = 3x^2 - 12x - 1$ in vertex form. Show your work in a table like **Figure 5.43**.

Algebra	Sketch of Tiles
a) Group the x-terms using parentheses.	
b) Factor out a 3 from both the x^2 and x-terms.	
c) Find $\frac{1}{2}$ of the coefficient of x. Square this value, then add it to your x-terms inside the parentheses.	
d) You want to convert the equation, not change its value. So, take the value from step c) that you added, multiply it by the 3 that you factored out (distribute, distribute!), and subtract it outside the parentheses. Factor your trinomial, and combine like terms to write the vertex form.	

FIGURE 5.43.
Table for Question 3.

4. Use completing the square to write $f(x) = \frac{1}{2}x^2 + 2x + 3$ in vertex form.

For Questions 5–7, solve by completing the square and inverse operations.

5. $x^2 + 4x = 12$

6 $x^2 + 8x - 11 = 0$

7. $5x^2 - 10x - 30 = 0$

8. A football is kicked into the air. It hits the ground 60 yards downfield. At its highest point, the ball is 7 yards above the ground. Assume the relationship between height and distance is quadratic.

a) Sketch a graph of the relationship between height and distance.

b) Label the vertex with its coordinates. What does this point mean here?

c) Label the x-intercepts with their coordinates. What do these points mean here?

d) Substitute the coordinates of the vertex into the vertex form $y = a(x - h)^2 + k$.

e) To find the value of a, choose a known point on the parabola and substitute the values for x and y. Solve for a.

f) Write a function rule to model the relationship between the height of the ball and the distance downfield.

g) Assume the ball is being kicked for a 51-yard field goal. The height of the field goal crossbar is 10 feet. Will the ball make it through the goal posts? Show how you found your answer.

The Quadratic Safety Net

You have seen three ways to write a quadratic function:

Polynomial Form: $y = ax^2 + bx + c$

Factored Form: $y = a(x - r_1)(x - r_2)$

Vertex Form: $y = a(x - h)^2 + k$

You have also seen two ways to solve quadratic equations. The first method, factoring, involves applying the Zero-Product Property to solve each factor for x. If the quadratic is in polynomial form, it must first be factored in order to use this method. The second method, using inverse operations, involves applying the inverse operations to the equation to isolate x, which includes taking the square root to "undo" the quadratic nature of the equation. If the quadratic is in polynomial form, it must first be expressed as a binomial squared, which usually involves completing the square.

1. Consider $y = x^2 + 4x + 3.75$.

   a) Does the polynomial $x^2 + 4x + 3.75$ factor? How can you tell? If so, what are the factors?

   b) Use your graphing calculator to graph the function $y = x^2 + 4x + 3.75$. Sketch your graph and record your window.

   c) What are the zeros of the function? How do you know?

   d) Based on your answers so far, what is the solution to $x^2 + 4x + 3.75 = 0$?

   e) Now, consider $y = x^2 + 5x - 3$. Can the function be written in factored form? If so, write it in factored form. If not, why not?

   f) Graph $y = x^2 + 5x - 3$ on your graphing calculator. Sketch your graph and record your window.

g) Find the zeros of $y = x^2 + 5x - 3$. Explain your process.

h) Based on your answers, what is the solution to $x^2 + 5x - 3 = 0$? Are these exact or approximate? How do you know?

i) Solve $x^2 + 5x - 3 = 0$ without using a graphing calculator. Leave your answer in exact form.

Solving Quadratic Equations Using the Quadratic Formula

The method of completing the square, followed by applying inverse operations, can be used to solve any quadratic equation. However, the arithmetic and algebra are time comsuming. So, mathematicians have developed a formula to solve any quadratic equation.

Instead of solving quadratic equations one at a time by completing the square, you can solve $ax^2 + bx + c = 0$ for x. This will give you a formula that can be used every time. You can complete the square once and for all!

Begin with $ax^2 + bx + c = 0$. Your goal is to solve this equation for x using completing the square, followed by inverse operations to isolate x.

Step 1: Factor out the a from the x-terms:

$$a\left(x^2 + \frac{b}{a}x\right) + c = 0$$

Step 2: Complete the square:

Take half of the coefficient of x. Square this value.

$$\left(\frac{1}{2}\cdot\frac{b}{a}\right)^2 = \frac{b^2}{4a^2}$$

Add this to the left side of the equation.

$$a\left(x^2 + \frac{b}{a}x + \frac{b^2}{4a^2}\right) + c = 0 + \boxed{???}$$

But wait! You cannot add something to one side of an equation without adding it to the other! To balance this equation, we have to add the same thing to the right side.

Look closely at the left side of the equation. We added $\frac{b^2}{4a^2}$ to the left side but we placed it inside the parentheses. When the a is distributed to each term in the parentheses, we actually added $\frac{b^2}{4a^2}$ to the left side. So we need to add this term to the right side: $a\left(x^2 + \frac{b}{a}x + \frac{b^2}{4a^2}\right) + c = 0 + \frac{a}{1}\cdot\frac{b^2}{4a^2}$.

Simplify the right side.

$$a\left(x^2 + \frac{b}{a}x + \frac{b^2}{4a^2}\right) + c = 0 + \frac{\cancel{a}}{1} \cdot \frac{b^2}{4a \cdot \cancel{a}}$$

$$a\left(x^2 + \frac{b}{a}x + \frac{b^2}{4a^2}\right) + c = \frac{b^2}{4a}$$

Factor the left side.

$$a\left(x + \frac{b}{2a}\right)^2 + c = \frac{b^2}{4a}$$

Step 3: Subtract c from both sides.

$$a\left(x + \frac{b}{2a}\right)^2 = \frac{b^2}{4a} - c$$

Step 4: Divide by a (or, multiply by the reciprocal of a).

$$\frac{1}{\cancel{a}} \cdot \cancel{a}\left(x + \frac{b}{2a}\right)^2 = \frac{1}{a} \cdot \left(\frac{b^2}{4a} - c\right)$$

$$\left(x + \frac{b}{2a}\right)^2 = \frac{b^2}{4a^2} - \frac{c}{a}$$

Step 5: Find a common denominator on the right side:

The common denominator can be found by identifying the least common multiple of $\frac{b^2}{4a^2}$ and $\frac{c}{a}$. In other words, we need to multiply the second fraction by $\frac{4a}{4a}$ in order to have a common denominator of $4a^2$.

$$\left(x + \frac{b}{2a}\right)^2 = \frac{b^2}{4a^2} - \frac{4a}{4a} \cdot \frac{c}{a}$$

$$\left(x + \frac{b}{2a}\right)^2 = \frac{b^2}{4a} - \frac{4ac}{4a^2}$$

$$\left(x + \frac{b}{2a}\right)^2 = \frac{b^2 - 4ac}{4a^2}$$

Step 6: Take the square root of both sides. (Don't forget that there are a positive and a negative square root!)

$$\sqrt{\left(x + \frac{b}{2a}\right)^2} = \sqrt{\frac{b^2 - 4ac}{4a^2}}$$

$$x + \frac{b}{2a} = \pm\sqrt{\frac{b^2 - 4ac}{4a^2}}$$

$$x + \frac{b}{2a} = \pm\frac{\sqrt{b^2 - 4ac}}{\sqrt{4a^2}}$$

$$x + \frac{b}{2a} = \pm\frac{\sqrt{b^2 - 4ac}}{2a}$$

Step 7: Solve for x.

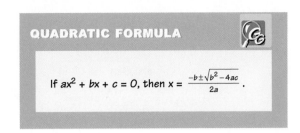

QUADRATIC FORMULA

If $ax^2 + bx + c = 0$, then $x = \frac{-b \pm \sqrt{b^2 - 4ac}}{2a}$.

Subtract $\frac{b}{2a}$ from both sides.

$$x = -\frac{b}{2a} \pm \frac{\sqrt{b^2 - 4ac}}{2a}$$

$$x = \frac{-b \pm \sqrt{b^2 - 4ac}}{2a}$$

Now we can use the Quadratic Formula to solve any quadratic equation.

EXAMPLE 1:

Use the Quadratic Formula to solve $3x^2 + 4x + 1 = 0$ for x.

Step 1: Define a, b, and c.

Let $a = 3$, $b = 4$, and $c = 1$.

Step 2: Substitute the values for a, b, and c into the Quadratic Formula.

$$x = \frac{-b \pm \sqrt{b^2 - 4ac}}{2a} \qquad x = \frac{-4 \pm \sqrt{4^2 - 4(3)(1)}}{2(3)}$$

Step 3: Simplify using the order of operations (the square root sign acts as a grouping symbol).

$$x = \frac{-4 \pm \sqrt{4^2 - 4(3)(1)}}{2(3)} \qquad x = \frac{-4 \pm 2}{6}$$

$$x = \frac{-4 \pm \sqrt{16 - 12}}{6} \qquad x = \frac{-4 + 2}{6} \qquad x = \frac{-4 - 2}{6}$$

$$x = \frac{-4 \pm \sqrt{4}}{6} \qquad x = \frac{-2}{6} \qquad x = \frac{-6}{6}$$

$$x = -\frac{1}{3} \qquad x = -1$$

EXAMPLE 2:

Use the Quadratic Formula to solve $4x^2 - 6x - 9 = 0$ for x.

Step 1: Define a, b, and c.

Let $a = 4$, $b = -6$, and $c = -9$.

Step 2: Substitute the values for a, b, and c into the Quadratic Formula.

$$x = \frac{-b \pm \sqrt{b^2 - 4ac}}{2a}$$

$$x = \frac{-(-6) \pm \sqrt{(-6)^2 - 4(4)(-9)}}{2(4)}$$

Step 3: Simplify using the order of operations (the square root sign acts as a grouping symbol).

$$x = \frac{-(-6) \pm \sqrt{(-6)^2 - 4(4)(-9)}}{2(4)} \qquad x = \frac{6 \pm 6\sqrt{5}}{8}$$

$$x = \frac{6 \pm \sqrt{36 + 144}}{8} \qquad x = \frac{6\left(1 \pm \sqrt{5}\right)}{8}$$

$$x = \frac{6 \pm \sqrt{180}}{8} \qquad x = \frac{3\left(1 \pm \sqrt{5}\right)}{4}$$

$$x = \frac{6 \pm 6\sqrt{5}}{8} \qquad x = \frac{3}{4} \pm \frac{3\sqrt{5}}{4}$$

SUMMARY

A quadratic equation in the form $ax^2 + bx + c = 0$ can be solved with the Quadratic Formula.

❖ If $ax^2 + bx + c = 0$, then $x = \frac{-b \pm \sqrt{b^2 - 4ac}}{2a}$.

To solve an equation with this formula:

❖ First, identify the values of a, b, and c.

❖ Second, substitute the values for a, b, and c into the Quadratic Formula.

❖ Third, simplify using the order of operations.

✓ The square root sign acts as a grouping symbol.

✓ Simplify the square root completely.

✓ Reduce all fractions.

Assignment

1. Use the Quadratic Formula to solve each of these:

    a) $3x^2 + 10x + 7 = 0$

    b) $2x^2 + 5x - 9 = 0$

2. Solve each of these using the most appropriate method.

    a) $y = x^2 + 6x + 9$

    b) $y = x^2 - 2.5x - 6$

    c) $y = -2x^2 - x - 0.12$

    d) $y = x^2 - 21x + 14$

    e) $y = \frac{1}{2}x^2 - 2x + 2$

    f) $y = -3x^2 + 9x + 12$

3. A stone is thrown from a catapult. The function $h(x) = -16x^2 + 80x$ describes the height (h) of the stone as a function of time.

    a) How high is the stone at 3.5 seconds? Explain how you found your answer.

    b) When will the stone be 36 feet high? Explain how you found your answer.

In this chapter you have explored three ways to write quadratic functions:

Polynomial form: $y = ax^2 + bx + c$

Factored form: $y = a(x - r_1)(x - r_2)$

Vertex form: $y = a(x - h)^2 + k$

You have used three methods to solve quadratic equations:

✓ Factor the polynomial, then apply the Zero-Product Property.

✓ Complete the square, then apply inverse operations.

✓ Use the Quadratic Formula.

Which method is best? It depends on the situation. In this section you will look at problems involving motion. You will consider why one form of a quadratic function is better than others and why one method of solution is quicker than others.

THREE WAYS TO REPRESENT QUADRATIC FUNCTIONS

1. Enter each of these into your graphing calculator:

   $p(x) = 2x^2 - 12x + 16$

   $f(x) = 2(x - 2)(x - 4)$

   $v(x) = 2(x - 3)^2 - 2$

   a) Graph the functions on your calculator. Sketch the graphs and record your window.

   b) What can you conclude about the three functions? On what evidence do you base your conclusion?

   c) Prove that $p(x)$ and $f(x)$ are equivalent.

d) Prove that $p(x)$ and $v(x)$ are equivalent.

e) What is the vertex of this parabola? Which function most closely reveals the vertex?

f) What are the x-intercepts of this parabola? Which function most closely reveals them?

g) What is the y-intercept of this parabola? Which function most closely reveals it?

2. For each function in **Figure 5.44,** fill in the missing representations.

Graph	Factored Form	Vertex Form	Polynomial Form
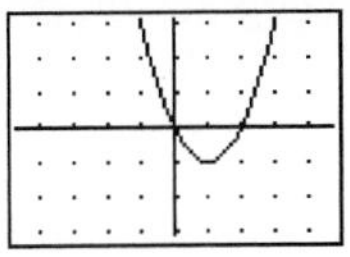	$y = x(x - 2)$	$y = (x - 1)^2 - 1$	$y = x^2 - 2x$
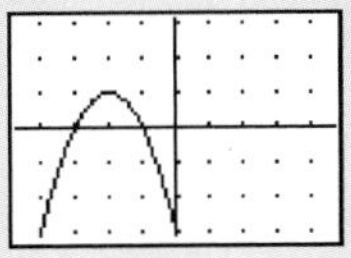			
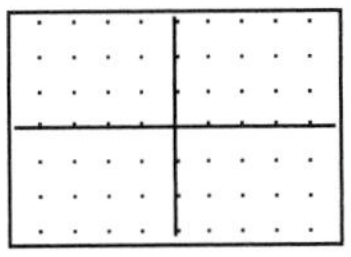	$y = (x - 2)(x + 1)$		
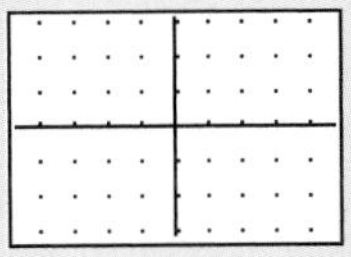		$y = 2(x - 2)^2 - 2$	
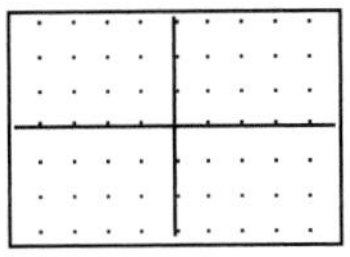			$y = -\frac{1}{2}x^2 + 2$

FIGURE 5.44.
Table for Question 2.

3. When is it best to solve equations with factoring and the Zero-Product Property?

4. When is it best to solve equations with inverse operations?

5. When is it best to solve equations with the Quadratic Formula?

6. Use the table in **Figure 5.45** to record your findings about the forms of quadratic functions and the methods to solve quadratic equations.

	General Form of Equation	Critical Attributes	Preferred Method of Solution
Factored Form			
Vertex Form			
Polynomial Form			

FIGURE 5.45.
Table for Question 6.

1. In **Figure 5.46,** fill in the blank columns.

Graph	Factored Form	Algebra Tiles	Vertex Form	Polynomial Form
				$y = -3x^2 + 6x$

FIGURE 5.46.
Table for Question 1.

2. Change each of these into the specified form.

   a) $y - 5 = 0.5(x + 2)^2$ into polynomial form

   b) $y = 2x^2 - 7x + 6$ into factored form

   c) $y = -2(x + 2)(x - 1)$ into polynomial form

   d) $y = 1.4(x - 1)(x - 5)$ into vertex form

   e) $y - 8 = -2(x + 4)^2$ into factored form

   f) $y = x^2 - 2x - 15$ into vertex form

3. Sketch the graph of each function without using a graphing calculator. In each case explain how the form of the function helped you determine the shape of the graph.

   a) $y = -0.5(x + 1)(x - 3)$

   b) $y = 2(x + 2)^2 - 4$

   c) $y = x^2 + 3x - 2$

4. A toy rocket blasts off from the ground and peaks in 3 seconds at 144 feet. Assume the relationship between height and time is quadratic, and let the value of a be −16.

a) Write a function in vertex form for the height of the rocket at time t.

b) Write a function in factored form for the height of the rocket at time t.

c) Write a function in polynomial form for the height of the rocket at time t.

d) Sketch a graph of the relationship between height and time. If you use a graphing calculator, record your window.

e) When does the rocket hit the ground? Why?

f) Find the coordinates of the vertex and the x-intercepts. Label these on your graph.

g) How high was the rocket 2 seconds after launch? Show how you found this.

h) When will the rocket again be the same height as your answer in g? Explain how you found this.

5. The freshman class is deciding whether or not to raise the price of tickets to the spring dance. $p = 75 + 100d − 20d^2$ can be used to estimate profit, p, by raising the ticket price d dollars.

a) What profit is expected if they do not raise the price (i.e., if $d = 0$)?

b) What is the least amount they can raise the price if they want to make $155?

c) What method did you use to answer b? Why did you choose that method?

Build a Better Sandbox . . .

A company builds sandboxes as shown in **Figure 5.47.** The series number determines the surface area covered with sand.

Series Number	Sandbox & Seating Tiles	Written Description
1		A 3 x 3 sandbox has 1 square unit of its surface area covered with sand.
2		A 4 x 4 sandbox has 4 square units of its surface area covered with sand.
3		A 5 x 5 sandbox has 9 square units of its surface area covered with sand.
4		A 6 x 6 sandbox has 16 square units of its surface area covered with sand.

FIGURE 5.47.
A series of sandboxes.

1. Find a function rule that describes the total surface area of the sandbox, including the part that is covered with sand and the surrounding seating tiles, in terms of the series number. Justify your answer.

2. Find the series number of the sandbox with 196 square units of surface area. Justify your answer.

Motion with a Pendulum

A **pendulum** is a string with a mass (called a "bob") attached at the end.

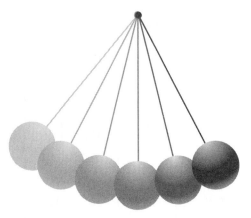

Galileo

One of the first people to investigate the motion of a pendulum was Galileo, a famous astronomer. He noticed a chandelier moving in a cathedral in Pisa, Italy. He explored the relationships among the length of a pendulum, the weight on the end of the pendulum, the angle of release, and the time it takes for one swing. Galileo's work led to the invention of the pendulum clock.

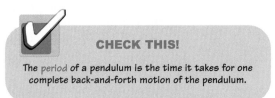

CHECK THIS!

The period of a pendulum is the time it takes for one complete back-and-forth motion of the pendulum.

1. What types of motion have you investigated so far in this chapter?

2. Give examples of events where the motion is like that of a pendulum.

Assignment

1. In **Figure 5.48**, a circle has been drawn with points evenly spaced around it. Straight lines connect all pairs of points. In order to predict the number of lines that have been drawn, you might want to start with a simpler problem and look for a pattern. Fill in the table in **Figure 5.49.** Hint: You may want to draw a circle, put the number of points on it, connect the points with lines, and then count them.

Number of Points on the Circle	Number of Lines Drawn
2	
3	
4	
5	
6	
7	

FIGURE 5.49.
Table for Question 1.

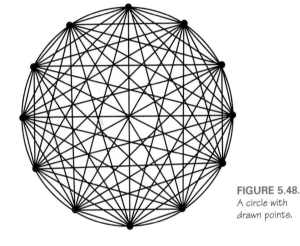

FIGURE 5.48.
A circle with drawn points.

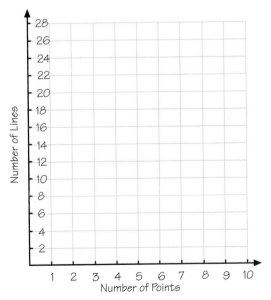

FIGURE 5.50. Graph for Question 2.

2. Graph the ordered pairs from the table on a graph like **Figure 5.50.**

3. Decide if the relationship between the number of lines and the number of points is linear or quadratic. Write a function rule that models this relationship.

4. Use your model to predict how many lines you would draw if there were 12 points on the circle.

5. Is your prediction close to the actual number of lines?

Reflections with Patty Paper

A REFLECTION WITH PATTY PAPER

1. Connect the points listed below in order on coordinate graph paper. When you see the word STOP, do not connect the point before the word STOP to the point after the word STOP.

   {(–8, 0), (–9, –2), (–3, –2), (–4, 0), (–8, 0), (–8, 8), (–4, 8), (–4, 0), STOP, (–8, 8), (–6, 10), (–4, 8), STOP}

2. Graph the line $y = x$ on the coordinate axes.

3. What do you notice about the coordinates of the points that lie on the graph of $y = x$?

4. Trace your drawing from Question 1, the line $y = x$, and the coordinate axes onto a sheet of patty paper.

5. Fold the patty paper on the line $y = x$.

6. Trace the reflection of the drawing onto the patty paper.

7. What do you notice about the reflection?

8. Trace the reflection of the drawing onto the coordinate grid. Draw the reflection with a different color pencil.

9. List the coordinates of the corresponding points of the reflected drawing in a table like **Figure 5.51**.

Coordinates of Points in Original Drawing	Corresponding Coordinates of Points in Reflected Drawing
(–8, 0)	
(–9, –2)	
(–3, –2)	
(–4, 0)	
(–4, 8)	
(–6, 10)	
(–8, 8)	

FIGURE 5.51.
Table for Question 9.

10. What do you notice about the coordinates?

11. What are the x-intercepts on the original drawing?

12. Which points in the reflection correspond to the x-intercepts in the original drawing?

13. What happens if you reflect the reflection back over the line $y = x$?

ANOTHER REFLECTION

14. Graph the line $y = 2x + 4$ on the coordinate axes.

15. Draw the graph of $y = x$.

16. Trace $y = 2x + 4$, $y = x$, and the coordinate axes onto a sheet of patty paper.

17. Fold the patty paper along the line $y = x$.

18. Trace the reflection of $y = 2x + 4$ onto the patty paper.

19. Place the patty paper back on the coordinate grid. Trace the reflection of $y = 2x + 4$ onto the coordinate grid. Draw the reflection with a different color pencil.

20. What do you notice about the reflection?

21. List three points that lie on $y = 2x + 4$ and list the coordinates of corresponding points on the reflected line.

22. What do you notice about the coordinates of corresponding points?

23. What is the slope of the line $y = 2x + 4$?

24. What is the slope of the reflected line?

25. What is the y-intercept of the line $y = 2x + 4$?

26. What is the x-intercept of the reflected line?

27. What is the x-intercept of $y = 2x + 4$?

28. What is the y-intercept of the reflected line?

29. What is the equation of the reflected line?

30. What happens if you reflect the reflection back over the line $y = x$?

A FINAL REFLECTION

31. Graph $y = x^2$ on a coordinate grid.

32. Graph $y = x$.

33. Trace $y = x^2$, $y = x$, and the coordinate axes onto a sheet of patty paper.

34. Fold the patty paper along the line $y = x$.

35. Trace the reflection of $y = x^2$ onto the patty paper.

36. Trace the reflection of $y = x^2$ onto the coordinate grid. Draw the reflection with a different color pencil.

37. What do you notice about the reflection?

38. Is the reflection a function? How do you know?

39. List three points that lie on $y = x^2$ and list the coordinates of corresponding points on the reflected graph.

40. What do you notice about the coordinates of corresponding points?

41. What are the domain and range of $y = x^2$?

42. What are the domain and range of the reflection of $y = x^2$?

43. What happens if you reflect the reflection back over the line $y = x$?

Assignment

1. Plot the following relation.

   {(2, 2), (2, 4), (–1, 4), (–3, 1), (–6, –1), (–5, –4), (–3, –6), (1, –5), (2, –2), (4.5, 0.5)}

2. What is the domain of this relation?

3. What is the range of this relation?

4. Plot the graph of $y = x$.

5. Plot the reflection of the relation from Question 1 over the line $y = x$.

6. List the coordinates of corresponding points as shown in the table in **Figure 5.52.**

Original Coordinates	Reflected Coordinates
(2, 2)	
(2, 4)	
(–1, 4)	
(–3, 1)	
(–6, –1)	
(–5, –4)	
(–3, –6)	
(1, –5)	
(2, –2)	
(4.5, 0.5)	

FIGURE 5.52.
Table for Question 6.

7. What is the domain of the reflected relation?

8. What is the range of the reflected relation?

9. What do you notice about the domain and range of the relation and its reflection?

10. If a point from the relation lies on the line $y = x$, where does the reflection of that point lie?

SECTION 5.17

Inverses

The line $y = x$ acts like a mirror for a relation and its reflection across the line. When you reflect a relation across the line $y = x$, the reflection is called the **inverse**. The x- and y-coordinates of each point in a relation are reversed when you graph the reflection.

1. You may also graph a relation and its inverse on the graphing calculator. Your teacher will give you instructions on how to graph the inverse with the graphing calculator. Enter the domain of the following data set (10 members) into List 1 of your graphing calculator. Enter the range (also 10 members) into List 2.

CHECK THIS!

A relation is a set of ordered pairs.

$\{(3, 2), (3, 3), (3, 4), (4, 1), (4, 5), (6, 1), (6, 5), (7, 2), (7, 3), (7, 4)\}$

2. Examine the scatterplot with List 1 serving as the domain and List 2 serving as the range.

3. Then activate a second scatterplot with List 1 acting as the range and List 2 acting as the domain.

4. Graph the line $y = x$. What do you observe?

CHECK THIS!

• The inverse of a relation is the reflection of the relation over the line $y = x$.

• When a relation is reflected over the line $y = x$, the x- and y-coordinates of corresponding points are reversed.

• The domain of the relation becomes the range of its inverse, and the range of the relation becomes the domain of its inverse.

• All relations have inverses.

• The term inverse is used two ways in mathematics: one to describe an opposite operation, the other to describe a type of relation.

An inverse reverses or *undoes* a previous action: A pitcher throws a ball; the catcher catches it and throws it back to the pitcher. Taking your shoes off and putting them back on is an example of an inverse. If you graph a relation, reflect it over the line $y = x$, and then reflect it back, you will have the same relation that you started with.

5. Where in mathematics have you used inverse operations?

6. Not every action can be reversed. List some examples of actions that cannot be reversed.

7. Graph $f(x) = x^2$. Graph the inverse. List five points from each graph in a table.

8. Is $f(x) = x^2$ a function?

9. What are the domain and range for $f(x) = x^2$?

10. Is the inverse of $f(x) = x^2$ a function?

11. What are the domain and range for the inverse?

Not all functions have inverses that are also functions. If you substitute $x = -2$ into $f(x) = x^2$ you get $f(-2) = (-2)^2 = 4$. If you try to undo 4 there are two possible values, 2 or -2 as shown in **Figure 5.53.**

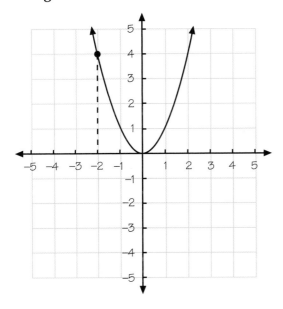
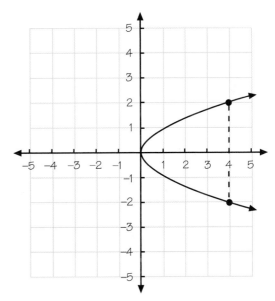

FIGURE 5.53. Inverse of a function.

In order for $f(x) = x^2$ to have an inverse that is also a function, mathematicians restrict the domain of $f(x) = x^2$ to $x \geq 0$. This gives you only the right branch of the parabola. Then the inverse is a function. The inverse of $f(x) = x^2$ needs to undo the action of squaring a number. The operation that reverses squaring is the square root. The mathematical notation for an inverse function is $f^{-1}(x)$.

If $f(x) = x^2$ for $x \geq 0$, then $f^{-1}(x) = \sqrt{x}$ for $x \geq 0$ (see **Figure 5.54**).

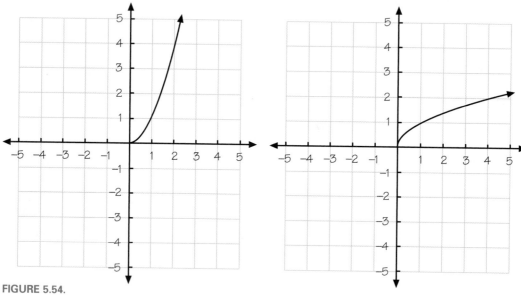

FIGURE 5.54.
Inverse of
$f(x) = x^2$ for x > 0.

$f(x) = x^2, x \geq 0, y \geq 0$

$f(x) = \sqrt{x}, x \geq 0, y \geq 0$

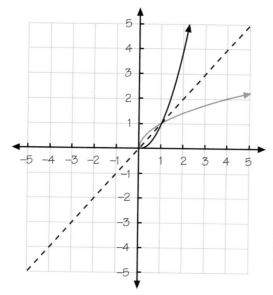

Graphing the function and its inverse on the same coordinate axes demonstrates the inverse relationship (see **Figure 5.55**).

FIGURE 5.55.
Plot of the inverse
relationship of Figure 5.54.

12. Use your graphing calculator to graph the parent function $y = \sqrt{x}$. Then graph the additional functions in **Figure 5.56** and fill in the blanks.

Function	Describe Changes to the Parent Function	Domain	Range
$y = \sqrt{x}$	Parent function	$x \geq 0$	$y \geq 0$
$y = \sqrt{x} + 2$		$x \geq 0$	$y \geq 2$
$y = \sqrt{x} - 3$		$x \geq 0$	$y \geq -3$

FIGURE 5.56.
Table for Question 12.

What do you observe?

13. Use your graphing calculator to graph the functions in **Figure 5.57** and fill in the blanks.

Function	Describe Changes to the Parent Function	Domain	Range
$y = \sqrt{x}$	Parent function	$x \geq 0$	$y \geq 0$
$y = \sqrt{x+3}$		$x \geq -3$	$y \geq 0$
$y = \sqrt{x-2}$		$x \geq 2$	$y \geq 0$

FIGURE 5.57.
Table for Question 13.

What do you observe?

14. Use your graphing calculator to graph the functions in **Figure 5.58** and fill in the blanks.

Function	Describe Changes to the Parent Function	Domain	Range
$y = \sqrt{x}$	Parent function	$x \geq 0$	$y \geq 0$
$y = 0.5\sqrt{x}$		$x \geq 0$	$y \geq 0$
$y = 2\sqrt{x}$		$x \geq 0$	$y \geq 0$

FIGURE 5.58.
Table for Question 14.

What do you observe?

15. Use your graphing calculator to graph the functions in **Figure 5.59** and fill in the blanks.

Function	Describe Changes to the Parent Function	Domain	Range
$y = \sqrt{x}$	Parent function	$x \geq 0$	$y \geq 0$
$y = -\sqrt{x}$		$x \geq 0$	$y \leq 0$
$y = -0.5\sqrt{x}$		$x \geq 0$	$y \leq 0$

FIGURE 5.59.
Table for Question 15.

What do you observe?

16. How are the transformations of $y = \sqrt{x}$ similar to what you learned about transformations of $y = x^2$?

17. Use a graphing calculator to make a table for $y = \sqrt{x-4}$, why does the word "ERROR" appear for values of x that are less than 4?

18. Graph $f(x) = x^2 + 2$ for $x \geq 0$. Graph its inverse. What is the equation of the inverse, $f^{-1}(x)$?

SUMMARY

The graph of the inverse of the graph of a function is the reflection of the function over the line $y = x$. The domain of the function becomes the range of its inverse. The range of the function becomes the domain of its inverse.

The inverse of the quadratic parent function where $x \geq 0$ is the square root function. Transformations of the square root function have effects like those of the quadratic parent function.

* $f(x) = \sqrt{x} + c$ results in a vertical translation.

* $f(x) = \sqrt{x+c}$ results in a horizontal translation.

* $f(x) = a\sqrt{x}$ results in a vertical compression, vertical stretch, or a reflection over the x-axis.

1. Plot this relation and its inverse on a grid. Use a different color to distinguish the function from its inverse.

   {(–8, 0), (–7, 1), (–4, 2), (1, 3), (5, 3.5), (8, 4)}

2. Graph the inverse of the relation in **Figure 5.60** on the same grid. (Hint: You can fold your paper along the line $y = x$ to check your answer.)

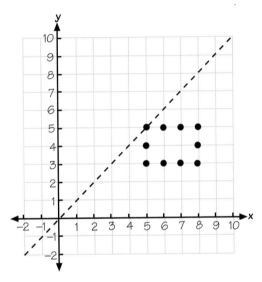

FIGURE 5.60.
Scatterplot for Question 2.

For Questions 3–8, is the statement true or false?

3. If the point (L, D) is a member of some function, then (D, L) is a member of its inverse.

4. If a function's graph intersects the graph of the line $y = x$ at a certain point, its inverse must intersect the line $y = x$ at the same point.

5. The inverse of any function is also a function.

6. A function could intersect the line $y = x$ at (5, 8).

7. The graphs of a function and its inverse are mirror images across the line $y = x$.

8. The domain of any set serves as the range of its inverse.

9. Graph $f(x) = \sqrt{x} + 3$ on a grid.

a) State the domain.

b) State the range.

c) Using your knowledge of inverses, state the domain and range of the inverse (f^{-1}) of $f(x) = \sqrt{x} + 3$.

d) Sketch the graph of the inverse on the same grid.

e) Now find the inverse function.

10. Graph the following transformations of the square root function.

a) $f(x) = -2\sqrt{x}$

b) $f(x) = \sqrt{x} + 2$

c) $f(x) = \sqrt{x+2}$

Aerial View

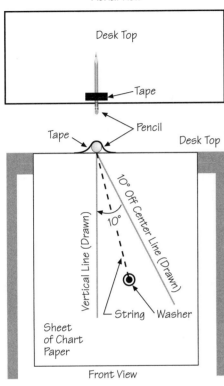

Desk Top

Tape

Pencil

Tape

Desk Top

10° Off Center Line (Drawn)

10°

Vertical Line (Drawn)

String   Washer

Sheet
of Chart
Paper

Front View

FIGURE 5.61.
Pendulum experiment.

Set up an experiment as shown in **Figure 5.61** and time the period of the pendulum for different lengths of string. Here are the steps.

Step 1: On a large sheet of chart paper draw a vertical line.

Step 2: Use a protractor to measure 10° off center of the vertical line and draw another line.

Step 3: Tape the sheet of chart paper to the edge of a desk or table.

Step 4: Tape a pencil to the surface of the desk at the point where the vertical line and the angled line intersect.

Step 5: Make sure you have enough space for a pendulum to swing freely.

Step 6: Tie a washer to the end of the string.

Step 7: Attach the string to the pencil so the length from the pencil to the center of the washer is 15 centimeters.

Step 8: Pull the washer back to the 10° line. Let the washer drop freely: do not push or throw it. The washer should swing back and forth in a plane. If it appears to be swinging in a circular fashion, repeat the trial.

Step 9: Using a stopwatch or a digital watch, find the time it takes for the pendulum to make 10 full swings. Record your measurement in a table like **Figure 5.62**.

FIGURE 5.62.
Table for pendulum experiment.

Length of Pendulum (cm)	Time for 10 Swings (sec)	Period (sec)
15		
25		
35		
45		
55		
65		

Step 10: Calculate the time for one period and record it in the table.

Step 11: Continue measuring the time required for 10 back-and-forth swings as you change the length of the string as shown in the table. You may try other lengths if you wish.

1. How does changing the length of the string affect the period of the pendulum?

2. Graph the results from your experiment on a coordinate grid, and then make a scatterplot on a graphing calculator.

3. Which parent function (linear, quadratic, or square root) best fits your data?

4. Write a function rule to fit your data. Test your function with the graphing calculator.

5. The relationship between the length of a simple pendulum and the period of the pendulum is described by

$$Period \ of \ a \ pendulum = \frac{2\pi}{\sqrt{Acceleration \ due \ to \ gravity}} \cdot \sqrt{Length \ of \ pendulum}.$$

If the acceleration due to gravity is 980 cm/sec², is this equation similar to the function you found?

$$P = \frac{2\pi}{\sqrt{\frac{980 \ cm}{sec^2}}} \cdot \sqrt{L} \ cm$$

$$P = \frac{2\pi}{\sqrt{980}} \sqrt{L} \ sec$$

$$P = 0.2\sqrt{L} \ sec$$

6. Use your function to predict the period for a pendulum that is 200 cm in length.

7. A clockmaker wants the swing of a pendulum to match a clock's ticking of the seconds. In other words, he wants the period of the pendulum to be 2 seconds. How long should the pendulum be?

8. Ava notices that her grandfather clock is slow. To adjust it should she make the pendulum wire longer or shorter?

Assignment

1. Sketch the inverse of each graph in **Figure 5.63.** The graph of $y = x$ is shown as a dotted line to help you sketch the inverse.

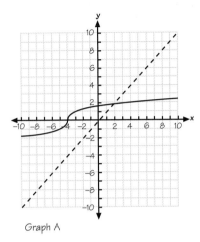

Graph A

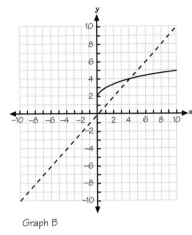

Graph B

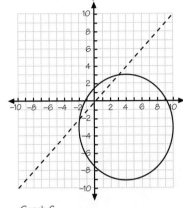

Graph C

FIGURE 5.63.
Three graphs for Question 1.

2. Look at the inverses you sketched in Question 1. Decide if each inverse is a function or a relation.

   Graph A: Function or relation?

   Graph B: Function or relation?

   Graph C: Function or relation?

3. The graph of $y = (x + 1)^2$ is shown in **Figure 5.64.** Answer these questions based on this graph.

   a) State the domain and range of the function.

   b) Sketch the graph of the inverse.

   c) State the domain and range of the inverse.

   d) What two functions can be entered into your calculator in order to see the graph of the inverse?

   e) Is the inverse of $y = (x + 1)^2$ a function? Justify your answer.

   f) How could you restrict the domain of $y = (x + 1)^2$ so its inverse would be a function?

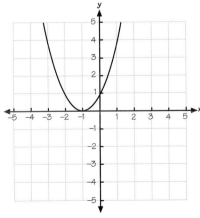

FIGURE 5.64. Graph of $y = (x + 1)^2$.

4. Graph each function.

    a) $f(x) = 2\sqrt{x}$

    b) $f(x) = \sqrt{x} - 3$

    c) $f(x) = \sqrt{x-3}$

The distance you can see from the top of a building depends on the building's height. If h is the height in meters of your viewing place, the distance you can see in kilometers on a clear day is modeled by

$S(h) = 3.532\sqrt{h}$.

5. From a height of 100 meters how far can you see?

6. Draw a graph for $S(h) = 3.532\sqrt{h}$. How can the graph of $S(h)$ be obtained from the graph of the square root parent function?

7. If you can see 14 km from the top of a ride at an amusement park, how tall is the ride?

    a) List examples of situations where this viewing model might not be accurate.

Rolling Stone

An object's acceleration on an incline is constant. The distance (d) moved from rest is related to the time (t) that a marble has been rolling by a constant k, which is unique to each incline (see **Figure 5.65**).

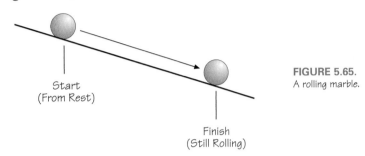

Start
(From Rest)

Finish
(Still Rolling)

FIGURE 5.65.
A rolling marble.

Trial	Distance (d)	Time (t)
A	0.25 m	1.77 sec
B	0.40 m	2.23 sec
C	0.55 m	2.65 sec
D	0.70 m	2.96 sec
E	0.85 m	3.25 sec
F	1.00 m	3.54 sec

FIGURE 5.66.
Math students' data.

1. A group of math students measured the distance, d, a marble rolled down a plane and the time, t, it took for the marble to roll each distance. They recorded their data in the table in **Figure 5.66.**

   a) Make a scatterplot of the group's data.

   b) Find a function that fits the data.

Trial	Distance (d)	Time (t)
G	0.25 m	1.50 sec
H	0.40 m	1.95 sec
I	0.55 m	2.25 sec
J	0.70 m	2.60 sec
K	0.85 m	2.87 sec
L	1.00 m	3.05 sec

FIGURE 5.67.
Data for Question 2.

2. Another group of students measured the distance and time for a marble to roll down a different plane. Their data are shown in **Figure 5.67.**

   a) Make a scatterplot of the data.

   b) Find a function that fits the data.

3. The two groups of students want to set up a marble stunt with their inclines as shown in **Figure 5.68.** At what distance should each marble be released for them to meet at the bottom exactly 4 seconds later?

FIGURE 5.68.
Inclines for Question 3.

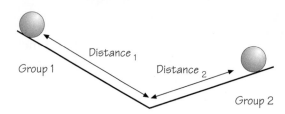

Group 1 · Distance$_1$ · Distance$_2$ · Group 2

Assignment

The questions below are based on the relationship $k = \frac{d}{t^2}$.

1. If your value for k is 0.5 m/s^2, write a function for distance (d) in terms of time (t).

2. Use your function from Question 1 to calculate the distance your marble would move during the times in the table in **Figure 5.69.** Complete the table.

FIGURE 5.69.
Data table for Question 2.

Time (t)	0 s	0.4 s	0.6 s	0.8 s	1.0 s	1.2 s
Distance (d)						

3. Let time (in seconds) serve as your domain and distance (in meters) as your range. Draw a scatterplot of your data from Question 2.

4. Graph the function you wrote in Question 1.

5. What is the parent function for this graph?

6. Now switch your domain and range so that distance (now in meters) serves as your domain and time (now in seconds) as your range. Draw a scatterplot using the data from Question 2.

7. What is the parent function for this graph?

8. Are the two graphs inverses of each other?

9. Find an equation for the inverse of your function from Question 1. Hint: Solve for t.

10. Graph the equation you found in Question 9.

Modeling Project Designing The Hero's Fall

In this project you will design a stunt for a western movie in which the hero falls off a rooftop onto the roof of a stagecoach being drawn by a team of horses. In order to make your plans, you'll need to choose the height of the building, the speed of the stagecoach, and the dimensions of the stagecoach. Make sure your choices are as realistic as possible.

In order to make your stunt work so that the hero does not break any bones, it is important for your hero to know when to jump.

CHECK THIS!

Recall that when an object falls freely, its height can be modeled by the function

$$h = \frac{1}{2}gt^2 + h_o$$

h is the height, g is the acceleration due to gravity (hint: remember this is a constant), and t is time

1. What function will you use to model the motion of the hero during the fall? Explain your reasoning.

2. How long will it take the hero to reach the height of the roof of the stagecoach? Explain your plan.

3. In this situation, height is calculated in feet and time in seconds. However, the speed of the stagecoach is 20 miles per hour. Convert the speed of the stagecoach to feet per second.

4. The hero should begin the fall the instant the center of the stagecoach (not including the horses) reaches a mark in the road. Where should you place this mark so that the hero will land safely on top of the stagecoach? (Show the distance between the mark and the hero's drop point.) Explain your answer.

Practice Problems

1. Answer these questions about the function
   $y - 3\sqrt{x - 2} + 1$.

   a) What is the domain?

   b) What is the range?

2. $h(t) = -5t^2 + 80$ describes the height in meters of a
   falling body after t seconds. Find the intercepts of the
   function's graph and interpret them in this context.
   Explain the role of the constants -5 and 80.

3. Using quadratic models sometimes requires solving
   quadratic equations. Among the methods you may
   have used to solve quadratic equations in previous
   courses are:

   A graphing calculator procedure such as graphing
   and zooming;

   A spreadsheet procedure such as zooming on a table;

   Factoring;

   The Quadratic Formula;

   Completing the square.

   The first two methods give approximate solutions and
   the last three give exact solutions. Solve $6x^2 + 17x = 10$
   by at least two methods.

4. Solve $2x^2 - 11x + 12 = 0$ using the factoring method.

5. The table in **Figure 5.70** contains data on the distance
   it takes a car to stop while traveling at various speeds.

Speed (mph)	20	30	40	50	60	70
Stopping Distance (ft)	42	74	116	173	248	343

   FIGURE 5.70.
   Stopping distances
   for different speeds.

   a) Create a mathematical model that predicts the
      stopping distance for any speed. Explain why you
      think your model is a good one.

b) A police officer at an accident scene estimates that a car needed 300 feet to stop. Write an equation for this situation, and use one of the methods you learned in this chapter to solve it. Describe the method you used.

6. Write each expression in the form stated:

a) $2(x + 4)(x - 3)$ as a general quadratic expression.

b) $3x^2 + 11x + 6$ in factored form.

7. Change the following general quadratic equations into the specified form:

a) $y = 0.5x^2 + 1.5x - 5.0$ in factored form.

b) $y = 2x^2 + 8x + 3$ in vertex form.

8. The graph of $y = 0.5x^2 - 1.5x - 2$ is shown in **Figure 5.71**.

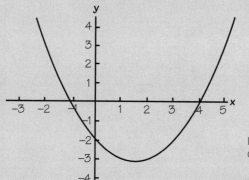

FIGURE 5.71.
Graph of $y = 0.5x^2 - 1.5x - 2$.

a) What are the x-intercepts?

b) What are the coordinates for the vertex?

c) How can you tell from looking at the original equation that the quadratic will look like a "U" (and not upside down)?

9. Given the function $y - 3 = 0.5(x + 1)^2$:

a) Explain the role that each of the three constants play in determining the graph of the function.

b) Sketch the graph of the function.

c) Stretch the graph vertically by a factor of 3 and move it 4 units to the right and 1 unit down. What equation describes the new graph?

10. Solve each of these equations. Explain the method you used.

a) $0.4(x + 3)(x - 2) = 0$

b) $1.6(x + 1)^2 - 3 = 10$

c) $2x^2 + 8x - 5 = 0$

11. A property of quadratic relationships that was explored in this chapter is that second differences are constant. You can prove that to be the case, as well as develop ways to construct the equation from patterns in the table of differences.

a) Substitute each value of x in **Figure 5.72** into
$y = ax^2 + bx + c$.

b) Record the *expression* obtained for y in the column labeled *y*-value.

x-value	y-value	First Differences	Second Differences
0			
1			
2			
3			
4			
5			

FIGURE 5.72.
Table for Question 11.

c) Calculate first and second differences the same way you would if there were numbers in the table. Record your answers in the table.

d) Explain how the work done shows that with quadratic functions, second differences are constant.

e) Explain how to find the values for a, b, and c from the table.

f) Use your answer to e to find a function that fits the data in **Figure 5.73**.

x	y
0	2
1	9
2	20
3	35
4	54
5	77

FIGURE 5.73.
Table for Question 11f.

12. From the top of the Eiffel Tower, which is 300 meters tall, how far can you see?

13. A tsunami, a series of waves created by an undersea earthquake, hit Asia in 2004. Hundreds of thousands of people died. Now monitoring stations determine the speed and force of these waves. A key variable in the damage a wave can do is ocean depth. Ocean depth can be found using the function $s = \sqrt{gd}$ where g is 9.8 meters per second squared (the force of gravity), s is the speed of the wave in meters per second, and d is the depth of the ocean. If a wave is moving at 99 meters per second, how deep is the ocean?

Glossary

KEY CONCEPTS

Binomials: Algebraic expressions involving two terms, usually with one of them containing a variable.

Coefficient: The real number in a mathematical expression. In $3x^2$ the 3 is the coefficient.

Completing the Square: The process of placing a quadratic in vertex form. Completing the square is used to solve quadratic equations.

Dilation (graph): A transformation that stretches or compresses a graph.

Equivalent: When two expressions produce identical values and solutions for all numbers.

Factored Form of a Quadratic: A quadratic of the form $y = a(x - r_1)(x - r_2)$.

First Differences: A pattern of numbers produced by subtracting successive values from a data table in which the x-values increase by a constant amount. If first differences are constant then the equation describing the data is linear.

General Form of a Quadratic: A quadratic of the form $y = ax^2 + bx + c$.

Inverse of a Relation: A reflection of the relation over the line $y = x$.

Inverse Operations: Pairs of operations that undo each other. Addition and subtraction are inverse operations.

Polynomial Form of a Quadratic: A quadratic of the form $y = ax^2 + bx + c$.

Quadratic Formula: A formula used to solve quadratic equations of the form $ax^2 + bx + c = 0$. The solutions are:

$$x = \frac{-b \pm \sqrt{b^2 - 4ac}}{2a} \; .$$

Quadratic Function : A function of the form $y = ax^2 + bx + c$. The highest power of x in a quadratic is 2.

Roots of an Equation: Values that make an equation true.

Second Differences: The pattern of numbers produced by subtracting successive first differences. If the second differences are constant, then the equation describing the data is quadratic.

Square Root: A factor of a number that, when squared, gives the original number.

$\sqrt{}$ (Square Root Symbol): The principal square root of a number. The principal square root of a positive number is the positive square root.

Square Root Function: A function in the form $f(x) = \sqrt{x}$ is called a square root function and is defined for all $x \geq 0$.

Transformation (graph): In this chapter, a rule that changes one graph into another.

Translation (graph): A transformation that affects either the x- or y-values of an equation by adding or subtracting a number. Translations produce a "shift" or "slide" effect on a graph.

Vertex Form of a Quadratic: A quadratic of the form $y = a(x - h)^2 + k$.

Vertex of a Parabola: A point on the graph of a quadratic where the left-side and right-side (symmetric) parts meet. The vertex is either the highest or lowest point of the parabola.

x-intercept: A point where the graph of a function crosses the x-axis. The x-coordinate of the intercept is called a zero of the function.

Zero-Product Property: If $ab = 0$, then $a = 0$ or $b = 0$ or both $a = 0$ and $b = 0$. For example, if $(x + 4)(x - 3) = 0$ the Zero-Product Property states that either $x + 4 = 0$ or $x - 3 = 0$ or both. Thus, $x = -4$ and $x = 3$ are solutions to this equation.

Zeros of a Function: Values for which a function is 0. They are also the x-coordinates of the points where the function's graph crosses the x-axis.

Growth & Decay: Exponential Functions

6
CHAPTER

"For the Benefit and Enjoyment of the People"

As American pioneers settled the West in the 1800s, many of them noticed the beauty of the Great Plains, Rocky Mountains, and Pacific coast. As early as the 1830s, some people began to express concern about the impact people were having on this land and the Native Americans who lived there.

In 1872, President Ulysses S. Grant signed legislation from Congress that set aside Yellowstone Country in Wyoming and Montana Territories "as a public park or pleasuring-ground for the benefit and enjoyment of the people." Yellowstone became the first piece of land in the world to become a national park and was the start of the United States National Park Service.

The National Park Service is now almost 400 parks, historic sites, and sanctuaries. States, counties, and cities also have park services that oversee state and local parks. These parks range from campgrounds and playgrounds to cultural sites.

Park managers must make many decisions about how to balance the needs of human park visitors with the need to preserve the natural environment. In a large park such as Yellowstone, park managers must be able to survey and define each part of the park. They must also keep a careful count of the population of wildlife such as bison, wolves, moose, and fish. Managing disease and giving medicine to sick animals is important in protecting endangered species.

On the human side, park managers must deal with resource management. Renewable resources such as forests must be managed in order to balance the needs of the lumber industry with the need to preserve the park's natural setting. Non-renewable resources such as coal and oil reserves must also be carefully managed.

In a popular park such as Yellowstone, there are many visitors in a year. Park visitors can have a large impact on the park from traffic and noise pollution to the large amounts of garbage that they leave behind.

In this chapter you'll examine some of the issues that park managers face. You'll use mathematical models to solve problems such as wildlife population control and resource management. You will explore and define exponential and logarithmic functions and use them to model real-world situations. Then you will use these models to solve practical problems.

Chapter 6 Growth & Decay: Exponential Functions

Paper Cutting

Just about everything in our world changes over time. Sometimes the change happens quickly, such as a teenager's growth spurt over a summer or an increase of 200,000 bacteria in a container of water. Other changes occur slowly, for example, the erosion of a mountain peak or an increase in Earth's surface temperature.

Patterns of growth can be used to classify behavior. In this chapter, patterns for describing change in quantities are explored. The challenge is to understand what kind of growth patterns a quantity is undergoing and how those patterns can be modeled. Once you determine the type of growth pattern, you can apply the models to a host of other situations such as wildlife populations, radioactive decay, resource depletion, and resource harvest.

1. Cut a piece of paper in half. Lay the two sheets on top of each other. Cut each of the two sheets of paper in half again. Continue cutting, counting the number of layers of paper when the pieces are stacked up, and filling in the table in **Figure 6.1.**

Number of Cuts	Process	Number of Layers of Paper
0		
1		
2		
3		
4		
5		
6		

FIGURE 6.1.
A table for paper cutting.

CHECK THIS!

Repeated multiplication can be written with a base and an exponent. For example, $5 \times 5 \times 5$ can be rewritten as 5^3. The factor 5 is the base. The number that tells how many times the base is multiplied is the exponent.

2. Use a copy of the grid in **Figure 6.2** to make a scatterplot of your data.

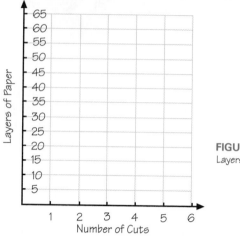

FIGURE 6.2.
Layers of paper versus number of cuts.

3. Write a function for the number of layers of paper if you cut the paper *n* times.

4. Is this function continuous or discrete? Why?

5. What are the domain and range of this function?

6. Use a graphing calculator to make a scatterplot of the data. Then sketch the graph of the function on the scatterplot and record your window.

7. A ream of paper is almost 2 inches thick. If a ream is 500 sheets of paper, about how thick is a piece of paper?

8. Complete the table in **Figure 6.3.**

Number of Cuts	Process	Thickness (in inches)
0		
1		
2		
3		
4		
5		
6		

FIGURE 6.3.
Paper cuts, process, and thicknesses.

9. Write a function rule for the thickness of the stack in inches if you cut the paper n times.

10. How are the function that models the thickness of the stack and the function you found for the number of layers after n cuts related?

11. Sketch a graph for the thickness of the stack, and label the axes.

12. How is this graph transformed from the graph that you drew in Question 6?

13. Complete the tables in **Figures 6.4 and 6.5** using the data from Questions 1 through 12.

Number of Cuts	Number of Layers
0	
1	
2	
3	
4	
5	
6	

FIGURE 6.4. Paper cuts and layers.

Number of Cuts	Thickness (inches)
0	
1	
2	
3	
4	
5	
6	

FIGURE 6.5. Paper cuts and thicknesses.

14. Recall that linear functions can be written from patterns that show repeated addition. The two functions in this section are exponential functions. Exponential functions can be written from patterns that show repeated multiplication or division. Show how the tables in Figures 6.4 and 6.5 display a pattern of repeated multiplication.

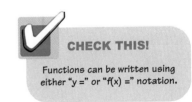

CHECK THIS!

Functions can be written using either "y =" or "f(x) =" notation.

15. A general form of an exponential equation is $f(x) = a \cdot b^x$. From the tables in Figures 6.4 and 6.5, what determines the value of a? How is a related to the graph of the function?

16. What determines the value of b?

Assignment

1. If you cut the paper 22 times, how many layers do you have? Write an equation to help you solve this problem.

2. If you cut the paper 22 times and stack it up, how tall is the stack? Write an equation to help you solve this problem.

3. A box of paper is 5 reams deep. A ream has 500 sheets of paper. About how many cuts would you need to make a stack at least as thick as a box of paper?

4. The Eiffel Tower is approximately 1050 feet tall. If you had a big enough piece of paper, how many cuts would you need to have a stack that equals or exceeds this height?

Moose Population

When animals move to a new place, it is called migration. In 1988, Adirondack State Park in upstate New York had a moose population of about 17 moose. The park is not fenced, and moose migrate into the park from other regions. Assume that the only change in population is due to migration and that the migration rate is two moose per year.

1. Make a table like **Figure 6.6** and use it to predict the number of moose in the year 2000 (12 years later).

Year	Year Number	Number of Moose
1988	0	
1989	1	
1990	2	
1991	3	
1992	4	
1993	5	
1994	6	
1995	7	
1996	8	
1997	9	
1998	10	
1999	11	
2000	12	

FIGURE 6.6.
Table for predicting the number of moose in the park.

2. Make a scatterplot of your data.

3. Find a function rule that models the moose migration. Check it with the graphing calculator. Explain why your function is a good model for the moose migration.

4. In Questions 1–3, the only reason for population change is the migration of moose into the park. Growth also occurs due to reproduction. Babies eventually have their own babies. If you assume that the moose population is 17 in 1988 and doubles each year due to migration and reproduction, make a table like **Figure 6.7** and use it to predict the moose population in 2000.

Year	Year Number	Process Column	Number of Moose
1988	0		
1989	1		
1990	2		
1991	3		
1992	4		
1993	5		
1994	6		
1995	7		
1996	8		
1997	9		
1998	10		
1999	11		
2000	12		

FIGURE 6.7.
Table for predicting the moose population in 2000.

5. Make a scatterplot of the data in Question 4.

6. Find a function rule that models the moose migration in Question 4. Check your function with a graphing calculator. Explain why your function is a good model.

SECTION

6.2

Assignment

For each of these tables, decide if the data show exponential growth, linear growth, or neither. If the growth set is linear or exponential, write a function rule for the pattern. Make a scatterplot with a graphing calculator and check your function to see if it fits.

1.

Year	Population
0	3
1	12
2	48
3	192
4	768

2.

Year	Population
0	3
1	12
2	21
3	30
4	39

3.

Year	Population
0	3
1	12
2	27
3	48
4	75

4.

Year	Population
0	3
1	6
2	12
3	24
4	48

Exponential Functions

Here are the types of function models that you have seen so far in this text:

❖ A linear function is an addition function. As the *x*-values in the domain increase by a constant amount, the *y*-values increase by a constant amount. The first differences of function values are the same. An example of linear data and a function rule is shown in **Figure 6.8.**

x	f(x)
0	5
1	7
2	9
3	11
4	13

+1 → +2 (between each row)

FIGURE 6.8.
A linear set of data and its function rule.

The rate of change is 2. The linear function is $f(x) = 2x + 5$.

❖ In a quadratic function the second differences are constant. An example of quadratic data and a function rule is shown in **Figure 6.9.**

FIGURE 6.9.
A quadratic set of data and its function rule.

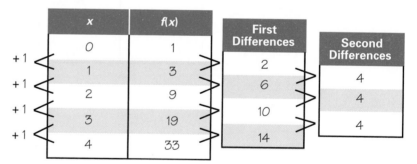

x	f(x)	First Differences	Second Differences
0	1		
1	3	2	
2	9	6	4
3	19	10	4
4	33	14	4

The quadratic function is $f(x) = 2x^2 + 1$.

❖ An exponential function is a multiplication function. As the x-values in the domain increase by a constant amount, the y-values are multiplied by a common multiplier. An example of exponential data and a function rule is shown in **Figure 6.10.**

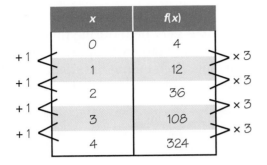

FIGURE 6.10.
An exponential set of data and its function rule.

The common multiplier is 3. The exponential function is $f(x) = 4 \cdot 3^x$.

In this same set of exponential data, notice that the first and second differences are not constants, as shown in **Figure 6.11.**

x	f(x)	First Differences	Second Differences
0	4		
		8	
1	12		16
		24	
2	36		48
		72	
3	108		144
		216	
4	324		

FIGURE 6.11.
An exponential set of data with first and second differences.

CHECK THIS!

The exponential parent function is written in the form $f(x) = b^x$. The constant b is the common multiplier, the value by which you repeatedly multiply or divide. In the paper cutting in Section 6.1, you started with 1 layer when there were 0 cuts. In your investigation of the thickness of the stack after a certain number of cuts, you started with a thickness of 0.004 inches when there were 0 cuts. In the modeling situations in this chapter, the initial value occurs when the independent variable is 0. Thus, in the exponential function $f(x) = a \cdot b^x$ that is used to model these situations, the constant a is the initial value of the dependent variable.

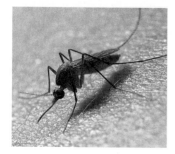

In previous sections you learned how to find a common multiplier. Look at the problem below and find the common multiplier. To find the common multiplier, you will use successive quotients. A successive quotient is the ratio of a y-value to the previous y-value.

A mosquito control scientist collected the data in **Figure 6.12** over a five-month period. Rainfall was very high during this summer, and the county did not have enough money to spray in order to control the mosquito population. Each month the scientist collected 5000 mosquitoes and tested them for West Nile Virus.

Month	Month Number	Number of Mosquitoes Found Testing Positive for West Nile Virus	Successive Quotients
May	0	400	—
June	1	600	$\frac{600}{400} = 1.5$
July	2	900	$\frac{900}{600} = 1.5$
August	3	1350	$\frac{1350}{900} = 1.5$
September	4	2025	$\frac{2025}{1350} = 1.5$

FIGURE 6.12.
Mosquito data.

1. Are the successive quotients the same?

2. How does finding the successive quotients help you find the common multiplier?

3. Find a function rule for the data in the table.

4. Check your function with a graphing calculator.

5. How are the three exponential functions listed below alike, and how do they differ?

$$f(x) = 1 \cdot 2^x \qquad f(x) = 3 \cdot 2^x \qquad f(x) = 0.5 \cdot 2^x$$

6. Use a graphing calculator to graph all three at the same time. Sketch the result on your paper.

7. How does changing a in $f(x) = a \cdot b^x$ change the graph?

CHECK THIS!

In the general exponential function $f(x) = a \cdot b^x$, if $b > 0$ and $a > 1$, the parent graph of $f(x) = b^x$ is stretched vertically. If the value of a is between 0 and 1, the graph is compressed vertically.

Assignment

Decide if a linear, quadratic, or exponential function best describes the data in each table. Find the function.

1.

x	f(x)
0	0.75
1	1.875
2	4.6875
3	11.719
4	29.297

2.

x	f(x)
−2	8
−1	2
0	0
1	2
2	8

3.

x	f(x)
0	2
1	2.44
2	2.88
3	3.456
4	4.1472

4.

x	f(x)
−3	−11
−2	−8
−1	−5
0	−2
1	1

Answer Questions 5–10 for $f(x) = 3(2.25)^x$.

5. Graph the function.

6. Fill in the table in **Figure 6.13** with 5 values for the function.

x	f(x)
0	
1	
2	
3	
4	

FIGURE 6.13.
Table of values.

7. What is the *y*-intercept of the graph?

8. What is the common multiplier?

9. Is $f(x)$ a continuous or discrete function?

10. Is $f(x) = 3(2.25)^x$ a vertical stretch or vertical compression of $f(x) = 2.25^x$?

Use **Figure 6.14** to answer Questions 11–15.

x	f(x)	Successive Quotients
0	0.5	
1	1.75	
2	6.125	
3	21.438	
4	75.031	

FIGURE 6.14.
Table for Questions 11–15.

11. Find the successive quotients for the set of data.

12. Why does an exponential function fit the data?

13. What is the initial value of the function?

14. What is the common multiplier?

15. Find an exponential function that fits the data.

M&M's® Growth

First and second differences and successive quotients are not always exactly equal in real-world data. In this activity you will gather data and fit a function.

Start with 4 M&M's in a cup. Shake the cup, roll out the M&M's, and then count the number with "m" showing. Multiply that number by 2 and add that many M&M's to the cup. Keep a record of the population and repeat the process 5 times. Before you start, answer Question 1.

1. What would you expect the population to be at each roll? Write your predictions in a table like **Figure 6.15.**

Trial	Predicted Population
0	
1	
2	
3	
4	
5	

FIGURE 6.15.
Table for predicting the M&M's population.

2. Add a column for the actual population to your table. Conduct the experiment, and fill in your actual population column.

3. How close are your results to what you predicted?

4. Combine the data with the rest of the class, and enter these data into lists in a calculator. Use transformations to create a function to fit the data.

5. Are the results more reasonable with all of the data? Take successive quotients to confirm your answer.

6. Find a function to fit the data. Check it by graphing the function on a scatterplot. Sketch the graph.

Assignment

In 1995 there were about 60,000 centenarians in the United States. A centenarian is a person at least 100 years old. Imagine that a researcher continued to collect data through 2002 as shown in **Figure 6.16.**

Year	Year Number	Number of Centenarians	Successive Quotients
1995	0	60,000	
1996	1	64,200	1.07
1997	2	68,694	1.07
1998	3	73,503	1.07
1999	4	78,648	1.07
2000	5	84,153	1.07
2001	6	90,044	1.07
2002	7	96,346	1.07

FIGURE 6.16.
Centenarian data from
1995–2002.

1. Make a scatterplot of the data.

2. Explain which type of function (linear, quadratic, or exponential) should fit these data.

3. Find a function that fits the data well.

4. A researcher claimed that the number of centenarians would reach 232,000 by the year 2015. Verify that the researcher is correct based on the function you found in Question 3.

5. When will the number of centenarians reach 1,000,000 based on your model?

6. A scientist is trying to project deer population in a forest. The sample data are shown in **Figure 6.17.** Find successive differences, and then find the successive quotients. Write them in a copy of the table. Round quotients to the nearest hundredth.

Year	Year Number	Population	Successive Differences	Successive Quotients
2000	0	1500		
2001	1	1545		
2002	2	1591		
2003	3	1639		
2004	4	1688		
2005	5	1739		
2006	6	1791		
2007	7	1845		

FIGURE 6.17.
Sample data for a deer population.

7. Sketch a scatterplot of the data.

8. Are the data linear or exponential? Why?

9. Determine the growth rate.

10. Find a function that models the growth.

11. Use your function to predict the population in the year 2010.

Moose Population in National Parks

Let's apply to a new situation what you have learned thus far about exponential functions.

Managers from four parks in New York, New England, and Canada are meeting to discuss how they can protect the moose population. Each of them has been recording the population of moose in their parks. The tables in **Figure 6.18** contain data that they brought to the meeting.

Adirondack Park (New York)		
Year	Year Number	Population
2000		50
2001		55
2002		61
2003		67
2004		73

Green Mountain National Forest (Vermont)		
Year	Year Number	Population
2000		2
2001		4
2002		8
2003		16
2004		32

Algonquin Provincial Park (Ontario, Canada)		
Year	Year Number	Population
2000		10
2001		15
2002		23
2003		34
2004		51

White Mountain National Forest (New Hampshire)		
Year	Year Number	Population
2000		50
2001		54
2002		58
2003		62
2004		66

FIGURE 6.18.
Moose population data from four parks.

If the moose population in each park continues to grow at the same rate, which park will have the most moose in 2015? Explain your answer.

SUMMARY

In this lesson you learned that exponential functions are good mathematical models for certain growth situations. You also reviewed other mathematical models:

❖ A linear model is an addition model.

❖ A quadratic model has second differences that are constant.

❖ An exponential model is a multiplication model.

Exponential models have these properties:

❖ Successive quotients are constant.

❖ The successive quotient is the common multiplier or the base of the exponential function.

❖ The parent function form for the exponential family is $f(x) = b^x$.

❖ A vertical compression or vertical stretch occurs when the parent function $f(x) = b^x$ is multiplied by a constant, a, $f(x) = a \cdot b^x$. This affects the y-intercept, the initial value, or a, in many modeling situations.

Fractional Parts and Area

Park managers often have to divide land into sections. One way to divide land is the rectangular survey system. This system uses rectangles or squares that are divided into equal parts. Each part is then given a name based on its location in relationship to the original rectangle or square.

In this section you will model the rectangular survey system by examining how the visible area of a sheet of paper changes as the paper is folded. You will look for patterns and make predictions.

First let's look at the part of a paper that is formed when a sheet of paper is folded.

You need an 8.5-inch by 11-inch sheet of paper. Before any folds are made, the fraction of the sheet of paper is 1 (the whole sheet).

1. Fold the sheet in half once. This fold splits the paper into regions bounded by the fold lines. Unfold the paper and count the number of regions. What fraction of the original sheet of paper is each new region?

2. Continue folding the paper and complete a table like **Figure 6.19.**

Number of Folds	Fraction of the Original Sheet of Paper
0	
1	
2	
3	
4	
5	
6	

FIGURE 6.19.
Table for paper folding data.

3. What patterns do you see in the table?

4. Based on the table values, does this relationship appear to be linear? How do you know?

5. Based on the table values, does this relationship appear to be quadratic? How do you know?

6. What is a reasonable domain here? Why?

7. What is a reasonable range here? Why?

8. Sketch what you would expect a graph of the fraction of the original sheet of paper versus the number of folds to look like. Why did you sketch the graph the way you did?

Now let's look at the area of the part of a paper that is formed when a sheet of paper is folded.

You will need a new 8.5-inch by 11-inch sheet of paper to fold. Before any folds are made, calculate the area in square inches of the sheet of paper.

9. What is the area of your sheet of paper?

10. Fold the sheet of paper one time; this fold splits the paper into regions bounded by the fold lines. What is the area of each new region?

11. Continue folding the paper and complete a table like the one in **Figure 6.20.**

Number of Folds	Area of a Region
0	
1	
2	
3	
4	
5	
6	

FIGURE 6.20.
Table for region area data.

12. What patterns do you see in the table?

13. Does the relationship between area and number of folds appear to be linear? How do you know?

14. Does the relationship appear to be quadratic? How do you know?

15. What is a reasonable domain here? Why?

16. What is a reasonable range here? Why?

17. Sketch what you expect a graph of the area of a region versus the number of folds to look like. Why did you sketch the graph the way you did?

Area of a Region

In the last section you used paper folding to break down a piece of paper into smaller regions, then you found the area of each region.

Examine the first table that you made in Section 6.6 (see **Figure 6.21**).

Number of Folds	Fraction of the Original Sheet of Paper
0	1
1	$\frac{1}{2}$
2	$\frac{1}{4}$
3	$\frac{1}{8}$
4	$\frac{1}{16}$
5	$\frac{1}{32}$
6	$\frac{1}{64}$

FIGURE 6.21. Table of values.

Number of Folds (n)	Process	Fraction of the Piece of Paper (f)
0	1	1
1	$1 \cdot \frac{1}{2}$	$\frac{1}{2}$
2	$1 \cdot \frac{1}{2} \cdot \frac{1}{2} = \frac{1}{2^2}$	$\frac{1}{4}$
3	$1 \cdot - \cdot - \cdot - = -$	$\frac{1}{8}$
4	$1 \cdot - \cdot - \cdot - \cdot - = -$	$\frac{1}{16}$
5	$1 \cdot - \cdot - \cdot - \cdot - \cdot - = -$	$\frac{1}{32}$
6	$1 \cdot - \cdot - \cdot - \cdot - \cdot - \cdot - = -$	$\frac{1}{64}$

FIGURE 6.22. Table of values with process column.

1. Use a graphing calculator to make a fraction of the original sheet versus the number of folds scatterplot. Record your window and sketch your graph.

2. What do you notice about the scatterplot?

3. How does each fraction of a sheet relate to the one before it?

4. Using the answer to Question 3, copy the table and fill in the blanks in the process column (see **Figure 6.22**).

5. What stays the same in the process column?

6. What changes in the process column?

7. Write a function rule that models this situation.

8. In the context of the paper folding activity, what does each number mean in your rule from Question 7?

9. Graph your function rule on your scatterplot. If the graph is not a good fit, change your rule. Sketch your graph.

CHECK THIS!

To find successive quotients, divide each value by the one before it.

10. What will be the fraction of the piece of paper after 9 folds?

11. If the fraction of the sheet of paper is $\frac{1}{256}$, how many times has it been folded?

12. Find the successive quotients for fraction of the sheet of paper and record them in a table like **Figure 6.23**.

Number of Folds	Fraction of the Original Sheet of Paper	Successive Quotients
0	1	
1	$\frac{1}{2}$	
2	$\frac{1}{4}$	
3	$\frac{1}{8}$	
4	$\frac{1}{16}$	
5	$\frac{1}{32}$	
6	$\frac{1}{64}$	

FIGURE 6.23.
Table of values with column for successive quotients.

13. What do you notice about the successive quotients?

14. Where do you see the successive quotients in your function from Question 7?

Now look at the second table you made in Section 6.6 (see **Figure 6.24**).

Number of Folds	Area of a Region
0	93.5 in^2
1	46.75 in^2
2	23.375 in^2
3	11.6875 in^2
4	5.84375 in^2
5	2.921875 in^2
6	1.4609375 in^2

FIGURE 6.24.
Table of values from Section 6.6.

15. Use a graphing calculator to make an area versus number of folds scatterplot. Record your window and sketch your graph.

16. What do you notice about the scatterplot?

17. How does each area relate to the area of the region before it?

Let's use a process column to help find a function.

18. Copy the table in **Figure 6.25** and fill in the blanks in the process column.

Number of Folds	Process	Area of a Region
0	93.5	93.5 in^2
1	$93.5 \cdot \frac{1}{2}$	46.75 in^2
2	$93.5 \cdot \frac{1}{2} \cdot \frac{1}{2} = 93.5 \cdot \frac{1}{2^2}$	23.375 in^2
3	$93.5 \cdot - \cdot - \cdot - = 93.5 \cdot -$	11.6875 in^2
4	$93.5 \cdot - \cdot - \cdot - \cdot - = 93.5 \cdot -$	5.84375 in^2
5	$93.5 \cdot - \cdot - \cdot - \cdot - \cdot - = 93.5 \cdot -$	2.921875 in^2
6	$93.5 \cdot - \cdot - \cdot - \cdot - \cdot - \cdot - = 93.5 \cdot -$	1.4609375 in^2

FIGURE 6.25.
Table of values with blanks in the process column.

19. What stays the same in the process column?

20. What changes in the process column?

21. Write a function rule that models this situation.

22. Graph your function rule on your scatterplot. If the graph is not a good fit, use transformations to change your rule. Sketch your graph.

23. What will be the area of the region of paper after 10 folds?

24. If the area of the region is 0.365234375 square inches, how many times has the paper been folded?

25. Calculate the successive quotients for area of a region and record them in a table like **Figure 6.26.**

Number of Folds	Area of a Region	Successive Quotients
0	93.5 in^2	
1	46.75 in^2	
2	23.375 in^2	
3	11.6875 in^2	
4	5.84375 in^2	
5	2.921875 in^2	
6	1.4609375 in^2	

FIGURE 6.26.
Table of values for calculating the successive quotients.

26. What do you notice about the successive quotients?

27. Where do you see the successive quotients in your function?

28. Write a sentence about the relationship of successive quotients to the function rule for an exponential function.

Assignment

1. Rhonda is drawing a design in art class. She started by drawing a triangle. Next she drew a new triangle by connecting the midpoints of the sides of the first triangle. She shaded a new triangle. Next she found the midpoints of the sides of the unshaded triangles and connected them. Once again she shaded some new triangles. Her first three drawings are shown. Based on this pattern, complete the term value column in a table like **Figure 6.27.**

Term Number	Picture	Process Column	Term Value (Number of *Unshaded* Triangles)
0		1	1
1		1 · 3	3
2			
3			
4			
n			

FIGURE 6.27.
Table for finding the number of unshaded triangles.

2. Use your graphing calculator to create a term value versus term number scatterplot. Record your window and sketch your graph.

3. What do you notice about the scatterplot?

4. How does each term value relate to the one before it?

5. Fill in the process column of your table.

6. What stays the same in the process column?

7. What changes in the process column?

8. Write a function rule that models this situation.

9. Graph your function rule on your scatterplot. If the graph is not a good fit, change your rule. Sketch your graph.

10. What will be the term value for the 9th term?

11. If there are 6561 new shaded triangles, what term number is this?

12. Next, Rhonda decided to examine the area of the smallest triangles she drew. She started by defining the area of the triangle for term number 0 as one square unit. So the area of the smallest triangle in her second figure is $\frac{1}{4}$ square units. Based on this pattern, complete the term value column in a table like **Figure 6.28.**

13. Use a graphing calculator to make a term value versus term number scatterplot. Record your window and sketch your graph.

14. What do you notice about the scatterplot?

15. How does each term value relate to the one before it?

16. Fill in the process column of your table.

Term Number	Picture	Process Column	Term Value (Area of the Smallest Triangle)
0	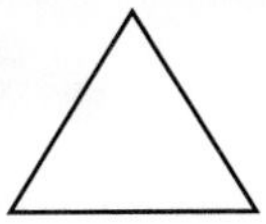	1	1
1		$1 \cdot \frac{1}{4}$	$\frac{1}{4}$
2	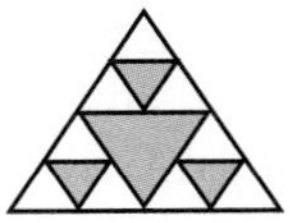		
3			
4			
n			

FIGURE 6.28.
Table for finding the area of the smallest triangle.

17. What stays the same in the process column?

18. What changes in the process column?

19. Write a function rule that models this situation.

20. Graph your function rule on your scatterplot. If the graph is not a good fit, change your rule. Sketch your graph.

21. What will be the term value for the 8th term in this pattern?

22. If the area of the smallest triangle is 0.000061 square units, what term number is this?

Characteristics of Exponential Functions

In the last sections you began to explore exponential functions. In this section you will see how and why those functions work the way they do.

Recall the paper cutting in Section 6.1. Each time the paper was cut, the number of layers of paper increased. This is an example of **exponential growth**.

1. How did the table show exponential growth?

2. How did the graph show exponential growth?

Recall the paper folding in Sections 6.6–6.7. Each time the paper was folded, the size of the region decreased. The pattern you observed is an example of **exponential decay**.

3. How did the table of data show exponential decay?

4. How did the graph show exponential decay?

5. How are the function rules for the relationship between the fraction of the paper and number of folds and the relationship between area and number of folds alike?

6. How do the function rules for the relationship between the fraction of the paper and number of folds and the relationship between area and number of folds differ?

Plant Height	
Number of Days	Height (cm)
0	2
1	5
2	8
3	11
4	14
5	17

+1
+1
+1
+1
+1
ΔX

_ +3 _
_ +3 _
_ +3 _
_ +3 _
_ +3 _
ΔY

Recall that linear relationships have a constant rate of change. In **Figure 6.29**, the height of a plant increases at a constant rate of 3 centimeters per day.

FIGURE 6.29.
Plant height.

7. In this linear relationship, what is the repeated operation?

8. What is a function rule that fits these data?

Let's compare the table for a linear relationship to the table for an exponential relationship. The table in **Figure 6.30** shows an exponential relationship between time and the amount of dissolved carbon dioxide (CO_2) in a can of soda.

Number of Hours	Dissolved CO_2
0	21.87
1	7.29
2	2.43
3	0.81
4	0.27
5	0.09

FIGURE 6.30.
Table of values for an exponential relationship.

9. How do these data show that this relationship is not linear or quadratic?

10. Is this relationship an example of exponential growth or exponential decay? How do you know?

11. Find successive quotients for the dissolved gas and record them in a table like **Figure 6.31.**

Number of Hours	Dissolved CO_2	Successive Quotients
0	21.87	
1	7.29	
2	2.43	
3	0.81	
4	0.27	
5	0.09	

FIGURE 6.31.
Table for recording successive quotients.

If a relationship is exponential and the rate of change of the independent variable is 1, then successive quotients of the dependent variable are constant.

12. What operation is repeated in an exponential relationship?

13. What is the function rule that fits these data?

The general function rule that models an exponential relationship is

y = (initial value) · (common multiplier)^x

$$y = a \cdot b^x$$

CHECK THIS!

The common multiplier in an exponential function is also called a common ratio because it is the ratio of successive values.

14. What must be the value of b, the common multiplier (successive quotient), in an exponential decay function?

15. What must be the value of b, the common multiplier (successive quotient), in an exponential growth function?

16. How can you find the base (b) of an exponential function in the form $y = a \cdot b^x$ from a table of values?

17. How can you find the initial value (a) for $y = a \cdot b^x$ from a table of values?

Wildlife scientists use models to predict animal populations. The data in **Figure 6.32** are for a bison population in a park.

Year	Year Number	Bison Population
1998	0	600
1999	1	570
2000	2	542
2001	3	514
2002	4	489
2003	5	464
2004	6	441
2005	7	419

FIGURE 6.32.
Bison population data.

18. Do these data show a linear relationship? How do you know?

19. Do these data show an exponential relationship? How do you know?

20. Is this relationship an example of exponential growth or exponential decay? How do you know?

21. Use your graphing calculator to make a population versus year number scatterplot. Record your window and sketch your graph.

22. Write a function rule that models this situation.

23. Graph your function rule on your scatterplot. If the graph is not a good fit, change your rule. Sketch your graph.

24. Predict the bison population in 2007.

1. State if each of these tables shows a linear, quadratic, or exponential relationship. If the pattern is exponential, state if it is growth or decay.

a)

x	y
0	100
1	50
2	25
3	12.5
4	6.25
5	3.125

b)

x	y
0	1
1	3
2	9
3	27
4	81
5	243

c)

x	y
0	1250
1	1350
2	1458
3	1574.6
4	1700.6
5	1836.7

d)

x	y
0	0
1	2
2	8
3	18
4	32
5	50

e)

x	y
0	1
1	$\frac{1}{3}$
2	$\frac{1}{9}$
3	$\frac{1}{27}$
4	$\frac{1}{81}$
5	$\frac{1}{243}$

f)

x	y
0	2
1	5
2	8
3	11
4	14
5	17

2. Match each window and graph to one of the tables in
Question 1, and write the function rule that fits the data.

a)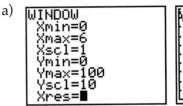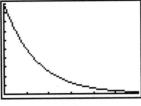

b)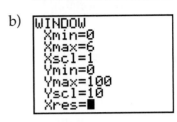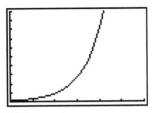

c)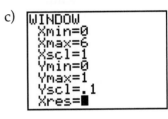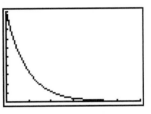

d)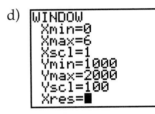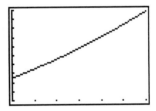

e)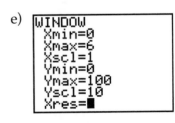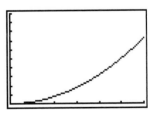

f)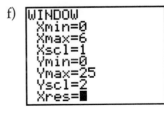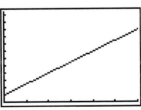

Hour	Milligrams
0	100
1	75
2	56.25
3	42.1875
4	31.6406
5	23.7304
6	17.7979
7	13.3484

FIGURE 6.33.
Medicine concentration data.

3. Park animals may need to be given medicine. The concentration of medicine in an animal's bloodstream decreases over time. **Figure 6.33** shows the concentration of a certain medicine in milligrams for each hour after an initial dose.

a) Do these data show a linear relationship? How do you know?

b) Do these data show an exponential relationship? How do you know?

c) Is this relationship an example of exponential growth or exponential decay?

d) Use your graphing calculator to make a milligrams versus number of hours scatterplot. Record your window and sketch your graph.

e) Write a function rule that models this situation.

f) Graph your function rule on your scatterplot. If the rule is not a good fit use transformations to change it. Sketch your graph.

g) How many milligrams of medication should remain in the bloodstream after 10 hours?

Newton's Law of Cooling

A national or state park is generally located around some natural attraction. For example, Yellowstone National Park contains geological interests such as the Old Faithful geyser. Crater Lake National Park in Oregon has a dormant volcano whose crater has filled with rainwater and become a deep lake.

Most lakes and pools fed by springs have a constant temperature. For example, Barton Springs in Austin, Texas, feeds a manmade pool in Zilker Park near downtown Austin. The water from the springs is always 68°F making the pool it feeds cool and popular during hot summer afternoons.

Since the temperature of the water is constant at 68°F, objects that are put in the pool cool to that temperature. Park managers can calculate how long it takes an object in the pool to reach 68°F using Newton's Law of Cooling.

To investigate Newton's Law of Cooling, we will perform two experiments. The first experiment will gather temperature data by putting a sensor in a cup of ice. The second experiment will gather temperature data by first putting a sensor in a cup of hot liquid then pulling it out and letting it cool to room temperature.

EXPERIMENT 1: CUP OF ICE

Before the start of the experiment, use a probe to record room temperature. This measurement is the temperature at time 0.

At the moment instructions tell you to PRESS ENTER TO START COLLECTING DATA, place the probe into a cup of ice. The calculator will collect 99 seconds of data.

1. Sketch a scatterplot of the data. Label your axes with units and record the window.

2. Copy the data for the first 7 seconds into a table like **Figure 6.34,** calculate the successive quotients and record them in the table.

Hour	Temperature (degrees Celsius)	Successive Quotients
0		
1		
2		
3		
4		
5		
6		
7		

FIGURE 6.34.
Table for data.

3. Do the data appear to be exponential? How do you know?

4. Use a graphing calculator to make a scatterplot of temperature versus seconds for the first 7 seconds. Sketch your graph.

5. If we continued this experiment, would the temperatures approach 0? Why or why not?

6. Write a function rule to model this situation.

7. Graph your function rule on the scatterplot. Sketch your graph and record your window.

EXPERIMENT 2: IT'S GETTING COLDER

For the second experiment, place the probe in a cup of hot water for 30 seconds. This time, at the moment instructions tell you to PRESS ENTER TO START COLLECTING DATA, press ENTER, take the probe out of the hot water, and hold it in the air.

8. Sketch a scatterplot of the data. Label your axes with units.

9. Copy the data for the first 7 seconds into a table like Figure 6.34. Calculate the successive quotients and record them in your table.

10. Do the data appear to be exponential? How do you know?

11. Use your graphing calculator to make a scatterplot of temperature versus seconds for the first 7 seconds of experiment 2. Sketch your graph and record your window.

12. If we continued this experiment, would the temperatures approach 0? Why or why not?

13. Transform your data from experiment 2 by subtracting room temperature from each temperature reading. Record the transformed temperatures in a table like **Figure 6.35.** Calculate the successive quotients and record them in the table.

Seconds	Temperature – Room Temperature	Successive Quotients
0		
1		
2		
3		
4		
5		
6		
7		

FIGURE 6.35.
Table for data and for successive quotients.

14. Do the data appear to be exponential? How do you know?

15. Use a graphing calculator to make a scatterplot of temperature – room temperature versus seconds. Sketch your graph and record your window.

16. Write a function rule to model the transformed data.

17. Graph your function rule on the scatterplot. Sketch your graph and record your window.

18. Use your knowledge of transformations to find a function rule for your original data.

19. Graph your function rule on the original scatterplot.

20. Graph both rules in the same window. What effect did the transformation have on the graph of your function rule from Question 16?

21. What does this transformation represent here?

1. Use transformations to describe the graph of each function as it relates to the function $y = (0.93)^x$.

   a) $y = (0.93)^x + 2$

   b) $y = (0.93)^x - 4$

   c) $y = 5(0.93)^x$

   d) $y = 0.3(0.93)^x$

   e) $y = 8.3(0.93)^x + 3$

   f) $y = 0.5(0.93)^x - 6$

2. In each of these, the table represents a transformation of the function $y = (2)^x$. Write the function rule that fits the transformed data.

a)

x	y
0	6
1	7
2	9
3	13
4	21

b)

x	y
0	3
1	6
2	12
3	24
4	48

Bounce Number	Height of the Bounce (feet)
0	
1	5.231
2	4.715
3	4.436
4	4.015
5	3.788
6	3.600

FIGURE 6.36.
Rubber ball data.

3. Students in a math class collected the data in **Figure 6.36.** They dropped a rubber ball and measured the maximum height of each bounce.

   a) Do the data appear to be exponential? How do you know?

   b) Use your graphing calculator to make a scatterplot of height versus bounce number. Sketch your graph.

Chapter 6 Growth & Decay: Exponential Functions

c) What is the average of the successive quotients?

d) Use the average of the successive quotients to estimate an initial value.

e) Write a function rule to fit the transformed data.

f) Graph your function rule on the scatterplot. Sketch your graph and record your window.

g) What would be the height of the 8th bounce?

h) Suppose the same ball is dropped from an initial height of 10 feet instead of 5.624 feet. What is a function rule that models this situation?

i) Predict how the graph of the function rule in h will differ from the graph of the function rule in e.

j) Graph both rules in the same window. Sketch your graph and record your window. What effect does changing the initial height to 10 feet have on the graph?

k) Suppose a different ball is dropped from an initial height of 4624 feet and that every bounce height is 1 foot less that the heights in the original data. What is a function rule that models this?

l) Predict how the graph of the function rule in k will differ from the graph of the function rule in e.

m) Graph both rules in the same window. Sketch your graph. What effect does subtracting 1 have on the graph?

No More Foam

Cindy, Jamal, and Juanita are discussing sodas. Jamal likes Blue Cow soda best because when he pours it in a glass the foam disappears in less time than any other soda. Juanita thinks the foam in her favorite soda, Dr. Salt, disappears in less time than any other soda. Cindy thinks the same thing about her favorite soda, Diet Dr. Salt.

The three friends decide to do an experiment on the relationship between time and the amount of foam in a glass. Their data are shown in **Figure 6.37** with foam height measured in centimeters.

Time in Seconds	Blue Cow	Dr. Salt	Diet Dr. Salt
0	17.0	14.0	14.0
1	16.1	11.8	12.1
2	14.9	10.5	10.9
3	14.0	9.3	10.0
4	13.2	8.5	9.3
5	12.5	7.7	8.6
6	11.9	7.1	8.0
7	11.2	6.5	7.5
8	10.7	6.0	7.0
9	9.7	5.3	6.2

FIGURE 6.37.
Foam data on three brands of sodas.

1. If each person drinks his or her favorite soda exactly 20 seconds after it is poured, who will be have the least foam?

2. Write a function rule that models each relationship.

SUMMARY

The general function rule that models an exponential relationship is

$y =$ (initial value) $\cdot$ (common multiplier)^x

$y = a \cdot b^x$.

If $b > 1$, then the exponential function models exponential growth where successive y-values increase by a constant factor.

If $b = 1$, then the exponential function reduces to $y = a \cdot 1^x = a$, which is a constant function.

If $0 < b < 1$, then the exponential function models exponential decay where successive y-values decrease by a constant factor.

The common multiplier b can be found from a table using successive quotients as shown in **Figure 6.38.**

Number of Hours	Dissolved CO_2	Successive Quotients
0	21.87	
1	7.29	$7.29 \div 21.87 = \frac{1}{3}$
2	2.43	$2.43 \div 7.29 = \frac{1}{3}$
3	0.81	$0.81 \div 2.43 = \frac{1}{3}$
4	0.27	$0.27 \div 0.81 = \frac{1}{3}$
5	0.09	$0.09 \div 0.27 = \frac{1}{3}$

FIGURE 6.38.
Table using successive quotients.

The value of a is the y-value when $x = 0$ if there is no vertical shift.

In Figure 6.38 the value for a is 21.87 because that is the amount of dissolved carbon dioxide (CO_2) present at 0 hours. Since the successive quotients are all $\frac{1}{3}$, $b = \frac{1}{3}$; therefore this situation can be modeled using the function $y = 21.87\left(\frac{1}{3}\right)^x$.

In the context of a problem, the value for a is the initial amount of a thing that is decaying exponentially.

For example, if the initial dosage of medication is 500 milligrams and the amount of medication in the bloodstream decreases by 90% every hour, the initial value a is 500. The constant multiplier b is 0.90. The amount of medication y in the bloodstream at time t can be modeled with $y = 500(0.90)^t$.

Switching Directions

A LOOK AHEAD

In Chapter 5 you learned that the inverse of a function is the reflection of the function over the line $y = x$. The domain of the function becomes the range of its inverse; the range of the function becomes the domain of its inverse. In the case of the quadratic function $y = x^2$, the inverse is the square root function $y = \sqrt{x}$. Likewise, the inverse operation of raising a number to a power is to take the root of that number. In other words, to un-square a number such as 9, you take the square root of the number; for example, $\sqrt{9}$.

Let's explore some ways we might *un-power* an equation such as $3^x = 27$. We will also investigate the inverse of an exponential function.

1. Make a table for $f(x) = 3^x$ like **Figure 6.39,** then sketch a graph of the function on a grid like **Figure 6.40.**

x	−2	−1	0	1	2	3	4
f(x)							

FIGURE 6.39.
Table for $f(x) = 3^x$.

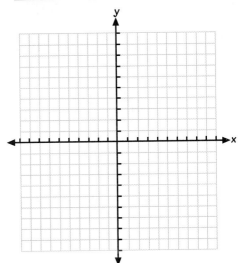

FIGURE 6.40.
Grid for graphing $f(x) = 3^x$.

2. What is the domain of this function?

3. What is the range of this function?

4. Trace the *x*-axis, *y*-axis, and the graph of *f*(*x*) onto a sheet of patty paper. Be sure to label the *x*- and *y*-axes. Reflect the graph of *f*(*x*) across the line *y* = *x* by holding the top-right and bottom-left corners of the patty paper in each hand and flipping the sheet of patty paper over (see **Figure 6.41**). Sketch what you see.

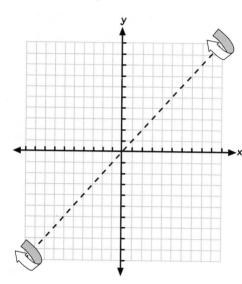

FIGURE 6.41.
Flipping over the patty paper.

CHECK THIS!

Recall that the inverse of a function is a reflection across the line *y* = *x*.

You can generate the inverse of a function by switching the domain and range values of the original function.

Now let's look at a function that undoes *f*(*x*). Let's call that function *g*(*x*). For example, when we evaluate *f*(*x*) at *x* = 0, we get a result (output) of 1. In *g*(*x*) when we evaluate *g*(*x*) at *x* = 1, we get a result of 0.

5. Complete a table like **Figure 6.42**, and sketch the graphs of both *y* = *f*(*x*) and *y* = *g*(*x*).

x							
g(x)	−2	−1	0	1	2	3	4

FIGURE 6.42.
Table for *g*(*x*).

6. Is the new curve that you sketched the graph of a function? How can you tell?

7. What is the domain of this function?

8. What is the range of this function?

The inverse of an exponential function has a special name; it is called a **logarithmic function**. In the next few sections you will explore and find some properties and uses for logarithmic functions.

It All Adds Up

In the last section you developed the inverse of an exponential function, a **logarithmic function**. This function is written as **log(x)** where x is any positive number. In the next few sections you will explore some properties and uses for logarithmic functions.

You have studied transformations of graphs of functions. In this section you will explore transformations of graphs of logarithmic functions and make some conjectures about patterns that occur in logarithmic functions.

1. On your graphing calculator, graph $y = \log(x)$. Record your viewing window and sketch your results.

2. What is the x-intercept? What is the y-intercept? Are there any other key points? Explain how you found these points.

3. In the same viewing window, graph $y = \log(x) + 2$ with $y = \log(x)$. Sketch your graph.

4. Now graph the parent function $y = \log(x)$ and $y = \log(x) - 2$ in the same window. Sketch your graph.

5. Compare the graphs of $y = \log(x)$, $y = \log(x) + 2$, and $y = \log(x) - 2$. How are they alike? How do they differ?

6. In terms of transformations, what will the graph of $y = \log(x) + k$ look like?

7. In a new viewing window, graph the parent function $y = \log(x)$ and $y = 2 \log(x)$. Sketch your graph and record your window.

8. Now graph the parent function $y = \log(x)$ and $y = \frac{1}{2} \log(x)$ in the same window. Sketch your graph.

9. Compare the graphs of $y = \log(x)$, $y = 2 \log(x)$, and $y = \frac{1}{2} \log(x)$. How are they alike? How do they differ?

10. In terms of transformations, what will the graph of $y = a \log(x)$ will look like?

11. Now enter Y1 = log(3x) and Y2 = log(3) + log(x) into the function list of your graphing calculator. Change the plot style for Y2 to a tracing ball then graph the two functions. Record your window and sketch the graphs. Explain what you saw while the calculator graphed both functions.

12. Now look at a table for these two functions. What patterns between the values for Y1 and Y2 do you see? What do these patterns suggest about the relationship between Y1 = log(3x) and Y2 = log(3) + log(x)?

13. Repeat Question 11 with Y1 = log(6x) and Y2 = log(3) + log(2x). Sketch your graphs and look at a table. What patterns between the values for Y1 and Y2 do you see? What do these patterns suggest about the relationship between Y1 = log(6x) and Y2 = log(3) + log(2x)?

14. Repeat Question 11 with Y1 = log(6x) and Y2 = log(2) + log(3x). Sketch your graphs and look at a table. What patterns between the values for Y1 and Y2 do you see? What do these patterns suggest about the relationship between Y1 = log(6x) and Y2 = log(2) + log(3x)?

15. Use your answers to Questions 12, 13, and 14 to write a sentence about log(xy).

16. Now enter Y1 = log(3x) and Y2 = log(6x) – log(2) into the function list of your graphing calculator. Change the plot style for Y2 to a tracing ball then graph the two functions. Record your window and sketch the graphs. Explain what you saw while the calculator graphed both functions.

17. Now look at a table for these two functions. What patterns between the values for Y1 and Y2 do you see? What do these patterns suggest about the relationship between Y1 = log(3x) and Y2 = log(6x) – log(2)?

18. Repeat Question 16 for Y1 = log(2x) and Y2 = log(8x) – log(4). Sketch your graphs and look at a table. What patterns between the values for Y1 and Y2 do you see? What do these patterns suggest about the relationship between Y1 = log(2x) and Y2 = log(8x) – log(4)?

19. Use your answers to Questions 17 and 18 to make a generalization about log$\left(\frac{x}{y}\right)$.

20. Now enter Y1 = log(x^3) and Y2 = 3 log(x) into the function list of your graphing calculator. Change the plot style for Y2 to a tracing ball then graph the two functions. Record your window and sketch the graphs. Explain what you saw while the calculator graphed both functions.

21. Now, look at a table for these two functions. What patterns between the values for Y1 and Y2 do you see? What do these patterns suggest about the relationship between Y1 = log(x^3) and Y2 = 3 log(x)?

22. Repeat Question 20 for Y1 = log(x^5) and Y2 = 5 log(x). Sketch your graphs and look at a table. What patterns between the values for Y1 and Y2 do you see? What do these patterns suggest about the relationship between Y1 = log(x^5) and Y2 = 5 log(x)?

23. Use your answers to Questions 21 and 22 to make a sentence about about log(x^n).

Assignment

1. Sketch a graph of $f(x) = 5^x$.

2. What are the domain and range of this function?

3. Sketch the graph of $g(x) = 5^x - 3$. How is this graph of $g(x)$ related to the graph of $f(x) = 5^x$?

4. What are the domain and range of $g(x) = 5^x - 3$? How do they compare to the domain and range of $f(x) = 5^x$?

5. Sketch a graph of $f(x) = \log(x)$.

6. What are the domain and the range of this function?

7. Sketch the graph of $g(x) = \log(x - 2)$. How is this graph of $g(x)$ related to the graph of $f(x) = \log(x)$?

8. What are the domain and range of $g(x)$? How do they compare to the domain and range of $f(x) = \log(x)$?

9. Find the value of each of these:

    a) $\log(3) + \log(7) =$

    b) $\log(4) + \log(10.5) =$

    c) $\log(48) - \log(16) =$

    d) $\log(60) - \log(8) =$

It's Not What You Say, But How You Say It

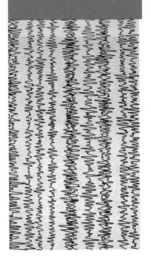

Logarithms were developed to help scientists work with large numbers in the days before calculators and computers. John Napier (1550–1617), a Scottish mathematician, developed logarithms. It has been said that science advanced quickly in the 1700s because logarithms made calculations easier.

Today, calculators and computers make calculations easy. On the other hand, logarithmic functions can still be used for modeling in science. Exponential relationships, such as energy or sound, can be expressed inversely using logarithms. For example, the Richter scale is a logarithmic scale that is used to measure the intensity of an earthquake, and the decibel scale is a logarithmic scale that is used to measure the intensity of sound.

An exponential equation has three parts as shown in **Figure 6.43**.

FIGURE 6.43.
An exponential equation.

$$y = 3^x$$

Power, Base, Exponent

As you recall, a function that undoes another function is called the **inverse** of that function. One way to find an inverse of a function is to switch y and x in its equation then solve for y. A name for a function that is the inverse of $f(x) = 3^x$ is **logarithm, base 3, of x**. It is written $f(x) = \log_3 x$. The abbreviation for logarithm is log.

To write an exponential equation as a logarithm, the base of the power becomes the base of the logarithm. The power is the quantity of which you are taking the logarithm; the logarithm of that quantity is equal to the exponent (see **Figure 6.44**).

FIGURE 6.44.
Writing an exponential
equation as a logarithm.

$$y = 3^x \qquad x = \log_3 y$$

Exponent, Base, Power

Every exponential equation can be rewritten in logarithmic form and vice-versa. For example, to write $y = \log_9 81$ in exponential form.

Step 1: Identify the base as 9 and the exponent as y.

Step 2: Using the base and exponent, write an exponential expression, 9^y.

Step 3: Set the expression equal to the power in the logarithmic equation, $9^y = 81$.

To evaluate a logarithm, set it equal to x and then rewrite it in exponential form. For example, evaluate $\log_7 49$.

Step 1: Set the expression equal to x: $\log_7 49 = x$.

Step 2: Rewrite the expression in exponential form: $7^x = 49$.

Step 3: Rewrite 49 as a power of 7: $7^x = 7^2$.

Step 4: Evaluate $x = 2$.

Because logarithms can be written with any base, it is important to state the base. Our number system is base-10 (that is, place value is based on powers of 10: 1, 10, 100, etc.). Base-10 logarithms are often used; thus they are called **common logarithms**. To save time, $\log_{10}$ can be written without the 10 simply as log.

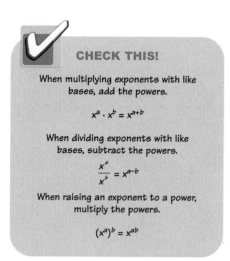

CHECK THIS!

When multiplying exponents with like bases, add the powers.

$$x^a \cdot x^b = x^{a+b}$$

When dividing exponents with like bases, subtract the powers.

$$\frac{x^a}{x^b} = x^{a-b}$$

When raising an exponent to a power, multiply the powers.

$$(x^a)^b = x^{ab}$$

There are three key properties of logarithms that you explored in Section 6.12. First, you explored $\log(xy)$.

1. Use the log key on your calculator to find the values in **Figure 6.45**. Keep a copy of your table because you will use it later in this section.

log 2	
log 5	
log 10	
log 50	
log 100	

FIGURE 6.45.
Logarithms for Question 1.

2. Use your table to evaluate and compare the quantities in **Figure 6.46.**

log 5 + log 10		log 50	
log 2 + log 5		log 10	
log 10 + log 10		log 100	

FIGURE 6.46.
Quantities for Question 2.

3. What patterns do you notice? Explain these patterns. What can you conclude?

4. Do you think this might be true for all logarithms of all bases? Explain.

Second, you explored $\log\left(\frac{x}{y}\right)$.

5. Now use your table to evaluate and compare the quantities in **Figure 6.47.**

log 10 – log 5		log 2	
log 50 – log 10		log 5	
log 100 – log 10		log 10	

FIGURE 6.47.
Quantities for Question 5.

6. What patterns do you notice? Explain these patterns.

7. Do you think this might be true for all logarithms of all bases? Explain.

Third, you explored $\log(x^b)$.

8. Use the log key on your calculator to evaluate the logarithms in **Figure 6.48.**

9. Now use your table to evaluate and compare the quantities in **Figure 6.49.**

log 2	
log 5	
log 10	
log 25	
log 100	
log 125	

FIGURE 6.48.
Logarithms for Question 8.

2 log 5		log 25	
2 log 10		log 100	
3 log 5		log 125	

FIGURE 6.49. Quantities for Question 9.

10. What patterns do you notice? Explain these patterns.

11. Do you think this might be true for all logarithms of all bases? Explain.

Assignment

1. Write each of these in logarithmic form.

   a) $x = 8^y$

   b) $x = 4^y$

   c) $x = 3^{2y}$

   d) $y = 5^x$

2. Find the value of each of these logarithms.

   a) $\log_3 27$

   b) $\log_{10} 100$

   c) $\log_2 64$

   d) $\log_8 64$

3. Simplify each of these. Leave your answers written as logarithms.

   a) $\log_3 9 + \log_3 5$

   b) $\log_2 8 + \log_2 7$

   c) $\log_2 35 - \log_2 7$

   d) $\log_7 144 - \log_7 4$

   e) $\log_{10} 125$

   f) $\log_3 64$

4. Simplify each of these as much as you can. Evaluate the logarithm when possible.

   a) $\log_2 8 + \log_2 4$

   b) $\log_2 32 - \log_2 16$

   c) $\log_2 32 - \log_2 4$

d) $\log_3 81 - \log_3 9$

e) $\log_5 125$

f) $\log_2 64$

5. Can you take the logarithm of 0? Why or why not?

6. Solve each of these for x.

a) $\log_3 x = 5$

b) $\log_3 5 + \log_3 x = \log_3 30$

c) $\log_5 x - \log_5 8 = 1$

d) $\log_7 5 + x \log_7 6 = \log_7 1080$

I Feel the Earth Move

Mount St. Helens National Volcanic Monument in Washington State has an active volcano. Active volcanoes can cause earthquakes.

Park managers work closely with geologists to monitor earthquakes. Scientists who study volcanoes use earthquake data to make predictions about volcanic eruptions. Stronger or more frequent shakes can mean that magma underground is moving toward the surface and an eruption may happen soon. By measuring the magnitude or strength of an earthquake, scientists can make decisions about an impending volcanic eruption. Such decisions could include whether or not to evacuate a park or to relocate wildlife.

In Section 6.13, you learned three properties of logarithms.

❖ $\log_b m + \log_b n = \log_b m \cdot n$

❖ $\log_b m - \log_b n = \log_b \frac{m}{n}$

❖ $\log_b mn = n \log_b m$

These properties are useful in solving exponential equations. We will solve exponential equations by taking the base 10 logarithm of each side. (We will use base 10 since this is one of two bases that the calculator uses.)

We will use $3^x = 14$ as an example.

We need a way to isolate or *undo* the exponent. To do this we need an inverse operation for exponentiation. This inverse operation is called a **logarithm**. If we take the logarithm of both sides, the logarithm undoes the exponent.

$$3^x = 14$$
$$\log(3^x) = \log(14)$$

Now we can use the properties of logarithms to solve for x. Recall that $m^n = n \log_b m$. So,

$$\log(3^x) = \log(14)$$
$$x \log(3) = \log(14)$$

Remember that log(3) and log(14) are real numbers. Therefore, we can apply division to both sides of the equation in order to solve for x. We will divide both sides of the equation by log(3) in order to undo the multiplication.

$$x \log(3) = \log(14)$$

$$\frac{x\log(3)}{\log(3)} = \frac{\log(14)}{\log(3)}$$

$$x = \frac{\log(14)}{\log(3)}$$

$$x \approx 2.402$$

The values for logarithms are often rounded to the nearest thousandth. However, sometimes you will be asked for an exact answer. In that case leave your answer in terms of logarithms.

In Section 6.12 you saw that logarithmic functions can be graphed on a graphing calculator. Most scientific and graphing calculators have two logarithm keys: LOG, which is the base-10 logarithm, and LN, which is the **natural logarithm** having a base e. The number e is similar to pi that recurs throughout mathematics and science. You can graph base-10 and base-e logarithmic functions on most graphing calculators using these keys. You can also graph other logarithmic functions, but first you must change the base to either 10 or e.

Consider the function $y = 3^x$.

1. Find the inverse function of $y = 3^x$ by switching x and y. Solve for y by taking the logarithm of both sides and using the properties of logarithms.

2. How can you write the inverse of $y = 3^x$ as a logarithmic function?

3. Based on your answer to Questions 1 and 2, what is $\log_3 x$ equal to?

4. Based on your findings in Questions 1–3, what is $\log_b x$ equal to?

The formula you found in Question 4 is called a change of base formula. It allows you to convert a logarithm of any base to a base 10. You can also use the change of base formula to convert a logarithm to any other base. In this section we are interested in changing logarithms to base 10 so that we can use the graphing calculator.

5. Use the change of base formula to graph $y = \log_5 x$ on your calculator. Record your window and sketch your results.

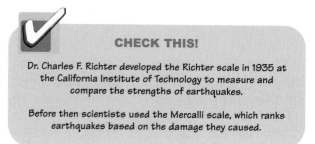

CHECK THIS!

Dr. Charles F. Richter developed the Richter scale in 1935 at the California Institute of Technology to measure and compare the strengths of earthquakes.

Before then scientists used the Mercalli scale, which ranks earthquakes based on the damage they caused.

Dr. Charles Richter

6. Use the Trace or Table feature to calculate the value of $\log_5 3$.

Logarithms can be used to compare large and small numbers. Dr. Charles F. Richter was the first scientist to recognize that the strength of a seismic wave could be used to describe the magnitude of the earthquake that caused it.

One of Richter's formulas relates the amount of energy released by an earthquake to its Richter magnitude:

$$\log E = 11.8 + 1.5M$$

In this formula, E is the seismic energy released by the earthquake and M is the Richter magnitude of the earthquake.

7. Solve Richter's formula for E in terms of M.

8. Use your graphing calculator to graph E as a function of M. Record your window and sketch your graph.

9. How useful is the graph in interpreting the relationship between energy and magnitude?

10. Use the Table feature of your graphing calculator and Richter's formula to compute the energy released for the magnitudes in **Figure 6.50.** Then find the successive ratios.

Magnitude	Energy	Successive Ratios
1		–
2		
3		
4		
5		
6		
7		

FIGURE 6.50.
Energy released by earthquake magnitudes.

11. What patterns do you notice in the ratios?

12. If the magnitude of an earthquake increases by one, by what factor does the energy increase?

13. If the magnitude of an earthquake increases by two, by what factor does the energy increase?

14. If the magnitude of an earthquake increases by four, by what factor does the energy increase?

15. The Northern Sumatra earthquake on December 26, 2004, which caused a tsunami that killed hundreds of thousands of people, had a magnitude of 9.0. In 1992, an earthquake of magnitude 4.5 struck Midland, Texas. Was the Northern Sumatra earthquake twice as strong as the Midland earthquake? Explain your answer.

1. Solve each of these equations for x. Round to the nearest thousandth.

    a) $6^x = 8$

    b) $4^{2x} = 23$

    c) $5^x + 2 = 26$

    d) $2^x = -4$

    e) $5^x = 10$

    f) $7^x = 95$

    g) $6^x = 38$

2. Graph each of these on your graphing calculator. Record your window and sketch the graph.

    a) $y = 2 \log x + 4$

    b) $y = \log_6 x$

    c) $y = 3 \log_6 x - 2$

For Questions 3–5, use $y_1 = \log x$ and $y_2 = \log_2 x$.

3. In the same window graph $y_1 = \log x$ and $y_2 = \log_2 x$. Record your window and sketch the graph.

4. Compare the domain and range of $y_1 = \log x$ and $y_2 = \log_2 x$. How are they alike? How do they differ?

5. Use your knowledge of transformations to describe how the graphs of $y_1 = \log x$ and $y_2 = \log_2 x$ are related.

Use $10^{(11.8+1.5M)} = E$ for Questions 6 and 7.

6. On July 20, an earthquake of magnitude 2.0 was recorded on the west-southwest face of Mount St. Helens. How much energy was released by this earthquake?

7. On July 21, an earthquake of magnitude 3.4 was recorded on the same face. Compare the strength of this earthquake to the one on July 20.

8. Chemists use the pH scale to measure acidity. The pH of a solution is measured on a scale of 1–14 and can be found using the function $y = -\log H$ where y is the pH and H is the concentration of hydrogen ions. The concentration of hydrogen ions in a solution determines how acidic the solution is; the greater the concentration, the greater the acidity.

It is important for park managers to know the pH of lakes and streams. This information helps them decide how much air or water pollution is in the park. It also helps them assess the risk to wildlife.

a) Use your graphing calculator to generate a pH table for the concentrations of hydrogen ions shown in **Figure 6.51.** Round your answers to the nearest tenth.

Concentration	pH
1×10^{-13}	
1×10^{-5}	
2×10^{-13}	
6.2×10^{-10}	
7.4×10^{-5}	

FIGURE 6.51.
pH table.

b) Bunsen has a solution with a hydrogen ion concentration of 0.00001. Use the properties of logarithms to find the pH of his solution. Check your answer with a calculator.

c) The solution in beaker A has a pH of 3. The solution in beaker B has a pH of 8. How many times greater is the concentration of hydrogen ions in beaker A than in beaker B?

What Did You Say?

The intensity of sound is described with the **decibel** (dB) scale. Music producers and sound engineers use this scale to mix recordings of music and to achieve the right balance between layers of sound.

Park visitors often complain about noise. So park managers can use the decibel scale to set noise level limits and use sound meters to measure the intensity of noise.

CHECK THIS!

The human ear can detect a sound wave that displaces air particles by one-billionth of a centimeter.

The intensity of a sound wave that matches this displacement is called the threshold of hearing.

Figure 6.52 shows typical decibel levels for certain sounds.

Sound	Intensity (dB)
Refrigerator	50
Normal conversation	60
Headphones on full volume	100
Chainsaw	120
Jet engine taking off	150

FIGURE 6.52.
Sound intensities.

The decibel scale is defined using $\beta = 10 \log \left(\frac{I}{1.0 \times 10^{-12}} \right)$ where β is the intensity of the sound in decibels and I is the intensity of the sound wave in watts per square meter.

1. How many times more intense is the sound produced by a chainsaw than the sound produced during normal conversation? Explain your answer.

2. In the data in Figure 6.52, how many times more intense is the sound produced by a jet engine taking off than the sound produced by a refrigerator? Explain your answer.

SUMMARY

A logarithmic function is the inverse of an exponential function. Exponential functions can be written in logarithmic form. For example, $y = 3^x$ can be written as $x = \log_3 y$ as shown in **Figure 6.53.**

$$y = 3^x \qquad x = \log_3 y$$

Exponent

Base

Power

When expressions or equations are written in logarithmic form, three properties of logarithms can be used to manipulate them or to solve for a variable.

* $\log_b m + \log_b n = \log_b m \cdot n$

* $\log_b m - \log_b n = \log_b \frac{m}{n}$

* $\log_b mn = n \log_b m$

Logarithms can be used to solve exponential equations for an unknown exponent of the expression.

You can convert between logarithm bases using the **change of base** formula:

$$\log_b x = \frac{\log x}{\log b}$$

Modeling Project Noise Pollution

Noise pollution is a major issue in cities and towns around the country. Noise pollution is associated with certain illnesses, fatigue, and even hearing loss. Many places have passed laws limiting noise levels.

There may be sources of noise pollution at your school, home, or community that could be distracting or unhealthy. For this chapter's modeling project you will develop a report to the principal about sources of noise pollution and ways to lessen its impact on your school. You will need to back up your arguments with data and research about the cause of the noise pollution and what levels of noise are distracting or unsafe.

Look (and listen) around your school for sources of noise pollution. These could be the cafeteria, music room, or traffic on a nearby street.

Create a list of the sources of noise pollution you think are the worst.

Use a decibel meter at various distances to measure the noise of these objects. Record the data.

Use your library or the Internet to research about safe, unsafe, and distracting noise levels. Find out how far from these noise sources you have to be for them to be safe and non-distracting.

Develop a report for your principal about noise pollution at your school. Your report should contain recommendations for limiting noise pollution that are supported by data in tables and graphs. Your report should also summarize the research you did.

Practice Problems

1. Recall the migration problem in Section 6.2. In 1988, the population estimate was between 15 and 20. In 1993, the population estimate was between 25 and 30.

   a) Explain how the migration rate of two moose per year was derived from these observations.

   b) The reason for using the average migration rate is that it predicts "typical" results. Concern about overpopulation might lead a mathematical modeler to use the largest migration rate instead. Based on these observations, what would be the largest rate?

   c) Estimate the largest possible moose population in the park in the year 2000 based on the observations. (We still assume that the only change in population is due to migration.)

2. In Section 6.2 a moose model was used to predict the population in the year 2000. Use the same initial population of 27 moose in 1993, and assume the growth rate is 10% per year on average.

   a) What would be the population for the year 2000 based on this set of assumptions?

   b) What would have been the yearly percentage increase if the same population reached 75 moose in the year 2000, as shown in **Figure 6.53?**

3. Reports show that satellite-based remote sensors measured a 6% decline in the area of ice covering the earth's surface during the period from 1978 to 1994.

   a) What is the overall rate of decay over the 16 years?

Year	Population
1993	27
1994	31
1995	36
1996	42
1997	48
1998	56
1999	65
2000	75

FIGURE 6.53.
Moose population data.

b) **Figure 6.54** shows the percentage of the Earth's surface covered with ice in 1978 that was still covered with ice in the given year. What is the annual rate of decay?

FIGURE 6.54.
Ice coverage data.

Year	Percentage of Ice Coverage	Year	Percentage of Ice Coverage
1978	100%	1986	97.0
1979	99.6	1987	96.6
1980	99.2	1988	96.2
1981	98.8	1989	95.8
1982	98.5	1990	95.5
1983	98.1	1991	95.1
1984	97.7	1992	94.7
1985	97.3	1993	94.4
		1994	94.0

c) Find the annual percent of decrease in the ice coverage.

d) Use 1978 as year zero and assume the amount of ice coverage to be 100% at that time. Write an equation for the percentage of ice coverage remaining after t years.

e) Graph your equation over a large domain (several centuries). If the model holds true well into the future, what is the long-term result of this decline in the ice coverage?

4. An article reported that the buffalo population in a national park grew as shown in **Figure 6.55.**

Year	Population
2000	125
2001	160
2002	205
2003	260
2004	340

FIGURE 6.55.
Buffalo population data.

a) Find the annual growth rate for the buffalo population over that time period. Explain what it means.

b) Write a function rule to predict the population of buffalo in the park over any time period. (Use the year 2000 to represent $t = 0$.)

c) Verify your function using the data from 2000 to 2004. How well does your function predict population?

d) Use the function developed in b to predict how many buffalo will be in the park in 2010.

5. The 2000 Census determined the population used to allocate seats in Congress. As of April 1, 2000, the population of the United States was around 281,422,000. That was a 13.2% increase over the 1990 census figure. This is an average annual increase of about 1.25%.

a) Write a function rule that could be used to predict the population of the United States after the year 2000 (assume a constant growth rate).

b) At the same rate of increase, what would be the population of the United States on April 1, 2003?

c) If the population increases at a yearly rate of 1.00% instead, what effect would that have on the population estimate from b?

6. A trout farm is estimated to have around 5000 fish when it opens. Each day approximately 2% of the fish in the lake are caught. At the end of each week the lake is restocked with 1200 fish.

a) How many fish are caught by the end of the first week?

b) Will the lake run out of fish? Explain.

7. Suppose the authorities at Adirondack State Park had counted 27 moose in 1988 and 42 moose in 1993.

a) If you assume that the growth is caused by migration, what is the migration rate (average number of moose migrating each year)?

b) At the same migration rate, how many moose would be in the park in the year 2013?

c) If you assume that the growth is caused by reproduction, then you can calculate an annual growth factor of 1.0924. What is the annual percentage increase?

d) Let t be the number of years since 1993. Assuming a constant growth rate write an exponential function to predict the number of moose in the park.

e) Based on your answer to d predict the number of moose in the park in the year 2013.

f) How many years would it take for the population of moose to reach 84 (double the original amount)?

8. As regional director of the Fish and Game Department you estimate that there are 6000 deer in the area, and their population grows at a rate of 9% per year. To best manage this resource you would like to keep the population constant. How many deer should you allow to be harvested by hunters next year?

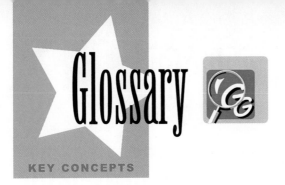

Glossary

KEY CONCEPTS

Base: In an exponential expression, the base is the number that is multiplied by itself. For example, in 2^3, 2 is the base because $2^3 = (2)(2)(2)$.

Common multiplier (common ratio): When multiplication by a constant is used to get successive values, the constant is called the common multiplier.

Exponent: Number that indicates how many times the base is multiplied. In 2^4 2 is multiplied four times, $2 \times 2 \times 2 \times 2 = 16$.

Exponential decay: Negative exponential growth; a growth factor in which $0 < b < 1$.

Exponential function: A function in the form $f(x) = a \cdot b^x$.

Exponential growth: A growth pattern in which multiplication by a constant larger than 1 generates successive values

Growth factor: The common multiplier in an exponential growth or decay situation.

Logarithm: A term used to describe an exponent. For example, in 2^3, 3 is an exponent and therefore a logarithm. 2^3 can be written as $\log_3(2)$.

Logarithmic function: A function of the form $y = \log(x)$. The inverse of an exponential function is a logarithmic function.

Migration: Process by which a population changes by moving from one location to another.

Natural logarithm: A logarithm with a base of e, which has a value of about 2.718.

pH: A scale used to measure the level of acidity in a substance.

Power: The value of an exponential expression. For example, in $2^3 = 8$, 8 is called the third power of 2.

Reproduction: Process by which a population (wildlife or people) increases by births.

Richter scale: A scale invented by Charles Richter used to measure and compare the strength of earthquakes.

7

CHAPTER

Money and You: Mathematics and Finance

Meet the Gonzales Family

Alfredo and Wilma Gonzales have been married for a long time. They both work and bring home paychecks.

Mr. and Mrs. Gonzales have three children. Their oldest, Maria, is a sophomore in college. She lives with her parents and commutes to a college in her town in order to save money.

Paco is a senior in high school. He plays sports, drives a car, and enjoys spending time with his friends on the weekend. He can't wait for graduation. He's planning to go to a community college for his basic coursework and then transfer to a university to major in business. He wants to go to law school after that.

Tico is a freshman in high school. He struggled in middle school last year, and his parents are grateful that he is going to a new school this year. Tico knows a lot about sports cars and can tell you how much a mint-condition 1968 Ford Mustang is worth. He doesn't know what he wants to do after high school. He is planning to spend the next four years having fun and learning what he can.

Mr. and Mrs. Gonzales have their hands full! With a daughter in college and two teenage sons, they must manage their money with care. Since they have been married for a long time, they are no strangers to careful budgeting. They make their decisions together. But sometimes, they need to consult financial planners to help them make decisions.

Financial planners spend their time studying money. They know what factors cause you to spend or lose money and what factors will help you save or earn more money. Financial planners can help a family save and plan for retirement. They can also help the family learn how to reduce expenses in order to make the most of their budget.

In this chapter you will act as a financial planner for the Gonzales family. You will follow the five members of the family as they encounter different situations where they earn or spend money. You will develop and use mathematical models to describe the patterns that occur in finance. Then you will use these models to help the Gonzales family make decisions.

Keep it Simple

In this section you will begin to examine personal finance. When it comes to finances, there are many types of accounts and a variety of programs for handling your money. The Gonzales family uses Lone Star Bank, which offers several options.

Financial advisors always suggest that people keep a little money in easy reach, in case of an emergency. For interested customers, banks offer a free safe-deposit box where money can be kept under lock and key.

1. Lone Star Bank offers safe-deposit boxes for its long-time customers. Mr. Gonzales decided to put $200 in a safe-deposit box and leave it there. How much money would be in the safe-deposit box after one week? 10 weeks? What assumptions are you making about the situation?

2. How much does the amount of money change each week?

Putting your money in a safe-deposit box will not earn you any interest. Let's see what happens when the assumptions change slightly.

To encourage customers to deposit money in a savings account, the Lone Star Bank promotes this New Saver's account:

For every $1000 deposited for an entire year, the bank will pay the customer $50 in interest.

CHECK THIS!

The assumption of the New Saver's account is that interest is paid at a rate of 5% per year and only on the principal, or amount deposited. For each $100 you deposit, the interest you will receive is 5% of $100 or $5 each year that the money is in the account. This interest accrues, which means is paid, usually at the end of each year. When interest is calculated only on the principal, it is called simple interest.

3. Suppose Mr. Gonzales deposits $1000 in a New Saver's account. (That is, the original balance, or **principal**, is $1000.) Complete a table like **Figure 7.1** to find the **balance**, or how much money is in the account, at the end of each year for five years.

Year Number	Interest	Total Balance
0	$0	$1000
1		
2		
3		
4		
5		

FIGURE 7.1.
Account balance for five years.

4. What is a reasonable domain for this situation? Why?

5. What is a reasonable range for this situation? Why?

6. Create a scatterplot of account balance versus number of years. Record your viewing window and sketch your graph.

7. Based on your graph, what type of relationship does this appear to be?

8. How can you use your table to verify your answer to Question 7?

9. Use the process column in a table like **Figure 7.2** to write a function rule that models this situation.

Number of Years (n)	Process	Account Balance (A)
0	$1000 = 1000 + 50(0)$	
1	$1000 + 50 = 1000 + 50(1)$	
2		
3		
4		
5		
n		A

FIGURE 7.2.
Table with process column.

10. Test your function rule by graphing the rule on your scatterplot. If the rule is not a good fit, use transformations to modify it. Sketch your graph.

11. In the context of this problem, what is the meaning of the *y*-intercept of your function?

12. In the context of this problem, what is the meaning of the slope of your function?

13. If there are no deposits or withdrawals, how much money will be in the account at the end of 10 years?

14. If there are no deposits or withdrawals, how many years will it take for Mr. Gonzales' money to triple?

15. If there are no deposits or withdrawals, how many years will it take to have $2250.00 in the account?

16. Suppose instead of earning $50 in interest, Mr. Gonzales earns $65 in interest for every $1000 deposited for an entire year. How will this increase affect the graph of the function?

Taxing Your Earnings

Paco took a job last summer to earn money to spend in his senior year. When he got his first paycheck, he noticed it was less than his total earnings for the week. Money was deducted (subtracted) from his earnings before he got his paycheck!

In this section you will examine some of these deductions and learn how they are calculated.

Let's calculate earnings for different pay periods.

DETERMINING EARNINGS

Hourly Earnings	Weekly Earnings	Monthly Earnings	Yearly Earnings
$6.50			
$10.75			
$17.50			
	$350.00		
	$715.00		
	$1400.00		
		$1250.00	
		$1855.00	
		$2000.00	
		$3333.33	
			$15,000.00
			$23,000.00
			$32,000.00
			$54,000.00

FIGURE 7.3.
Table for calculating earnings.

1. Complete the table in **Figure 7.3.** Be sure to use these conversions:

   40 hours = 1 work week

   52 weeks = 1 year

   12 months = 1 year

2. Explain the process you used to complete the table.

3. What is the mean of the hourly wages?

4. Are the above hourly wages a good indicator of how much the average worker makes in the United States? Why or why not?

5. Listed in **Figure 7.4** are the 2004 yearly earnings for some well-known people. Complete the table to find their hourly wages.

Person	2004 Yearly Earnings	Hourly Wages
George Lucas	$200 million	
Oprah Winfrey	$150 million	
Tiger Woods	$69 million	
Adam Sandler	$47 million	
Madonna	$43 million	
Britney Spears	$39.2 million	
Derek Jeter	$18.5 million	
Sammy Sosa	$18 million	

FIGURE 7.4.
Yearly salaries for some well-known people.

Source: Forbes Magazine on-line www.forbes.com

6. What is the mean hourly wage for the people in Figure 7.4?

FEDERAL WITHHOLDING TAX

According to the Internal Revenue Service (IRS), about 37% of the income for the United States government in fiscal year 2003 came from personal income taxes. The amount of income tax that an employee pays depends on the amount of money earned in a year.

It can be hard to find the cash to pay your annual taxes all at once. The IRS requires employers to withhold money from paychecks in order to spread the tax burden over time.

The amount of withholding, or the money that an employer takes out of an employee's paycheck, can be found using a mathematical formula or a **tax withholding table.**

Many factors go into determining the amount of taxes that someone must pay. The primary factor is personal income. Another factor is marital status.

Use the federal income tax withholding tables to answer these questions.

7. Determine the effect of marital status on federal income tax withheld. Using a withholding allowance of 1, find the amount withheld for a single person and a married person for each wage in **Figure 7.5.**

Monthly Income	Single Withholding	Married Withholding
$500		
$1000		
$1500		
$2000		
$2500		
$3000		
$3500		
$4000		
$4500		

FIGURE 7.5.
Table for Question 7.

8. What can you conclude?

9. On the same set of axes, make a scatterplot of single withholding versus monthly income and a scatterplot of married withholding versus monthly income. Record your window and sketch your graph.

10. What do you notice in the graph?

11. Who has less money withheld, a single person making $3000.00 a month or a married couple making a total of $3000.00 a month? Explain your thinking.

12. Mr. Gonzales earns $1500 per month and Mrs. Gonzales earns $3000. If they were both single, what would be their combined withholding? Since they are married, what is the withholding on their total earnings of $4500? How do these compare?

13. Make a general statement about withholding amounts for married people compared to single people. Explain your thinking.

THE EFFECT OF ALLOWANCES

Another factor that affects withholding is the number of dependents you have. The IRS allows you to reduce your withholding by a certain amount for each dependent. The IRS uses **allowances** to estimate your withholding. In general you can claim one allowance for yourself, one for your spouse, and one for each of your dependents.

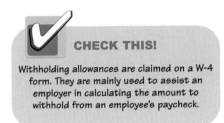

CHECK THIS!

Withholding allowances are claimed on a W-4 form. They are mainly used to assist an employer in calculating the amount to withhold from an employee's paycheck.

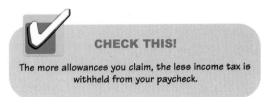

CHECK THIS!

The more allowances you claim, the less income tax is withheld from your paycheck.

Let's explore the effect of withholding allowances. Use federal withholding tables to find the amount withheld for married people in **Figure 7.6.**

Monthly Salary	Withholding for Number of Allowances					
	0	2	4	6	8	10
$500						
$1000						
$1500						
$2000						
$2500						
$3000						
$3500						
$4000						
$4500						
$5000						

FIGURE 7.6.
Monthly salaries for finding withholding tax.

14. What can you conclude?

15. Make scatterplots of withholding for zero allowances versus monthly salary, withholding for two allowances versus monthly salary, and withholding for four allowances versus monthly salary on the same set of axes. Record your window and sketch your graph.

16. Make scatterplots of withholding for six allowances versus monthly salary, withholding for eight allowances versus monthly salary, and withholding for 10 allowances versus monthly salary on the same set of axes. Record your window and sketch your graph.

17. What do you notice in the graphs?

18. What is an estimate of the withholding for a married person with an income of $6000 claiming 10 allowances? Explain how you got your answer.

19. What is an estimate of the withholding for a married person earning $5000 claiming 12 allowances? Explain how you got your answer.

20. Federal withholding is not the only thing deducted from your paycheck. What are some other things?

After all deductions are made, the amount of money that remains is called your **net pay.**

Maria earned $2695.40 in February of 2005. When she received her paycheck, her net pay was much less. Along with federal withholding tax based on zero allowances, the items listed in **Figure 7.7** were deducted.

21. How much was her net pay?

22. Medicare tax and social security tax are each a certain percent of a person's income. Use Figure 7.7 and Maria's income of $2695.40 in February of 2005 to find each of these.

Deduction	Amount
Federal Withholding	?
Medicare	39.08
Health Insurance	121.23
Social Security	167.11
Life Insurance	5.65

FIGURE 7.7.
Deductions from Maria's pay.

a) What percent of her income is deducted for medicare?

b) What percent of her income is deducted for social security?

1. In Section 7.1 the savings account plan earned $50 in interest for every $1000 kept in the account for an entire year. The assumption was that the interest was paid at the rate of 5% per year. This kind of savings account pays **simple interest** because the interest calculation is based on the principal only.

   a) Use the 5% interest rate to find the interest earned when $1000 is deposited for one year.

   b) How much interest would the bank have paid for one year, at that same rate, if only $600 was put into the account?

   c) At that same rate, if $600 were put into the account for five years, how much interest would the bank pay?

   d) Suppose the $600 was put into an account for five years, but the simple interest rate was 4% per year. How much interest would the bank pay? At the end of the five years, what would be the balance in the account?

2. Translate the work done in Question 1 into equations that represent simple interest calculations. Let P be the principal. Let t be the number of years the money is in the account, and let r be the rate (expressed as the decimal).

   a) Write an equation to calculate the amount of interest, I, that the bank must pay.

   b) Write an equation to calculate the balance, B, in the bank at the end of that time period. (Assume interest is deposited into the account.)

3. If you deposit $1250 into a savings account that pays at a simple interest rate of 6.2%, what will be the balance after three years?

4. The Lone Star Bank features a Christmas savings account that pays a 6% simple interest rate per year, but the account can only stay open for one year. If you open the account with $500 and deposit $300 more half a year later, how much money will be in the account when it closes?

5. According to the U.S. Bureau of Labor Statistics, the average hourly wage in the United States for 2003 was $17.75. If you earn this amount find each of these:

   a) Weekly earnings

   b) Yearly earnings

   c) Monthly earnings

6. Shirley earned a total $24,750 in 2005.

   a) What was her monthly salary?

   b) If Shirley claims one allowance, what is the amount withheld each month according to the tax tables for single persons?

   c) If Shirley claims four allowances, what is the amount withheld from her pay each month?

7. The amount of income tax withheld from Ricardo's paycheck each month is $82. He claims five allowances. What is his approximate yearly income?

8. Martin earned $34,000 in 2005. He claimed no allowances and had money withheld according to the 2005 tax tables. At the end of the year when he filled out his tax return he calculated his tax to be $3840. How much money should he expect as a tax refund?

9. Huang earned $25,000 in 2005. He claimed five allowances and had money withheld according to the 2005 tax tables. At the end of the year when he filled out his tax return he calculated his tax to be $2544. How much more money does he owe in taxes?

You Earned It!

In Section 7.2 you were given hourly, weekly, monthly, or yearly earnings. Explain how you found each of these.

1. Weekly earnings when you were given hourly earnings

2. Yearly earnings when you were given hourly earnings

3. Weekly earnings when you were given yearly earnings

4. Hourly earnings when you were given monthly earnings

5. Monthly earnings when you were given yearly earnings

Figure 7.8 lists the yearly salaries for six people in a book study group.

Name	Occupation	Yearly Salary
Duc Lee	Chemical Engineer	87,000
Samantha Rivers	Cashier	15,246
Silvia Rodriquez	Medical Doctor	150,000
Tammera Jones	Veterinarian	130,000
Jose Rubio	Teacher	45,000
Oprah Winfrey	Talk Show Host	50 million

FIGURE 7.8.
Yearly salaries for people in a book study group.

6. What is the mean of the yearly salaries?

7. Is the mean salary representative of this group? Why or why not?

8. What is the median of the yearly salaries?

9. What is the mode of the yearly salaries?

10. Of mean, median, and mode, which gives the best "typical" yearly salary for this group? Why?

When you calculated earnings in Section 7.2, you were calculating what is called **gross pay.**

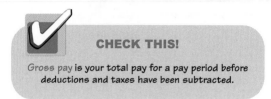

CHECK THIS!

Gross pay is your total pay for a pay period before deductions and taxes have been subtracted.

Federal withholding tax is deducted (subtracted) from your gross pay. In Section 7.2 you used tables to find the amount withheld based on monthly earnings. Let's examine a different way to calculate withholding.

Figure 7.9 explains how to calculate federal withholding tax for single persons.

11. What patterns do you observe in the table?

FIGURE 7.9.
Federal withholding tax table for wages paid January 1, 2005, and after.

MONTHLY PAYROLL PERIOD				
Single Persons				
If the Taxable Income is:		Computed Tax is:		
Over	**But Not Over**	**The total of this column PLUS>>**	**The % of taxable income**	**The amount over** (Taxable income less the amount in this column)
0	221	NO TAX WITHHELD		
221	817	0	10%	221
817	2625	59.60	15%	817
2625	5813	330.80	25%	2625
5813	12,663	1127.80	28%	5813
12,663	27,354	3045.80	33%	12,663
27,354	OVER	7893.83	35%	27,354

12. Explain in your own words how to use the table.

Let's look at an example.

A single person who earns $3285.00 in a month falls between $2625 and $5813 (see **Figure 7.10).**

FIGURE 7.10.
Withholding tax for a single person earning $3285 a month.

Over	**But Not Over**	**The total of this column PLUS>>**	**The % in this column TIMES>>**	**The amount over** (Taxable income less the amount in this column)
0	221	NO TAX WITHHELD		
221	817	0	10%	221
817	2625	59.60	15%	817
2625	5813	330.80	25%	2625
5813	12,663	1127.80	28%	5813

This means the withholding tax will be $330.80 plus 25% of the amount over $2625. Calculations are shown in **Figure 7.11.**

Over	But Not Over	The total of this column PLUS>>	The % in this column TIMES>>	The amount over (Taxable income less the amount in this column)
0	221	NO TAX WITHHELD		
221	817	0	10%	221
817	2625	59.60	15%	817
2625	5813	330.80 +	25% ($3285 –	2625) =
5813	12,663	1127.80	28%	5813

FIGURE 7.11. Calculations for withholding tax.

Tax = initial tax + tax rate(income − limit of previous bracket)
Tax = 330.80 + 0.25(3285.00 − 2625.00)
Tax = 330.80 + 0.25(660)
Tax = 330.80 + 165
Tax = 495.80

13. Use the information in Figure 7.11 to write a linear function that models the relationship between a person's monthly income, x, and computed tax, T, in the 25% tax bracket.

14. What are the domain and range for this situation?

15. How does $495.80 compare to the withholding for $3285 from the tables? Does $495.80 make sense? Why?

16. Use the method in Figure 7.11 to find the withholding for a single person who claims zero allowances and earns $2500.00 per month.

17. Compare your answer to Question 16 to the tax tables. What do you notice?

18. In Section 7.2 we learned that Oprah earned $150 million in 2004. If she claims zero allowances and earns the same amount in 2005, what is her monthly federal withholding tax?

SOCIAL SECURITY TAX AND MEDICARE TAX

When you work for someone, 6.2% of your wages is withheld as social security tax. Your employer also pays a 6.2% matching amount for social security programs. In 2004 the employee tax stopped after $87,900 of wages. Also, 1.45% of your wages is withheld, and the employer pays a matching 1.45% amount to the medicare program. All wages are subject to the medicare tax; there is no ceiling.

19. Assume you claim zero allowances, and your gross pay for one month is $5234.00. The cost of your health insurance is $87.00 per month and is deducted from your earnings. Federal withholding, social security, and medicare taxes are also deducted. Use your knowledge of these deductions to calculate your net pay.

20. Assume you are single and claim zero allowances; your net pay for one month is $1738. The cost of your health insurance is $26 per month and is deducted from your earnings. Federal withholding tax is $264; social security and medicare taxes are also deducted. What is your gross pay?

Assignment

The Internal Revenue Ser
document other than th

1. Calculate each of these.

   a) Weekly earnings when you make $12.75 an hour

   b) Yearly earnings when you make $525 a week

   c) Weekly earnings when you make $26,568 a year

   d) Hourly earnings when you make $2500 a month

   e) Monthly earnings when you make $75,300 a year

2. How is gross pay different from net pay?

Figure 7.12 shows how to calculate federal withholding tax for married persons.

MONTHLY PAYROLL PERIOD				
Married Persons				
If the Taxable Income is:		Computed Tax is:		
Over	**But Not Over**	**The total of this column PLUS>>**	**The % of taxable income**	**The amount over** (Taxable income less the amount in this column)
0	667	NO TAX WITHHELD		
667	1883	0	10%	667
1883	5517	121.60	15%	1883
5517	10,063	666.70	25%	5517
10,063	15,800	1803.20	28%	10,063
15,800	27,771	3409.56	33%	15,800
27,771	OVER	7359.99	35%	27,771

FIGURE 7.12.
Federal withholding tax for wages paid January 1, 2005, and after.

3. What patterns do you observe in the table?

4. Explain in your own words how to use the table.

5. Use the method shown in Figure 7.12 to find the amount withheld for a married couple who claim zero allowances and make $6352 a month.

6. What is the federal withholding tax for a married couple who claim zero allowances and earn $3500 per month?

7. Mrs. Schroeder earned gross pay of $4695.40 in February 2005. When she received her paycheck, her net pay was much less. Along with federal withholding tax based on zero allowances and being married, the items listed in **Figure 7.13** were deducted.

Deduction	Amount
Federal Withholding	
Medicare	68.08
Health Insurance	201.23
Social Security	291.09
Life Insurance	15.65

FIGURE 7.13.
Mrs. Schroeder's deductions.

a) How much was her net pay?

8. Assume you are married, claim zero allowances, and your gross pay for one month is $8274. The cost of your health insurance is $127 per month and is deducted from your earnings. Federal withholding, social security, and medicare taxes are also deducted. Find your net pay.

9. Assume you are single and claim zero allowances; your gross pay for one year is $25,740. The cost of your health insurance is $133 per month and is deducted from your earnings. Federal withholding, social security, and medicare taxes are also deducted. Find your net pay for one month.

10. Assume you are married and claim zero allowances; your net pay for one month is $3296.37. The cost of your health insurance is $97 per month and is deducted from your earnings. Federal withholding tax is $780.85; social security and medicare taxes are also deducted. Find your gross pay.

11. Assume you are single and claim zero allowances; your net pay for one year is $16,653.24. The cost of your health insurance is $27 per month and is deducted from your earnings. Federal withholding is $97; social security and medicare taxes are also deducted. Find your gross pay for one month.

SECTION 7.4

The Certainty of Taxes

PART A

Recall from Section 7.3 that in 2004 each person who contributed to social security and made $87,900 or less per year had 6.20% withheld for social security. They also had 1.45% withheld for medicare.

Use the above information to complete a table like **Figure 7.14.**

Base Salary (in dollars)	Social Security Taxes Withheld (in dollars)	Medicare Taxes Withheld (in dollars)	Total Taxes Withheld (in dollars)
10,000			
20,000			
30,000			
40,000			
50,000			
60,000			
70,000			
80,000			
87,900			

FIGURE 7.14.
Table for Part A.

1. Use a graphing calculator to make each scatterplot using the same window. Record your window and sketch your graphs.

   a) Amount of social security taxes versus base salary

   b) Amount of medicare taxes versus base salary

   c) Total of taxes versus base salary

   d) Graph all three relationships on the same set of axes.

2. Do each of these graphs appear to describe direct variation? Why or why not?

3. a) What is a function rule that can be used to calculate the amount of social security taxes for a given salary?

b) Graph your function rule on the scatterplot from Question 1b. Sketch your graph.

4. a) What is a function rule that can be used to calculate the amount of medicare taxes for a given salary?

b) Graph your function rule on the scatterplot from Question 1b. Sketch your graph.

5. a) What is a function rule that can be used to calculate the total amount of taxes for a given salary?

b) Graph your function rule on the scatterplot from Question 1b. If the function is not a good fit, use transformations to adjust it. Sketch your graph.

6. Complete a table like **Figure 7.15** to find the money remaining after taxes are deducted. Use the values from your first table to fill in the total taxes withheld column, then compute the money remaining column.

Base Salary (in dollars)	Total Taxes Withheld (in dollars)	Money Remaining (Base Salary – Total Taxes Withheld)
10,000		
20,000		
30,000		
40,000		
50,000		
60,000		
70,000		
80,000		
87,900		

FIGURE 7.15.
Table for Question 6.

7. Make a new scatterplot of money remaining versus base salary. Record your window and sketch your graph.

8. Find a function rule for the relationship between money remaining and base salary. Explain how you got it.

9. Graph your function rule on the scatterplot from Question 7. Sketch your graph.

10. Is this a direct variation function? Why or why not?

PART B

For people who make more than $87,900 per year, $5449.80 is withheld for social security, plus 1.45% of the base income is withheld for medicare. Complete a table like **Figure 7.16** based on this information.

Base Salary (in dollars)	Social Security Taxes Withheld (in dollars)	Medicare Taxes Withheld (in dollars)	Total Taxes Withheld (in dollars)
90,000			
100,000			
150,000			
200,000			
250,000			
300,000			
350,000			
400,000			
450,000			

FIGURE 7.16.
Table for Part B.

11. Use a graphing calculator to make each scatterplot using the same window. Record your window and sketch your graphs.

   a) Amount of social security taxes versus base salary

   b) Amount of medicare taxes versus base salary

   c) Total taxes versus base salary

   d) Graph all three relationships on the same set of axes.

12. a) What is a function rule that can be used to find the social security taxes for a given salary greater than $87,900?

   b) Graph your function rule on the scatterplot from Question 11d. If the function is not a good fit, use transformations to adjust it. Sketch your graph.

13. a) What is a function rule that can be used to find the medicare taxes for a given salary greater than $87,900?

   b) Graph your function rule on the scatterplot from Question 11d. If the function is not a good fit, use transformations to adjust it. Sketch your graph.

14. a) What is a function rule that can be used to find the total taxes for a given salary greater than $87,900?

   b) Graph your function rule on the scatterplot from Question 11d. If the function is not a good fit, use transformations to adjust it. Sketch your graph.

15. Is this a direct variation function? Why or why not?

Assignment

Employers often contribute a percentage of a person's salary to a retirement account. In **Figure 7.16** are graphs that show the retirement contributions of two companies as a function of salary.

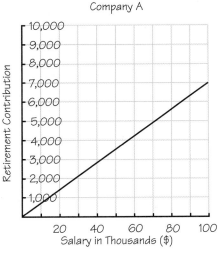

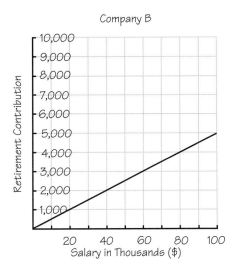

FIGURE 7.16.
Retirement contributions
of two companies.

1. Find a constant of proportionality for each graph. What is the real-world meaning of this number?

2. How much money would be contributed to the retirement account of a person making $48,000 under each plan?

3. How much money would need to be made to have $48,000 contributed to a retirement account under each plan?

4. Write function rules that model the contribution for each company for any salary.

5. Use your functions to find the amount each company would contribute to the retirement plan of a person making $60,000. Explain your solution.

6. Use your functions to find the amount a person would have to make to have $10,000 contributed to a retirement account under each plan. Explain your solution.

A common example of direct variation is sales tax. The current tax rate for San Antonio, Texas has these parts:

State sales tax: $6\frac{1}{4}\%$

City tax: $1\frac{1}{8}\%$

Metropolitan Transit Authority tax: $\frac{1}{2}\%$

7. Write the equation (y_1) that gives the total sales tax for a given purchase (x). Then write an equation (y_2) that gives the final price (after taxes) for a given purchase price (x).

8. Find the sales tax and final price for each item:

   a) A DVD player priced at $149.95

   b) A pair of tennis shoes priced at $79.99

   c) A diamond bracelet priced at $5000

Which Checking Account?

Let's apply what you have learned thus far about finances and simple interest to a new situation.

Mr. and Mrs. Gonzales want to open a checking account at Lone Star Bank. The bank offers three options: Personal Checking 1, Personal Checking 2, or Personal Checking 3. The Gonzaleses will deposit $2000 in the account and leave it there for four years. Also they plan to use the ATM machine about 10 times every month.

The details for each account are in **Figure 7.17.**

FIGURE 7.17.
Account details.

Account	Monthly Interest Rate	ATM Fee	Monthly Service Fee
Personal Checking 1	0%	$1.25 each use	$12.00
Personal Checking 2	0.1%	$1.25 each use over 2	$15.00
Personal Checking 3	0.3%	No Charge	$25.00

Note: For each account, the interest is simple interest. The interest is calculated only on the initial deposit and is deposited into the account. For each account there is no charge for writing checks.

1. Complete each of these tables.

a) Personal Checking 1

Month	Beginning Balance	Interest	ATM Fee	Service Fee	Ending Balance
1	2000				
2					
3					

b) Personal Checking 2

Month	Beginning Balance	Interest	ATM Fee	Service Fee	Ending Balance
1	2000				
2					
3					

c) Personal Checking 3

Month	Beginning Balance	Interest	ATM Fee	Service Fee	Ending Balance
1	2000				
2					
3					

2. Write a function rule for Personal Checking 1 where x is the month and y is the Ending Balance.

3. Write a function rule for Personal Checking 2 where x is the month and y is the Ending Balance.

4. Write a function rule for Personal Checking 3 where x is the month and y is the Ending Balance.

5. Keep in mind that Mr. and Mrs. Gonzales will keep this account for four years. Which account is the best deal for them?

What's It Worth? A Look at the CPI

In this section you will examine the Consumer Price Index (CPI). The CPI is a measure of the change in price over time of a fixed basket of goods and services. It is based on items that cost 100 dollars in 1982. If the inflation rate is 3%, it means the CPI increased by 3% from a year earlier. You can use the CPI to find changes over time in how much you can buy with a certain amount of money. For example, something that cost about 10 cents in 1913 would cost about $1.85 in 2003.

1. Imagine that a news report stated that from 1990 to 1996 the CPI had an annual percent increase of about 3%. Based on this report, if you bought $10.00 worth of groceries in 1990, how much do you think the same groceries would cost in 1996?

To help get a better answer to Question 1, let's examine the CPI for 1990 to 1996.

2. Use the information from the news story and the actual CPI for 1990 (130.7) to complete the CPI column in a table like **Figure 7.18.**

Year	Year #	CPI
1990	0	130.7
1991	1	
1992	2	
1993	3	
1994	4	
1995	5	
1996	6	

FIGURE 7.18.
Table for CPI data.

3. Make a scatterplot of CPI versus year number. Record your window and sketch your graph.

4. Use the process column to write a function rule for this situation.

5. Test your function rule by graphing it on your scatterplot. If the function rule is not a good fit, use transformations to modify it. Sketch your graph.

6. Use a function similar to the function you found in Question 4 to find out how much $10.00 worth of groceries purchased in 1990 cost in 1996.

7. Assuming this trend continues, use your function rule to predict the CPI in 2002.

8. If this trend continues, when will the CPI reach 200?

To see if the news story is accurate, let's look at real data. **Figure 7.19** shows the CPI from 1990–1996.

Year	Year #	CPI
1990	0	130.7
1991	1	136.2
1992	2	140.3
1993	3	144.5
1994	4	148.2
1995	5	152.4
1996	6	156.9

FIGURE 7.19.
CPI data from 1990–1996.

9. Make a scatterplot of CPI versus year number. Record your window and sketch your graph.

10. Use successive quotients as shown in **Figure 7.20** to write a function rule.

Year	Year #	CPI	Successive Quotients
1990	0	130.7	
1991	1	136.2	1.042
1992	2	140.3	1.030
1993	3	144.5	1.029
1994	4	148.2	1.026
1995	5	152.4	1.028
1996	6	156.9	1.030
Mean			**1.031**

FIGURE 7.20.
CPI data with successive quotients.

11. Test your function rule by graphing it on your scatterplot. If the function rule is not a good fit, use transformations to modify it. Sketch your graph.

12. Use a function rule similar to the function you found in Question 10 to determine how much $10.00 worth of groceries purchased in 1990 cost in 1996. How does this value compare to your answer to Question 6?

13. Use your function rule to predict the CPI in 2002. How does this value compare to your answer to Question 7?

14. If this trend continues, when will the CPI reach 200? How does this value compare to your answer to Question 8?

15. Based on your answers to Questions 12–13, is the news story accurate? Why or why not?

16. Based on the actual data, the CPI will most likely reach 200 in 2006. Were you able to conclude this using your function rule? Why or why not?

17. The CPI for 2002 was 179.9. Were you able to conclude this using your function rule? Why or why not?

Compound Interest

In Lone Star Bank's New Saver's account described in Section 7.1 the interest is simple interest. There is another form of interest that earns money differently. This type of interest is called **compound interest.** In this section you will explore compound interest.

CHECK THIS!

Recall that when interest is calculated only on the principal, it is called simple interest. Compound interest accounts receive interest on both the principal and any interest previously earned. The interest is compounded during a specified time period, usually daily, monthly, quarterly, or yearly. For example, if you deposit $100 at 5% interest compounded annually:

Year	Beginning Balance	Ending Balance
1	$100.00	$105.00
2	$105.00	$110.25
3	$110.25	$115.76

Looking a little closer at the New Saver's account raises an interesting question. Suppose you have $1000 in a New Saver's account. At the end of the year you should be paid $50. The bank doesn't hand over this money; they put it in your account. That means the principal should change for the next year. Let's see what happens with an account that combines the interest earned with the principal.

When interest is calculated once a year on both principal and prior interest earned we say the interest is **compounded annually.**

1. Suppose that you deposit $1000 in an account that pays 5% interest on the balance you have at the beginning of each year instead of just on the principal.

   a) How much money is in the account at the end of the first year?

   b) How much money is in the account at the end of the second year?

c) How much money is in the account at the end of the third year?

d) Continue this process to complete a table like **Figure 7.21.**

2. What kind of function do you think best models this relationship? Use your table to support your answer.

3. Make a scatterplot of account balance versus year number. Record your window and sketch your graph.

4. Does your graph confirm your answer to Question 2? Why or why not?

5. At the end of year one, what percent of the account balance is interest? Write your answer as a percent and a decimal.

6. At the end of one year, what percent of the original balance is the new balance? Write your answer as a percent and a decimal.

7. Write a function rule using the answers to Questions 5 and 6 to relate the original balance and the balance at the end of the first year.

8. When we find the balance at the end of the first year, we are taking the starting balance plus 5% of the starting balance, which can be written as $1000 + 1000 \cdot (0.05)$. Factoring this expression gives $1000(1 + 0.05)$. Use your function rule for Question 7 to complete the process column in a table like **Figure 7.22** and write an exponential function for this situation.

Year #	Account Balance
0	1000.00
1	
2	
3	
4	
5	

FIGURE 7.21.
Account balance after five years.

Year #	Process	Account Balance
0	$1000(1 + 0.05)^0$	1000.00
1	$1000(1 + 0.05) = 1000(1 + 0.05)^1$	1050.00
2		1102.50
3		1157.63
4		1215.51
5		1276.28
x		$f(x)$

FIGURE 7.22. Account data with process column.

9. Test your function rule by graphing it on your scatterplot. If the rule is not a good fit, use transformations to modify it. Sketch your graph.

10. If you make no deposits or withdrawals, how much money will you have in the account at the end of 10 years?

11. If you make no deposits or withdrawals, how many years will it take for your money to triple?

12. If you make no deposits or withdrawals, how many years will it take to have $2292 in the account?

Just as you are about to open your account with Lone Star Savings, you see an ad for USA Savings Bank. They are offering a Super Savings account that pays 5% annual interest that is compounded four times a year. When interest is compounded four times a year we say the interest is **compounded quarterly.**

CHECK THIS!

When interest is compounded quarterly, it is compounded every three months.

Let's develop a function that can be used to find an account balance when interest is compounded quarterly. Once we have a function we can use it to compare the two offers.

When interest is compounded quarterly, $\frac{1}{4}$ of the interest is added to the previous quarter's balance every three months.

13. If the interest rate is 5% per year, what percent is compounded each quarter? How did you get your answer?

14. Complete a table like **Figure 7.23** to find your balance at the end of one year.

FIGURE 7.23.
Finding a bank balance after one year.

Quarter	Beginning Balance	Process	Ending Balance
1	1000.00	1000(1 + 0.0125)	1012.50
2	1012.50		
3	1025.16		
4	1037.97		

15. How does the balance at the end of one year when interest is compounded quarterly compare to the balance at the end of one year when interest is compounded annually? What is the reason for the difference?

Using the process in Question 14 is a valid way to find your ending balance for any number of years, but it takes a lot of time. A way to help speed up this process is to use a **recursive routine** on your calculator.

16. Follow the steps below to develop a recursive routine.

CHECK THIS!

A recursive routine generates a sequence of values by applying a rule to an initial value and then applying the rule to the resulting value. Continuing this process generates a sequence of values.

a) Start with the ordered pair that represents your starting point. In this case, at time 0 you have $1000 in the account. What is the ordered pair?

b) Enter the ordered pair into your calculator as shown in **Figure 7.24**, then press ENTER.

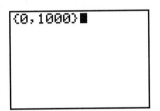
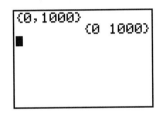

FIGURE 7.24.
Entering an ordered pair into a calculator.

Each time we add 1 to the number of quarters, we need to multiply the previous balance by (1 + 0.0125). Write (1 + 0.0125) as a single decimal number without parentheses.

c) Enter this into your calculator as shown in **Figure 7.25**, then press ENTER.

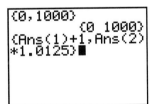
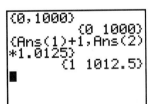

FIGURE 7.25.
Calculator screens from Question 16c.

CHECK THIS!

Since we want the calculator to work with two previous answers, we need to specify Ans (1) and Ans (2). The quarter number is Ans (1), and the account balance is Ans (2).

CHECK THIS!

You obtain "Ans" by pressing the 2nd key, and then the negative key (–).

Press ENTER a second time. What ordered pair results? What is the meaning of this ordered pair in this context?

d) Continue pressing ENTER until you reach the 4th quarter. How does this compare to the account balance at the end of the 4th quarter you found in Question 14?

e) How many quarters are there in 10 years?

f) Continue using your recursive routine to find the account balance at the end of 10 years. How does your answer compare to the answer for Question 10? Why are they different?

17. Write a recursive routine to calculate the balance in an account if you start with $500 and earn 8% interest compounded quarterly.

18. Use your recursive routine from Question 17 to find the account balance after eight years.

1. Suppose there is a constant 2% annual inflation for the next four years as a result of economic policy changes in 2001.

    a) How much will you have to pay in 2005 for an item that costs $100 in 2001?

    b) What is the cost of an item in 2001 that will cost $100 in 2005?

2. Samantha is depositing $25,000 that her uncle left her into an account with a 12% annual interest rate that is compounded six times a year. Complete a table like **Figure 7.26** showing her balance after the first three interest periods.

Interest Period	Balance
0	$25,000
1	
2	
3	

FIGURE 7.26.
Samantha's account.

3. Use this function for calculating the balance in a savings account that earns interest compounded quarterly:

$$A(t) = 15,000 \left(1 + \frac{0.06}{4}\right)^{4t}$$

    a) What is the annual interest rate?

    b) What growth factor can be used to make a sequence of yearly ending balances?

    c) Which term number corresponds to a time interval of two years, nine months?

    d) How much interest is earned during the second year?

4. Suppose you deposit $1800 in an account with an interest rate of 8% per year. What is the balance at the end of one year if the interest paid is:

a) Simple interest?

b) Compounded annually?

c) Compounded quarterly?

d) Compounded daily, assuming 365 days per year?

5. Each of these gives an investment in a savings account earning compound interest and the final balance after a certain time period. Find the rate that provides those earnings.

a) Investment: $900; final balance: $1000; 2 compounds per year; $t = 1$ year

b) Investment: $1100; final balance: $1250; 1 compound per year; $t = 3$ years

c) Investment: $2000; final balance: $2250; 6 compounds per year; $t = 1$ year

6. Suppose $1 is deposited in an account that pays 10% interest compounded annually. The plan is to keep the money in the account for just the first year. Later, the owner decides to keep the money in the account for the second year as well.

a) We would like to know how the individual growth factors compare with the overall one. Fill in a table like the one in **Figure 7.27** with the growth factors that correspond to the term lengths.

FIGURE 7.27. Term lengths.

1st Term Length	Growth Factor	2nd Term Length	Growth Factor	Total Term Length	Overall Growth Factor
2 years		5 years		7 years	
4 years		3 years		7 years	

b) How can you determine the overall growth factor if you know only the individual ones? Test your idea on the two examples in Figure 7.24. Does it work?

c) What do you notice about the overall growth factor in each case? Does that make sense in this problem?

Very Interesting

In Section 7.6 you used a news story and the actual CPI for 1990 (130.7) to complete the CPI column in a table like **Figure 7.28.** The news story reported that from 1990 to 1996 the CPI had an annual growth rate of about 3%. We can use a recursive routine to fill a table.

1. Write the recursive routine that can complete the table.

2. Use your recursive routine to compute the values in the CPI column in Figure 7.28.

Year	Year #	Process	CPI
1990	0		130.7
1991	1		
1992	2		
1993	3		
1994	4		
1995	5		
1996	6		

FIGURE 7.28.
Table for computing CPI values.

3. A function rule for the situation described in the news story in Section 7.6 is $f(x) = 130.7 \cdot 1.03^x$. How did you use this model to predict how much groceries purchased in 1990 cost in 1996?

4. In Section 7.7 you investigated Lone Star Bank's New Saver's account. For this type of account the bank pays 5% interest compounded annually. Say again that you deposit $1000 in this account. Write a recursive routine to calculate the balance at the end of any given year.

5. Use your recursive routine to verify the values in your table from Section 7.7. The table is shown in **Figure 7.29.**

Year #	Account Balance
0	1000.00
1	1050.00
2	1102.50
3	1157.63
4	1215.51
5	1276.28

FIGURE 7.29.
Table from Section 7.7.

6. Find the successive quotients for the account balances in **Figure 7.30.**

Year #	Account Balance	Successive Quotients
0	1000.00	
1	1050.00	
2	1102.50	
3	1157.63	
4	1215.51	
5	1276.28	

FIGURE 7.30.
Account balance data.

7. How can you use successive quotients and a starting balance of $1000 to write a function rule for this situation?

The function rule in Question 7 allows you to compound interest annually. We know that when interest is compounded annually there is a greater return than with simple interest.

8. What do you think will happen to the amount of interest earned over a period of years when interest is compounded quarterly?

9. Complete a table like the one in **Figure 7.31** and use the process column to write an expression that can be used to calculate the number of times interest is calculated for x years, if that interest is compounded quarterly.

Number of Years	Process	Number of Times Interest is Compounded
1	4·1	4
2		
3		
4		
5		
x		

FIGURE 7.31.
Number of times interest is calculated.

Recall that if the rate is 5% per year then $\frac{0.05}{4}$ or 1.25% is compounded each quarter. Substitute $\frac{0.05}{4}$ and the expression for the number of times interest is compounded from Question 9 into the function rule from Section 7.7, $f(x) = 1000(1 + 0.05)^x$. This gives $f(x) = 1000\left(1+\frac{0.05}{4}\right)^{4x}$, which can be used to calculate interest compounded quarterly when the initial balance is $1000.

10. Use $f(x) = 1000\left(1+\frac{0.05}{4}\right)^{4x}$ to complete a table like Figure 7.32.

| Year # | Account Balance | |
	Interest Compounded Annually	Interest Compounded Quarterly
0	1000.00	
1	1050.00	
2	1102.50	
3	1157.63	
4	1215.51	
5	1276.28	

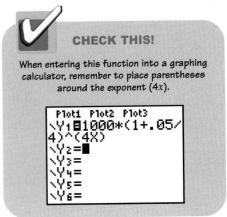

CHECK THIS!

When entering this function into a graphing calculator, remember to place parentheses around the exponent (4x).

FIGURE 7.32.
Table for Question 10.

11. Based on your answer to the questions above, would it be better to open a Lone Star Bank New Saver's account or a USA Savings Bank Super Savings account? Why?

12. Use the function rule from Question 10 and substitution to write the general formula for this type of savings plan. Assume these conditions.

❖ You start by depositing P dollars.

❖ The interest rate is r.

❖ The number of times the interest is compounded in a year is n.

❖ The number of years is t.

❖ The ending balance is B.

13. Paco deposited $4000 in an account that earns 6% interest compounded monthly. If he makes no deposits or withdrawals, what will be his balance after eight years?

1. Imagine you put $10,000 in a savings account for one year at 5% interest for the compounding periods in **Figure 7.33**. Determine how much money you will have at the end of the year for each compounding method.

$1000 at 5% Interest of One Year	
Compounded	Total at End of One Year
Annually	
Semiannually	
Quarterly	
Monthly	
Weekly	
Daily	
Hourly	

FIGURE 7.33.
Different compounding periods.

2. Suppose you deposit $1200 in an account with an annual rate of 4%. How much interest do you receive after one year:

   a) If the bank compounds daily using a 360-day year (called the "360 over 360" method)?

   b) If the bank compounds daily using a 365-day year (called the "365 over 365" method)?

   c) Why might a bank use the "360 over 360" method?

3. The exponential equation $A_t = A_0 \cdot b^t$ has four quantities where A_t is the amount of money in the account after t years, A_0 is the amount initially invested, and b is the yearly growth factor. Given any three of these quantities, you can solve for the missing one. In each of the following, three quantities are given. Show how to solve for the missing item, and then find its value.

   a) $A_0 = \$800$; $b = 1.08$; $t = 5.2$ years

   b) $A_t = \$1200$; $b = 1.12$; $t = 3.8$ years

   c) $A_0 = \$1500$; $A_t = \$2500$; $b = 1.05$

CHECK THIS!

A_t is an example of a subscript. The subscript is a method of showing different values of the same variable. Here it is the account balance over time.

Annuities: Mixed Growth

Making regular deposits into an account that earns compound interest creates what is called **mixed growth**. An annuity is an example of a mixed growth account. It can be described recursively as a repeating two-step calculation of multiplying and adding (in either order, but the result is affected by which is done first). Because the basic process is two operations, it is easy to find terms of such a sequence. However, finding the total of the terms is a little more involved.

CHECK THIS!

Annuities come in two forms:

The first is an asset, things such as stocks, bonds, or cash that pay a constant amount each year to the holder until the annuity expires.

The second is a savings vehicle in which regular deposits are made for a specified period to grow an asset. Tico is creating this kind of annuity.

1. Tico deposits $500 in an annuity that earns 5% compounded annually at the beginning of each year for a six-year time period. How is this situation different from those discussed so far?

2. Fill in a table like **Figure 7.34**. For each year calculate the amount in the account that earns interest (after the deposit is made), the amount of interest earned that year, and the ending balance.

Year (n)	Deposit Made	Beginning Balance	Total Amount Earning Interest	Interest Earned During Year	Ending Balance
1	$500.00	$500.00	$500.00	$25.00	$525.00
2	$500.00	$1025.00	$1025.00		
3	$500.00	$1576.25			
4	$500.00	$2155.06			
5	$500.00	$2762.81			
6	$500.00	$3400.95			

FIGURE 7.34.
Table for account calculation.

3. What calculation finds the beginning balance for year two from the beginning balance in year one?

4. Does that same process work for finding the beginning balance for year three? Write a recursive routine for the beginning balances.

5. Use your recursive routine to calculate the ending balance in the account after 10 years.

6. How does the answer to Question 5 relate to the recursive routine you found in Question 4?

Let's look at Tico's annuity in a different way. This time think of each deposit as a separate account with deposits made at different times and different amounts of interest earned based on how long the money is in the account.

7. Fill in a table like **Figure 7.35**. Use the number of years each deposit is in the account to calculate the ending balance and interest that portion of the account earns.

Year (n)	Deposit Made	Number of Years Deposit is in Account	Process	Ending Balance
1	$500.00	6	$500 \cdot (1.05)^6$	≈ $670.05
2	$500.00	5		
3	$500.00	4		
4	$500.00	3		
5	$500.00	2		
6	$500.00	1		
Total				

FIGURE 7.35.
Table for Question 7.

8. Write the expressions from the process column that you used to calculate the ending balance for each deposit.

9. Write *one* expression that calculates the sum of the ending balances at the end of the six-year period from all the deposits made.

10. Evaluate the expression you wrote in Question 9. How does this number compare to the total of ending balances you found in the table in Question 7?

11. The terms of the expression from Question 9 have two common factors since every payment earns interest for at least one year. Identify the common factors, and then factor using the distributive property so that the common factors are in front of a set of parentheses that group what is left of your original expression.

Choosing any number as a base and raising it to exponent powers that are consecutive integers forms a **geometric sequence**. When those sequence values are added together, that is called a **geometric series**. There is a shortcut to finding the sum of a geometric series when the first term is 1 and the common ratio is b:

$$1 + b^1 + b^2 + b^3 + \ldots + b^{(n-1)} = \frac{b^n - 1}{b - 1}.$$

12. Use the formula for summing a geometric series to find the sum of the geometric series you wrote in Question 9. The part in the bracket is the geometric series.

13. Use the same formula and your work from Question 12 to write the general formula for this type of savings plan. Assume these conditions.

   ❖ At the beginning of each time period you put d dollars into an account.

   ❖ The account has a growth factor of b during that time period.

   ❖ There is a total of n time periods.

14. Write an expression for the total of all the payments of d dollars made into the account n times.

15. To calculate the total interest, adjust the formula you found in Question 13 by subtracting the expression for the total of all the payments.

16. Check to see if this new formula correctly predicts the total interest earned after six years. How do you know this formula is correct?

One of the powerful features of mathematics is that a formula that required a lot of work to derive can be used to solve many problems of the same type. (If you are lucky, the formula can even describe situations that aren't from the same context but *do* have the same underlying process.) The **savings formula** is one of those friendly formulas. Now that we know where it comes from, we can use it to solve a lot of problems!

Recall that b is the growth rate so $b = (1 + r)$. Substituting $(1 + r)$ for b in the *total interest earned formula from Question 15 gives us the savings formula.*

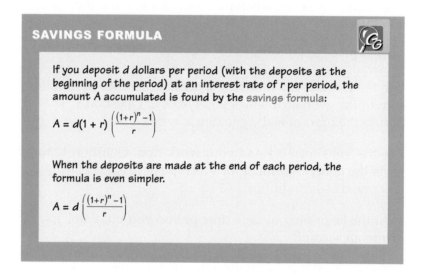

SAVINGS FORMULA

If you deposit d dollars per period (with the deposits at the beginning of the period) at an interest rate of r per period, the amount A accumulated is found by the savings formula:

$$A = d(1 + r)\left(\frac{(1+r)^n - 1}{r}\right)$$

When the deposits are made at the end of each period, the formula is even simpler.

$$A = d\left(\frac{(1+r)^n - 1}{r}\right)$$

17. Maria deposits $125 at the end of each quarter into an account that earns 5% compounded quarterly. Calculate the balance at the end of six years.

18. If Maria wants the account balance to reach $5000 in that same six-year period, how much would she have to deposit each quarter?

Assignment

1. Find the ending balance after one year for a savings account in which $1200 principal is deposited, an additional $200 is deposited on the last day of each month, and the interest rate is 8% per year compounded monthly.

2. A couple wishes to save for the college education of their child. They want to make regular deposits at the end of each month into an account that earns 6.5% interest per year, compounded daily. Using a 360-day year, find the rate of increase for the account for one month.

3. If you invest $5000 into a savings plan that pays 6% interest compounded quarterly and wish to have $9000 after five years, how much should you deposit at the end of each quarter?

4. You decide to invest in an annuity when you are 30 years old.

   a) You will invest $25 per month for 35 years at 5% interest. How much will you have after 35 years?

   b) What if you invest $50 per month for the same time and interest?

   c) What if you invest $100 per month for the same time and interest?

Which Investment?

Let's apply what you have learned thus far to a new situation.

Investing money is a great way to make money. By making some careful decisions you can have a comfortable financial future.

Say your parents give you a graduation party and most of your guests give you money as a gift. When you count all of it you realize you have $1500.

You want to deposit the $1500 into an account at Big Bank. The account offers 6.25% compounded monthly. Your father thinks you should buy a Certificate of Deposit (CD) for $1500 that offers 6.5% compounded annually. Your aunt thinks you should deposit the $1500 in a credit union account that earns 6% compounded quarterly.

Whose idea will give you the most money at the end of 10 years? Justify your answer.

I'm Feeling Down—Depreciation

The Gonzales family owns three cars. Mr. and Mrs. Gonzales each have a vehicle and Maria, their oldest daughter, has a car.

Maria has decided to use her old car as a trade-in on a new one. Maria and the dealer do not agree on the trade-in value of her old car, but neither feels it is worth the $30,000 it cost in 2000.

Maria learns that the value of a new car goes down as soon as it leaves the lot. Once a car has been titled in someone's name, the car's status goes from "new" to "used." Maria also learned that most cars lose value all the time, even if kept in good condition.

This loss of value is called **depreciation**, and mathematical rules determine how much value a car loses from aging.

> ✓ **CHECK THIS!**
>
> In accounting, depreciation is a tool used to determine the expense of replacing a capital, or expensive, item.

One method is to specify a number of years and decrease the value by the same amount of money each year until the car has no value left. The graph in **Figure 7.36** shows the change in value of Maria's car over time using this method.

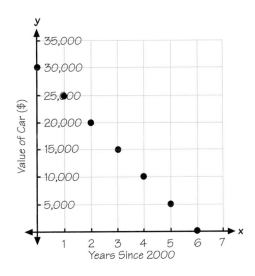

FIGURE 7.36.
Maria's car value over time.

1. What type of function does the graph seem to describe?

2. According to the graph, when will the value of Maria's car be $0?

This method of depreciation is called **linear depreciation**, since the car decreases in value by a constant rate each year. However, in reality, cars are rarely worth $0, even though we may joke about a worthless car.

Another method that Maria found is to have the value of the car go down by a certain percentage of its value *for that year*. Using this method, as the value of the car goes down so does the amount that the value decreases.

Maria wants to see how this method would work for a new car. She decides to use 20% as the rate of depreciation.

3. Calculate the value of the car at the end of the first year by taking away 20% of the Year 0 value. Fill in a table like **Figure 7.37** by taking away 20% of the value from the previous year. Show your calculations in the process column.

Year	Year Number (*t*)	Process	Value (*V*) in Dollars
2005	0		30,000.00
2006	1		
2007	2		
2008	3		
2009	4		
2010	5		
2011	6		

FIGURE 7.37.
Table for calculating car value.

4. Make a scatterplot of the value of the car versus time.

5. Write a function rule describing the depreciation of Maria's car.

6. Graph your function rule on your scatterplot. If necessary, use transformations to make the function rule fit the data.

7. What are the domain and range of your function rule?

8. What are the domain and range of the problem situation? How are these different from the domain and range for the function rule?

9. Using this method of depreciation, will the new car ever be worth nothing? How do you know?

Should I Buy or Lease?

Figuring out which car to buy is an important part of owning a car. There are many factors to consider. In Section 7.11 you investigated one factor, depreciation. Other factors may include fuel economy and insurance. However, the purchase is only the first step. You also have to *pay* for it! And if you're like Maria, you don't have $30,000.

Fortunately, there are several options.

OPTION 1: PURCHASING A CAR

One option is to look into businesses that **finance** car purchases. That means that they lend you the money to buy a car and you pay them back in monthly installments. For that privilege, your car will cost more than you thought it would. In researching her car purchase, Maria found out that the **annual percentage rate**, or **APR,** plays a big role.

Assume that Maria settles on a car whose selling price is $27,000. She considers having it financed by a four-year loan for the same amount with an APR of 6.8%.

1. Assume that Maria's loan will be computed with simple interest for four years. What is a reasonable estimate of the amount of interest Maria will pay?

2. Based on your answer for Question 1, what is the total amount of money including principal and interest that Maria will pay for her new car?

That is a nice way of thinking about APR, but that is not the way that interest and monthly payments are calculated. Each month you owe some of that interest, and the payment will pay for that *plus* some of the loan. As the amount you owe goes down, you make the same payment but you owe less interest; therefore, you repay more of the loan. By the end of the **loan term** (the amount of time it takes to pay off the loan) there will be nothing left to pay.

3. Maria is going to borrow $27,000 with a four-year 6.8% APR loan. Car loans are compounded monthly. What will be Maria's monthly interest rate?

4. Suppose Maria's payments are $640 per month. If her payments are due at the end of the month, interest is compounded from the original principal. What amount of interest does Maria pay in her first monthly payment?

5. If you begin with the original principal, add the first month's interest, then subtract the monthly payment, you can calculate the principal balance at the end of the first month. What is Maria's principal balance at the end of the first month?

6. Continue this process to fill in a table like **Figure 7.38**.

Payment Number	Process	Principal Balance
0		27,000.00
1	$27{,}000 \left(1 + \frac{0.068}{12}\right) - 640$	26,513.00
2		
3		
4		
5		

FIGURE 7.38.
Computing the principal balance.

7. How did you calculate each monthly payment?

We could continue this process for 48 months, but that is time-consuming. Instead, we can use technology. Some Websites have monthly payment calculators. You can also use a spreadsheet. You can also use the Time Value of Money, or TVM Solver on a graphing calculator.

Use the TVM Solver on your graphing calculator to find the exact terms of Maria's loan by using these steps.

Step 1: Press the APPS key and select the Finance application as shown in **Figure 7.39**.

FIGURE 7.39.
Finance application on a calculator.

FIGURE 7.40.
TVM Solver option.

Step 2: Select the TVM Solver as shown in **Figure 7.40**.

Step 3: Enter the known information into the list. If you do not know the value, enter 0.

Let's use the TVM Solver to solve Maria's problem. If she buys a car for $27,000, finances the purchase with a loan having a 6.8% APR for four years, what will she end up paying including interest for the car? Assume that interest is compounded monthly based on the amount of the principal that Maria has not yet paid.

Given four values, the TVM Solver solves for the fifth. Enter cash inflows as positive numbers and cash outflows as negative numbers. The variables are:

❖ N = Total number of payment periods

❖ I% = Annual interest rate (a percent, not a decimal value)

❖ PV = Present value (at the beginning of the loan)

❖ PMT = Payment amount

❖ FV = Future value (at the end of the loan)

❖ P/Y = Number of payment periods per year

❖ C/Y = Number of compounding periods per year

(When you store a value to P/Y, the value for C/Y automatically changes. To store a different value to C/Y, store the value to C/Y after you have stored a value to P/Y.)

---

8. What values would you enter into the TVM Solver for:

   a) N = Total number of payment periods?

   b) I% = Annual interest rate (a percent, not a decimal value)?

   c) PV = Present value (at the beginning of the loan)?

   d) PMT = Payment amount?

   e) FV = Future value (at the end of the loan or investment)?

   f) P/Y = Number of payment periods per year?

   g) C/Y = Number of compounding periods per year?

9. Enter these values into the TVM Solver.

10. Select the variable for which you want to solve. In Maria's case we want to find her monthly payment, so use the arrow keys to select PMT. Then press ALPHA-SOLVE. Record your results rounded to the nearest cent.

11. If Maria makes four years' worth of the monthly payments in Question 10, how much does she pay for her car including interest?

12. Based on your answer to Question 11, how much interest does Maria pay?

13. Maria decides to lower her monthly payment by financing the car over five years. Use the TVM Solver to calculate her new monthly payment.

14. If Maria finances the car over five years, how much interest does she pay?

15. How does the amount of interest Maria pays over five years compare to the amount of interest she pays if she finances the car for four years?

16. If Maria has budgeted $450 per month for a car payment, and the best interest rate she can get is 5.25% on a four-year loan, what is the most Maria can borrow?

Maria has found that three things affect the monthly payments and therefore the total amount that the loan will cost. These things are the amount of money borrowed, the APR, and the loan term. Using technology allows Maria to compare different combinations of the three in order to find the one that best fits her budget.

OPTION 2: LEASING A CAR

17. What are some expenses that you face when owning a car?

At most dealers you can **lease** a car. Leasing is like having a long-term rental agreement. The consumer pays a fixed amount each month to the leasing company (who owns the car), and also pays for insurance and operating costs. At the end of the lease agreement, the company gets the car.

Because leasing a car is different from purchasing one, there is a new set of terms used to describe the factors that influence the cost of the lease. For now, let's work with the numbers so that we can find the costs associated with leasing. We will explain what the terms are later.

Lease Conditions	Value
APR	0.9%
Term of Lease	48 months
Down Payment	$1000
Security Deposit	$800
Milage Allowance	45,000 miles

Leasing Figures	Amount
MSRP	25,810.00
Capitalized Cost	
Net Trade-in Allowance	
Adjusted Capitalized Cost	
Residual Value	
Money Factor	

FIGURE 7.41.
Conditions and cost of leasing an Explorer.

While shopping for her new car, Maria found this in an ad:

Lease a new Ford Explorer XLS with automatic transmission, 6-cylinder engine, anti-lock brakes, AM/FM radio, air conditioning, and power steering. MSRP is $25,810, and financing is available at a special APR rate of 0.9% for 48 months with a mileage allowance of 45,000 miles. $1000 down payment and $800 security deposit due on lease signing.

Maria and the dealer agreed to base the lease value of the car as $25,000. The car dealer agreed to buy Maria's old car for $5000 (this is commonly called **trading in** your old vehicle for a new one). Even though she thought it was worth more, she agreed to the offer of $5000.

Maria organized the information in the table in **Figure 7.41.**

18. Record the selling price in a table like Figure 7.41 under the amount next to capitalized cost. Record the trade-in allowance that the dealer gave Maria under the amount next to net trade-in allowance.

19. Consider the discounts and extra fees that Maria will either receive or pay. What is the final lease price of the vehicle? Record this value in the table under the amount next to adjusted capitalized cost.

20. The dealer thinks that the car will depreciate to 48.43% of its value over the term of the lease. To the nearest dollar, what will be the value of the car at the end of the lease?

21. Begin with the APR written as a decimal. Find the interest rate for one compounding period if the interest is compounded monthly.

22. Divide your answer from Question 21 by 2. Record this value in your table under the amount next to money factor.

Maria needs to know what her monthly payments are going to be. Assume that the sales tax rate is 6.25%.

23. Calculate the monthly depreciation by subtracting the **residual value**, the value of a car at the end of a lease, from the adjusted capitalized cost from Figure 7.41, then dividing by the length of the lease in months. Record that value in a table like **Figure 7.42.**

Leasing Calculations	Amount
Monthly Depreciation	
Monthly Interest Charge	
Monthly Sales Tax	
Monthly Payment	

FIGURE 7.42.
Monthly payment table.

24. Use your data from Figure 7.41 to calculate the monthly interest charge by adding the adjusted capitalized cost and residual value then multiplying that sum by the money factor. Record that value in your monthly payment table.

25. Calculate the monthly sales tax by multiplying the tax rate by the sum of the monthly depreciation and monthly interest charge. Record that value in your monthly payment table.

26. Calculate Maria's total monthly payment. Record that value in your monthly payment table.

In order to compare other kinds of financing, Maria needs to know how much the leasing agreement actually cost her. Assume that she will drive the car 52,000 miles during the four years, the surcharge for extra miles is $0.12 per mile, and she will be charged an additional $1000 for excess wear and tear.

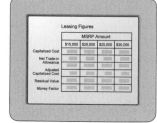

27. Calculate the total of the payments she will make.

28. Calculate the excess mileage charge.

29. Based on your answers to Questions 27 and 28, calculate the total cost of leasing the car for four years.

30. Add the trade-in allowance that Maria received to the total cost of the lease. This is the actual cost to Maria.

Maria wants to compare the leasing deal with what it would cost to buy the same car at the same selling price ($25,000), down payment ($1000), trade-in allowance ($5000), and finance rate (0.9% APR).

31. What would her monthly payments be for a four-year loan?

32. How much would it cost her to finance the car for four years?

33. Assuming that the new car will depreciate to 50% in four years, what will the new car be worth at the end of the loan?

34. Based on your answers so far, what will be the actual cost of owning the new car for the four years?

35. Based on Maria's experience, is it better to lease or buy a car? Explain your answer.

Assignment

1. Use the TVM Solver to find the missing values in **Figure 7.43.**

	Present Value	Interest Rate	Time	Future Value
a	5000	6%	10	
b	5000	6%		10,000
c	5000		10	10,000
d		6%	10	10,000

FIGURE 7.43.
Table with missing values.

2. Compute the monthly payment, total cost of the loan, and amount of interest paid for these car loans:

   a) Joanie can get an $11,000 loan for five years at 6% APR.

   b) Mitchell can get an $11,000 loan for five years at 7% APR.

   c) Marisela can get an $11,000 loan for five years at 8% APR.

At the end of Section 7.12, Maria had compared the total cost of leasing a new car for four years with the total cost of purchasing the car and financing it for the same amount of time. Her calculations showed that for the APR rate she was offered, the two strategies would cost her just about the same amount. By leasing a car she spent a lot of money without a car to show for it. Is this a good decision? She has all the tools she needs to compare options over a longer time period and see which strategy is best.

Maria has several options available now; exploring each one will allow you to answer her question about which one is the best. In all cases assume that operating costs are the same, and focus on the cost of buying versus leasing.

3. She could lease another car in four years. Assume that the same arrangements apply. How much will it cost her to lease a new car four years in the future for another four years?

4. How much will her total leasing costs be over the eight-year period?

5. At the end of the lease, Maria could buy the purchase option price and finance it over the next four years at the same APR (0.9%). Recall that the purchase option price was $11,500. How much would Maria's monthly payments be?

6. Assume that after eight years the resale value of the car will be only $5000. How much will it cost Maria to keep the car another four years?

7. How much did the lease and buy option cost her over the entire eight-year period?

8. Had Maria just bought the car, financed it over four years, and kept it for a total of eight years, how much would it cost her? (Hint: the value of the car is $5000 at the end of the eight years.)

Amortization of a Car Loan

In Section 7.12 you explored two ways that Maria could get a new car. One was to purchase the car and the other was to lease it. You also looked at ways Maria could borrow money to pay for her new car.

When you finance a purchase such as a new car, you borrow money and pay interest. Interest is often compounded monthly based on a yearly rate. As you pay off the loan, you pay less interest because the amount you owe is less. This process of computing the amount of interest and the amount of principal that you pay each month is called **amortization**.

Suppose you are buying a used car for $8000. You put $1600 down and finance the remainder at 9.5% interest for 36 months. The monthly payment is $205.01. Create an amortization table for the loan.

Handout 7.2 contains a copy of the table you just built. Use your table to answer these questions.

1. How is interest computed on a loan?

2. How much total interest do you pay?

3. How much do you end up paying for the car?

4. How much money do you save if you pay off the loan at 24 months?

5. How much money do you save if you pay off the loan at 12 months?

6. Compare your answers to Questions 3 and 4. Explain what you find.

7. Is it to your advantage to pay off the loan early? Why or why not?

Consider a leasing option.

Maria negotiates a $25,000 price for the car, and she chose $2000 in options. The **capitalized cost** is the combination of all expenses related to buying the car:

$25,000 base sticker price + $2000 options = $27,000.

Maria is prepared to give a **down payment** of $1000. She is also being given a **net trade-in allowance** of $3000 for her old car. To find the **adjusted capitalized cost**, subtract these credits from the selling price, or capitalized cost.

8. After applying the down payment and the net trade-in allowance, how much is Maria's adjusted capitalized cost?

Assume that the **lease term** (the loan term when purchasing) is 36 months. For that interval it is estimated that the residual value, or value of the car at the end of the lease term, is $13,955. Dealers calculate the residual value as a percentage of the MSRP (manufacturer's suggested retail price). This value is often between 50% and 60% of the MSRP, depending on the length of the lease.

Maria organized the information as a formula:

(capitalized cost) – (down payment) – (net trade-in allowance) = (adjusted capitalized cost)

There are three factors that determine the monthly payment for a lease:

❖ monthly depreciation

❖ finance charge or lease charge

❖ sales tax

The first step is to find the **monthly depreciation**, which measures the monthly decrease in the value of the car:

$$\text{monthly depreciation} = \frac{\text{adjusted capitalized cost} - \text{residual value}}{\text{lease term}}$$

9. In this case, what is the monthly depreciation?

The second step is to find the **monthly interest charge**, which is also called the rent charge or the lease charge. (This is like the finance charge in purchasing a car.) To find the monthly interest charge, you add the residual value to the adjusted capitalized cost and multiply that sum by something called the **money factor**. A money factor is used to find the finance charge, which is found as follows:

$$\text{finance charge} = \frac{(\text{adjusted capitalized cost} + \text{residual value})}{2} \times \frac{\text{interest rate}}{12}$$

$$\text{finance charge} = (\text{adjusted capitalized cost} + \text{residual value}) \times \left(\frac{\text{interest rate}}{2 \times 12}\right)$$

$$\text{finance charge} = (\text{adjusted capitalized cost} + \text{residual value}) \times \text{money factor}$$

The money factor includes dividing the interest rate by 12 (since there are 12 months in a year and interest is compounded monthly), then dividing by 2 (to get a mean value by averaging the adjusted capitalized cost and residual value). While this is a common method, the money factor is not calculated the same way by all lease companies.

10. Suppose, for example, that Maria can get an interest rate of 8.04%. What is the money factor?

11. In this case what is Maria's monthly interest charge?

The third part is sales tax. You pay sales tax on both the monthly depreciation and the monthly interest charge.

12. Assume that the tax rate is 6.25%. How much sales tax does Maria pay per month?

13. Based on these three parts, what is Maria's monthly payment?

14. Over the term of the lease, how much does Maria pay to the lease company?

15. List some reasons why leasing is popular.

16. List some reasons why leasing might not be a good idea.

17. Write a formula that describes how to calculate monthly depreciation.

18. Write a formula that describes how to calculate the monthly interest charge.

19. Write a formula that describes how to calculate the monthly sales tax.

20. Write a formula that describes how to calculate the monthly payment.

21. Summarize the vocabulary in this section by completing **Figure 7.44.**

FIGURE 7.44.
Vocabulary table.

Definition	Finance Term	Leasing Term
Price that the buyer and seller of a car agree upon after negotiation		Capitalized cost
Amount of money that the buyer pays toward the purchase or lease before the loan or lease begins	Down payment	
Amount of money that the dealer will pay for the buyer's old car		Trade-in allowance
Amount of money that the buyer will either finance or lease, after any credits and charges are applied to the negotiated price	Principal balance	
Length of time required to make payments		Lease term
Rate used to calculate the interest portion of the monthly payments	Annual percentage rate	
Loss in value of the car over time		Depreciation
Value of the car after the loan has matured	Equity	

Assignment

4% APR
with $293.52 monthly payment

Drive a little • Save a lot!

Daves' Auto Leasing

Best terms in the business

Jennifer is considering how to finance a Sport Utility Vehicle (SUV). She isn't sure if the Special Edition, or SE option package, is worth the extra money so she considers both. With sales tax the list price for a standard SUV (XE) is $25,095.19 and for a fully loaded SE the price is $26,786.69.

1. Use the rounded figure of $25,000 for the selling price of the XE.

   a) If the dealer offers financing at 7.4% APR over four years, what does the XE cost Jennifer?

   b) If the loan is financed over five years instead of four years, how much does the payment go down?

   c) How much more does the car cost Jennifer if the loan is for five years rather than four?

2. Use the rounded figure of $27,000 for the selling price of the fully loaded SE.

   a) A local bank offers five-year loans at 7.0% APR. How much does the car cost Jennifer with this financing?

   b) A local lending company offers five-year loans at 6.6% APR. How much does Jennifer save by borrowing the money from this company instead of the bank?

3. Jennifer *really* wants the SE. She has been saving money for the day when she can get one and has $1000 for a down payment. She also has a car that the dealer will take as a $3000 trade-in allowance. By shopping around she found the lowest financing of 6.4% APR for a four-year term. Assuming the rounded $27,000 price for the SE, if she applies both the down payment and the trade-in and finances the rest, how much does the car cost her?

4. A tip that consumer groups suggest for reducing costs is to purchase options like car stereos *after* you buy the car. Assume Maria buys the basic-model car; with sales tax included, the selling price is around $30,000. She will use her old car's trade-in value of $5000 and finance the rest with a five-year loan at 7.0% APR. She has three options for getting the sound system.

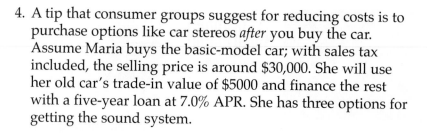

a) One option is to drive her new car to a store that sells car stereos and buy a system with 6-CD changer for $1000 (installation included). How much will that option cost her above the cost of the car?

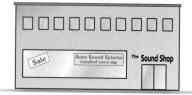

b) A second option is to have the dealer install the system and 6-CD changer as options. The selling price of the car will go up by $1000. How much extra will she pay as a result of having to borrow more money to buy the car?

c) A third option is to have the dealer install the sound system as an option, but to match the additional cost by making a $1000 down payment. How much will that option cost her above the cost of the car?

d) What assumptions is the consumer group making when they recommend that you buy the stereo *after* you buy the car?

5. Jennifer also found dealers who would lease an SE. One dealer offered her these terms: MSRP of $25,211; 36-month lease at 6.4% APR; 10,000 miles per year; excess mileage surcharge $0.18/mile; residual value of $14,500.

a) Under those terms, how much is Jennifer's monthly lease payment?

b) Excluding operating costs, how much does the car cost Jennifer to drive for the three years?

6. Jennifer really liked the car and negotiated a selling price of $24,500. She extended the lease term to 48 months. (The residual value was adjusted to be $12,200.) She agreed to make a $1000 down payment and applied the $3000 trade-in value of her old car.

a) Under these terms, how much is Jennifer's monthly lease payment?

b) Assume Jennifer put 53,000 miles on the car during the lease term. Assume she also paid $2100 for excess wear on the car. Excluding operating costs, how much does the car cost Jennifer to drive for four years?

c) If she had *bought* the car, and financed what she needed at the same APR, how much would the car have cost Jennifer during those four years? (Use the residual value to estimate the worth of the car after the lease term expired.)

d) Do you think leasing the car is a good option? Explain.

7. Jennifer decided that shopping around for the best lease made sense. She checked with two car dealers using the same assumptions: capitalized cost of $24,500, down payment of $2000 with financing of 6.9% APR. Her information is in **Figure 7.45.**

Lease Term	Leasing Company A			Leasing Company B		
	36 mo	48 mo	60 mo	36 mo	48 mo	60 mo
Monthly Payment	$402	$344	$316	$474	$412	$373
Money Factor	0.00308	0.00312	0.00319	0.00436	0.00431	0.00427
Mileage Penalty	$0.17/mi	$0.17/mi	$0.17/mi	$0.15/mi	$0.15/mi	$0.15/mi
Purchase Option	$12,849	$12,102	$10,857	$11,747	$10,486	$9,225
Residual Value	$12,699	$11,952	$10,707	$11,597	$10,337	$9,076

FIGURE 7.45.
Two leasing options.

a) Which company makes more money from monthly interest charges? Explain.

b) Which company makes more money from penalty charges? Explain.

c) Which company makes more money from re-selling leased vehicles as used cars? Explain.

d) Which company do you think Jennifer should select? Why?

e) For the company you picked in d, which lease term costs the least? Why?

There's No Place Like Home

Buying a home is like buying a car. Because it is a high-cost item, most people have to borrow money in order to buy a home. There are many types of home loans available for consumers.

All home loans calculate interest monthly based on the remaining principal, which is just like a car loan. Some home loans use a fixed rate for the entire life of the loan. These loans are called **conventional fixed-rate loans.**

Lenders offer fixed-rate loans of different term lengths. Common term lengths are 15 years and 30 years.

The Gonzales family is considering a conventional fixed-rate loan to buy a new home. After a down payment, they need to finance $150,000. They consider two options:

Option A: 30-year home loan for $150,000 at 7.25% interest

Option B: 15-year home loan for $150,000 at 7.25% interest

1. What is the monthly payment for each option?

2. How much money do they save per month with the 30-year loan?

3. How much money does the house cost, including principal and interest, with the 30-year loan?

4. How much money does the house cost, including principal and interest, with the 15-year loan?

5. Since the time doubles, does the cost of the house double, too? If not, then how much did the cost of the house change?

6. How long does it take to pay off half of the 30-year loan?

7. How long does it take to pay off half of the 15-year loan?

8. When would someone want to use a 15-year conventional loan?

9. When would someone want to use a 30-year conventional loan?

Mrs. Gonzales found this in a magazine:

WHAT A DEAL!!!

Move into your $150,000 home today for only $750 a month! Adjustable Rate Mortgages are available today for only 4.39% APR.

CALL NOW!

Mrs. Gonzales called the bank offering this deal. They told her that the ad was for a 30-year loan term. The APR for the adjustable rate loan is based on the six-month US Treasury Bill index, which is currently 3.11%. The bank calculates the APR by adding 1.27% to the index. Right now that means an APR of 4.39%.

10. If the Gonzales family chooses this type of loan, what is the principal balance in one year?

11. Suppose that after one year the index rises to 4.5%. What is the Gonzales family's new interest rate?

12. What is the new monthly payment? How does this compare to their original monthly payment?

13. What is the principal balance of their loan after three years?

14. After the third year of their mortgage, the index rises to 6.1%. What is their new interest rate?

15. What is the new monthly payment? How does this compare to their original monthly payment?

16. What is the percent change in the Gonzales' monthly payment from when they first got the loan?

17. When would someone want to use an Adjustable Rate Mortgage? When would someone NOT want to use one?

RENTING A HOUSE OR AN APARTMENT

As with cars, many people choose to rent a home. There are several reasons that someone might do this.

18. What costs are involved with owning a home?

19. Of these costs, which ones are fixed amounts each month? Which ones vary from month to month?

Use newspapers, magazines, or the Internet to find two houses in a similar neighborhood. One house should be for sale and one for rent.

20. Calculate the monthly payment for a 30-year loan at 6.25%.

21. How does this monthly payment compare to the monthly rent for a similar house?

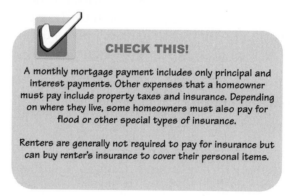

CHECK THIS!

A monthly mortgage payment includes only principal and interest payments. Other expenses that a homeowner must pay include property taxes and insurance. Depending on where they live, some homeowners must also pay for flood or other special types of insurance.

Renters are generally not required to pay for insurance but can buy renter's insurance to cover their personal items.

22. When is it better to own a home? When is it better to rent a home or apartment?

Assignment

1. Find the monthly payment, total amount of interest paid, and total cost of the loan for a 30-year home loan for $175,000 at 7.25%.

2. Find the monthly payment, total amount of interest paid, and total cost of the loan for a 15-year home loan for $175,000 at 7.25%.

3. The Washington family is buying a new home. They plan to get a 30-year adjustable rate mortgage beginning at 4.75% for $125,000.

   a) What is their initial monthly payment?

   b) At the end of three years, what is their principal balance?

   c) After three years the interest rate increases to 6.25%. What is the monthly payment now?

   d) At the end of the fourth year, what is the principal balance?

   e) At the end of the fourth year, the interest rate decreases to 5.75%. What is the monthly payment now?

   f) When the Washingtons bought their home, they could have gotten a 30-year conventional fixed-rate loan for 5.625%. What would their monthly payment have been?

   g) After four years of payments, what is the principal balance for this loan?

   h) Which option do you think is better for the Washingtons?

4. Bryan is looking for a townhouse in the city where he has a new job. He does some research and finds these offers:

MUST BUY!	Now Leasing!
2108 Elmcroft Drive— a bargain at $85,500. 2 bedrooms, 1-1/2 bathrooms. No money down.	2115 Maplecroft Drive— a great neighborhood! Lease for only $850 per month. 2-bedroom specials. Going Fast! Call now.
CALL TODAY!	

a) Bryan can get a 15-year fixed-rate mortgage at 6.25%. Which offer gives him a lower monthly payment?

b) Insurance will cost Bryan $1100 per year. Property taxes are $3200 per year. If Bryan buys the townhouse on Elmcroft Drive and pays both insurance and taxes monthly, what are his total monthly costs?

c) Bryan spoke to the leasing agent for the Maplecroft townhouse. The agent said that renter's insurance costs $450 per year, and there is a residents' fee of $25 per month. If Bryan leases the townhouse on Maplecroft Drive, what are his total monthly costs?

Gotta Have My Tunes!

Paco, Tico, and Maria have convinced their parents to buy a $2400 home theater system, but they only have $100 per month budgeted for the purchases. The Gonzales family has two options:

Option 1: Place $100 a month into a savings account that pays 5% annual interest compounded monthly and purchase the system when they have saved $2400.

Option 2: Use a credit card to purchase the system now. The credit card charges 11.5% annual interest compounded monthly. They plan to pay $100 a month until the stereo is paid off.

Which will happen first: Will they save enough money to buy the stereo with money in their savings account or will they be able to pay the credit card balance entirely? Explain your answer.

SUMMARY

Vehicles: Purchase or Lease?

There are two ways to acquire a vehicle.

First, you can buy one. If you need to borrow money you can finance the purchase. You will pay a certain amount per month depending on several factors: the principal, or amount of money you borrow; the annual percentage rate (APR), which is compounded monthly; and the length of the loan.

Second, you can lease a car. When you do so, you agree to pay for the depreciation in the car's value over the length of the lease. To calculate the monthly payment the following formulas are used:

* monthly depreciation $= \dfrac{\text{adjusted capitalized cost} - \text{residual value}}{\text{lease term}}$

* monthly finance charge = (adjusted capitalized cost + residual value) × money factor

- monthly sales tax = (monthly depreciation + finance charge) × tax rate

- monthly payment = monthly depreciation + monthly interest charge + monthly sales tax

Home: Purchase or Lease?

There are two options to pay for housing costs.

First, you can buy a home. Most people need to borrow money, or obtain a **mortgage,** to do so. Mortgages are long-term loans, often 15 or 30 years, and can have fixed interest rates as in a **conventional fixed-rate loan** or variable rates as in **adjusted rate mortgages (ARMs).**

Second, you can rent or lease a home. As with cars, leasing a home has advantages. Since you do not own the home, you are not responsible for major repairs or property taxes.

Modeling Project *Money for Life*

One problem that faces all people is managing income and planning for retirement. Do you buy a house or rent, and at what point? Do you keep your money secure and easily accessed, or do you put it in an account that pays higher interest, but with restrictions on its use?

The world of finances is complex, with interest rates that change almost daily, speculative investments that might lose all value, and a multitude of costs—some changing with the seasons, others constant. Even the decision to raise a family has implications for your finances. As good mathematical modelers, it is better to keep the problem reasonable and work on making it more realistic in the next version of the model.

For that reason, this Modeling Project includes only these assumptions:

❖ You just finished school, have no outstanding debts, and have $10,000 in a savings account.

❖ You just took a job at a firm where you will remain for the rest of your career. The starting salary is $32,000 and you receive pay raises of $2500 every two years.

❖ You are 25 years old and can work for up to 40 years. After that you have to retire.

❖ The selling price for a typical house is $180,000 and is going up by 2% per year (the value of the house goes up at the same rate).

❖ You need at least 20% of the price of a house as a down payment, and the closing costs are 5% of the price. The down payment must come from assets but the closing costs can be financed with the house, which is financed at 6.5% per year compounded monthly over 30 years.

❖ You will be in the 15% tax bracket until your salary goes above $44,000; then it will be 20%.

❖ You need $1400 a month in living expenses including the monthly rent on your apartment, which is $900 but goes up $100 every two years. Other expenses go up by 3% per year.

❖ Your savings account pays 5% interest compounded monthly.

❖ Certificates of Deposit (CDs) pay 7% interest compounded quarterly but you have to keep the money in the bank for two years.

❖ You may put up to $2000 in an Individual Retirement Account (IRA) every year, which pays 10% per year. Any contribution to an IRA is not taxed until the money is taken out of the account. However, you may not take out any money until you retire.

Prepare a written plan for your financial future. You know the mathematics. You can explore this problem with spreadsheets as another option. What investments will you make? At what point in time will you make them? When you retire what will your worth be? Try and make as much money as you can—but remember, *"Money doesn't buy happiness!"*

Practice Problems

1. Eduardo wants to buy a computer that costs $1200. The store offers a one-year financing plan that allows him to pay $100 as a down payment and then $100 per month for 12 months.

   a) What interest rate is he paying?

   b) Is he paying simple interest?

   c) He can get a loan for the same computer from his college at a rate of 5% but he can only finance 85% of it. Is this loan a better deal? Why?

2. Lucy buys a used car for $15,000. The dealer tells her she will make monthly payments of $360.94 for 48 months.

   a) What interest rate is she paying?

   b) Lucy can only afford $300 per month. If she extends the financing to 60 months, can she get her payment under $300?

   c) Lucy's parents lend her $1500 so that she can make a down payment. The new amount financed is $13,500. Also the dealer tells her that the down payment gets her a lower rate. Her payments are now $317.05 for 48 months. What is the new interest rate?

3. A principal amount of $3500 is deposited in an account that earns 4.5% interest. Calculate the ending balance after five years if the interest is:

   a) Simple interest

   b) Compounded annually

   c) Compounded monthly

   d) Compounded daily (using 360 days)

4. Find the missing values in **Figure 7.46.**

	APR	Compounding Condition	Time Interval	Growth Factor	APY
a)	5%	annually	3 years		
b)	5%	bi-monthly	6 months		
c)	5%	monthly	2.5 years		
d)	5%	continuously	1/4 year		
e)		bi-monthly	1 year	1.10	
f)		quarterly	1.5 years	1.25	
g)		monthly	5 years	2.00	
h)		semiannually	4 years		8%
i)		bi-monthly	3 years		10%

FIGURE 7.46.
Table for Question 4.

5. A credit card company offers this plan to new customers: no annual fee, an initial rate of 6% for the first year, and then 9% after the first year.

   a) Robert decides to take the deal and transfers a balance of $2000. He also charges $1000 more on the card. Assuming Robert pays $200 per month and does not charge anything more, how long will it take him to pay the bill in full?

   b) If after the first month he charges $25 per month, how long will it take him to pay the bill in full?

6. If you invest $5000 into a savings plan that pays 6% interest compounded quarterly, and you wish to have $9000 after five years, how much should you make in payments at the end of each quarter? (Hint: How much will the principal earn? The rest needs to come from the savings plan accumulation.)

7. You wish to buy a house for $220,000. Closing costs are $5600. You pay 20% of the total cost of the house as a down payment, and you want to finance the rest over 30 years with a loan that charges 7.6% interest per year compounded monthly.

a) What is the total cost for the house?

b) How much do you need for the down payment?

c) What will your monthly payments be?

8. a) Elena opened a savings account with a deposit of $50, and she deposits $25 per month from her earnings. The bank pays interest at the rate of 0.25% per month. That is, each month the bank adds 0.0025 × (current balance) to the account and then Elena adds $25 of her own money. Write a function rule describing her balance under this process.

b) Use a calculator or a spreadsheet program to simulate the growth of her account using this saving scheme. Round to the nearest cent.

9 a) Tia plans to retire in 20 years and expects inflation to average about 3% per year between now and then. If she retired now, she would want $1500 per month in income. How much income will she need in 20 years in order to maintain the same standard of living?

b) Tia can secure an investment that pays 1% per month. She wants to use the interest from this investment as her sole income during retirement. (She will remove each monthly interest payment as soon as it is put into the account; the principal will never be touched.) How much does she need in the account when she retires to meet the goal in a?

Glossary

KEY CONCEPTS

Accumulation: Combination of deposits made on a regular basis and the interest that those deposits make.

Adjusted Rate Mortgages (ARMs): A loan where the interest rate changes over the term of the loan.

Allowances: A means of determining your federal income tax withholding. Increasing the number of allowances decreases the amount of withholding taken from gross pay.

Amortization: Process in which a debt is repaid in regular installments.

Amortization formula: Formula for situations involving installment loans:

$P(1 + r)^n = d\ \dfrac{(1+r)^n + 1}{r}$, where P is the principal, r is the effective interest rate per compounding period, d is the payment amount made at the end of each period, and n is the number of payments required to pay off the loan.

Annual Percentage Rate (APR): The cost of a loan or the amount of interest received on an investment expressed as a yearly percentage rate.

Annuities: Assets that make periodic payments to the holder. Also can be an investment method.

Balance: Amount remaining in a savings account, or amount remaining to be repaid on a loan.

Capitalized cost: The purchase price of a capital item that forms the basis for leasing transactions. Net capitalized cost is the capital cost less any deductions, such as a trade-in.

Compound interest: Method of paying interest on both the principal amount and the interest previously earned. The compound interest formula is:

$A = P\left(1 + \dfrac{r}{n}\right)^{nt}$.

Compounding period: Time interval between interest calculations.

Consumer Price Index (CPI): A measure of the change in price over time of a fixed "basket " of goods and services. CPI is a measure of inflation. In general, things get more expensive over time.

Continuous compounding: Payment of interest based on using the limiting rate from compounding as frequently as possible.

Depreciation: The loss of value of an item over time.

Down payment: An initial payment that lowers the capitalized cost of a car, house, or other capital item.

Equity: The value of an individual's ownership of a capital item.

Fixed rate loan: A loan where the interest does not change over the term of the loan. Also known as conventional fixed rate loans.

Geometric sequence: Sequence of numbers that can be described recursively by a multiplication operation, or that is characterized by a repetitive multiplicative process.

Geometric series: A sum of consecutive terms (either some or all of them) of a geometric sequence.

Gross pay: Amount earned before deductions for taxes, benefits, and other reductions.

Inflation: Rate at which the cost of goods increases.

Installment: Amount paid on a regular basis to repay a loan.

Internal Revenue Service (IRS): The federal agency responsible for collecting federal taxes from individuals and businesses.

Lease: A rental agreement.

Lease term: The length of time (usually in months) that an item will be leased.

Medicare: A government health care program that serves people primarily 65 years of age and over. A medicare tax is deducted (and matched by the employer) from an individual's gross pay.

Mixed growth: Sequence of numbers that can be described recursively by a combination of one multiplication and one addition operation.

Net pay: Amount earned after taxes, benefits, and other reductions. Also known as take-home pay.

Present value: The value of money today that is to be received in the future.

Principal: Initial deposit or balance on a savings account or loan.

Quarterly: Four times per year.

Recursive: Description of change during a time interval, based upon the previous value.

Residual value: The value of a capital item after deducting depreciation.

Salary: The amount of money you receive for work.

Savings formula: Formula for the amount in an account in which regular deposits are made: $A = d\dfrac{(1+r)^n - 1}{r}$, where A is the amount accumulated, d is the amount deposited at the end of each compounding period, r is the interest rate over the compounding period, and n is the number of compounding periods (and payments).

Savings plan: Savings account with compound interest in which deposits are made regularly for the purpose of accumulating money.

Semiannual: Term used to describe two compounding periods per year.

Sequence: A list of variables.

Simple interest: Method of paying interest on only the initial balance in an account and not on any interest that is earned.

Social security: A government retirement and disability program. A social security tax is deducted (and matched by the employer) from an individual's gross pay.

Subscript: Method of showing different values for the same variable, such as an account balance (the variable) over time (the subscript). In x_t the subscript is t.

Tax: A payment to a federal, state, or local government that is based on earned income, unearned income, the value of a purchase, and other transactions.

Trade-in allowance: The dollar value realized from the sale of a vehicle to an automobile dealer for the purpose of lowering the net purchase price of another vehicle.

TVM Solver: A graphing calculator utility for solving financial problems.

US Treasury bill index: An index that measures the cost of buying the debt of the United States government. This index is often linked to variable interest loans such as ARM.

Withholding (tax): An amount deducted from your gross pay to cover anticipated taxes due. Withholding taxes are based on allowances.

8

CHAPTER

Cycles: Trigonometry and Sinusoidal Functions

It is Coming Around Again!

Certain things in nature follow a regular repeating pattern known as a cycle. Since these cycles occur again and again, they are predictable. One cycle you have seen is that of the moon, which is about a thirty-day cycle that begins and ends with a new moon. This process is constantly repeating. The National Weather Service can predict the dates of each phase of the moon for years to come. There are many cycles in nature, including tides, the position of planets in our solar system, and the populations of certain animals. In music, frequencies of a sound wave, or cycle, determine the sound, or note.

Cycles are not only in the natural world. For practical and social reasons cycles occur in many manmade things. For example, electricity use during the day, the purchase of items like heating oil over the course of a year, and, some people argue, public opinion on issues over time can be described as cycles. Sometimes the reasons for these cycles are easy to explain but sometimes they are a mystery.

The ability to make predictions about cycles is important. If local officials know that a storm will hit at high tide, they are more likely to worry about coastal flooding. If a retail store buys 10,000 bathing suits in October, they may not be able to sell them.

Mathematicians have created tools for describing patterns. You may find it surprising that these tools are based on similar triangles, like the ones you explored in Chapter 4. In this chapter we will expand these tools to explore cycles and understand the relationships that create them.

Right Triangle Relationships: Pythagorean Theorem and Similar Triangles

In Chapter 4 you saw pictures like **Figure 8.1**.

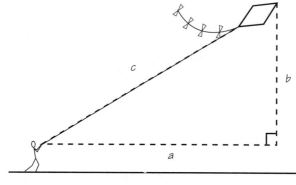

FIGURE 8.1.
Picture similar to
those in Chapter 4.

1. What type of problem does Figure 8.1 suggest?

2. What do you think the problem would ask you to find?

3. What information would you expect to be given?

Stand up and face the clock or some other object that is located near the top of your classroom wall. Look straight ahead and extend both arms out in front of your body. Your line of vision and your arms should be parallel to the floor. Pretend the clock on the wall is the kite; rotate your head and one arm upwards until the clock is directly in your line of vision. The angle you create with your arms is called the angle of elevation (**Figure 8.2**).

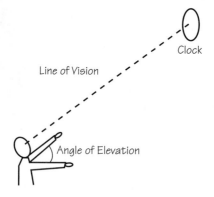

FIGURE 8.2.
Angle of elevation.

Once again think about the kite. Use this information.

The kite string is 100 yards long.

The kite's angle of elevation from the ground is 30° (**Figure 8.3**).

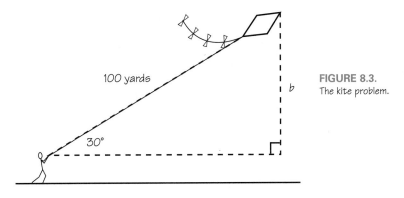

FIGURE 8.3.
The kite problem.

4. Use your knowledge of similar triangles from Chapter 4 and **Figure 8.4** to find the height of the kite.

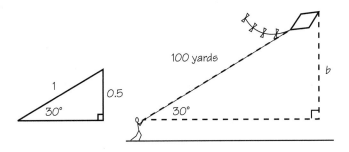

FIGURE 8.4.
Similar triangles.

5. 50 yards is the length of the missing leg of the right triangle. Is it actually the height of the kite? Why or why not?

6. Use two methods to find the length of the third side of each triangle in **Figure 8.5**.

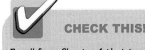

CHECK THIS!

Recall from Chapter 4 that two triangles are similar if their corresponding angles are congruent. Corresponding sides of similar triangles are proportional.

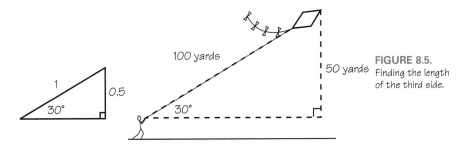

FIGURE 8.5.
Finding the length of the third side.

7. Use **Figure 8.6** to write a word problem with the answer of 86.6 yards.

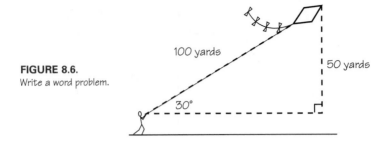

FIGURE 8.6.
Write a word problem.

100 yards

50 yards

30°

Right Triangle Relationships: Ratios

SAY WHAT YOU MEAN, MEAN WHAT YOU SAY

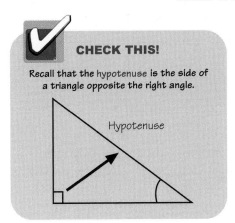

CHECK THIS!

Recall that the hypotenuse is the side of a triangle opposite the right angle.

Hypotenuse

Identifying a side of a right triangle can be awkward. If you say **hypotenuse**, your meaning is clear. On the other hand, when you say **leg**, there are two sides possible. To avoid confusion, there must be names for the sides of the triangle. The legs are named according to their relationship to a **reference angle**.

The **adjacent leg** is the leg that touches or is next to the reference angle (**Figure 8.7**).

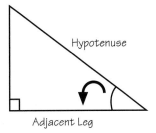

FIGURE 8.7.
Adjacent leg.

Hypotenuse

Adjacent Leg

The **opposite leg** is the side that is opposite the reference angle (**Figure 8.8**).

FIGURE 8.8.
Opposite leg.

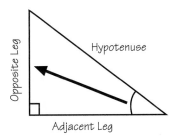

Opposite Leg

Hypotenuse

Adjacent Leg

1. Sketch two triangles like those in **Figure 8.9**. Label the hypotenuse of each, then label the side adjacent to the marked angle and the side opposite the marked angle.

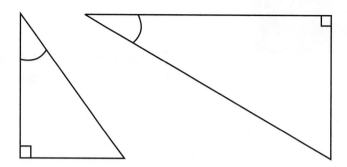

FIGURE 8.9.
Two triangles.

LET'S INVESTIGATE!

2. Use a protractor to draw a 55° angle on a piece of paper. Label it ∠A and extend the rays to the edges of the paper (**Figure 8.10**).

FIGURE 8.10.
Angle A.

3. In your drawing, add a small right triangle, △ABC, where ∠C is the right angle (**Figure 8.11**). Use a centimeter ruler to measure the lengths of the sides of the triangle to the nearest tenth of a centimeter. Record the measurements in a table like **Figure 8.12**.

FIGURE 8.11.
Triangle ABC.

Triangle	Hypotenuse	Opposite Side	Adjacent Side
△ABC	AB =	BC =	AC =
△ADE			
△AFG			

FIGURE 8.12. Measurements.

4. Add right triangles ADE (**Figure 8.13**) and AFG (not shown) in a way similar to how you drew △ABC in Question 3. Measure the lengths of the sides of both triangles and record the measurements in your table.

FIGURE 8.13.
Triangle ADE.

5. Are all three triangles, △ABC, △ADE, and △AFG, similar to each other? Explain.

6. Use the measurements from your table in Figure 8.12 to find and compare these ratios. Record your results in a table like **Figure 8.14**.

CHECK THIS!

Mathematicians tire of writing the word *leg* when they work with right triangles. So they agree to use just *adjacent* when they mean *adjacent leg* and *opposite* when they mean *opposite leg*.

Ratio	△ABC	△ADE	△AFG
$\dfrac{\text{opposite}}{\text{hypotenuse}}$			
$\dfrac{\text{adjacent}}{\text{hypotenuse}}$			
$\dfrac{\text{opposite}}{\text{adjacent}}$			

FIGURE 8.14.
Recorded results.

7. What patterns do you observe in the table?

8. What can you conclude?

9. Create another right △AHI with a hypotenuse of 10 cm. Use the ratios above to predict the length of the opposite leg.

$$\frac{\text{opposite}}{\text{hypotenuse}} = \underline{\qquad} \rightarrow \frac{\text{opposite}}{10} = \underline{\qquad} \rightarrow \text{opposite} = \underline{\qquad}$$

1. In **Figure 8.15**, ΔABC is similar to ΔEFG. Write three proportions based on the similarity.

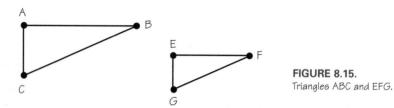

FIGURE 8.15.
Triangles ABC and EFG.

2. In relationship to the marked angles in **Figure 8.16**, label the hypotenuse, opposite leg, and adjacent leg.

FIGURE 8.16.
Three triangles.

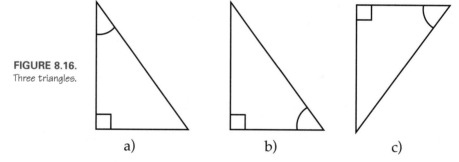

a)   b)   c)

3. Set up proportions and find the length of the hypotenuse and adjacent side of the larger triangle in **Figure 8.17**.

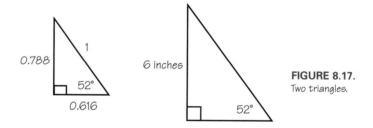

FIGURE 8.17.
Two triangles.

4. Use **Figure 8.18** to answer these questions.

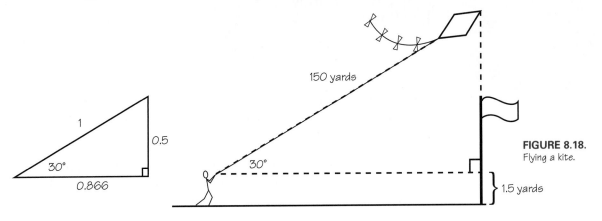

FIGURE 8.18.
Flying a kite.

a) About how high is the kite?

b) The kite is flying directly over a flagpole. How far is the
   kite flyer from the pole?

5. In **Figure 8.19**, find the width of the stream.

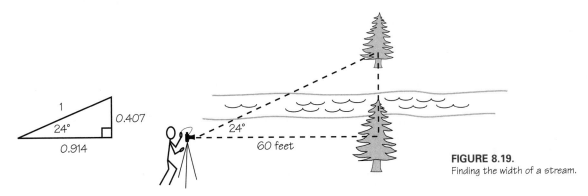

FIGURE 8.19.
Finding the width of a stream.

Trigonometric Ratios

1. In Section 8.1 you saw problems that cannot be solved with the Pythagorean theorem. Why was it not possible to solve these problems?

2. Many of the problems that cannot be solved with the Pythagorean theorem can be solved with similar triangles. How did you use similar triangles to solve problems?

3. In Section 8.2 you explored three ratios that are the same for similar right triangles. What were these ratios?

These right triangle ratios are called **Trigonometric Ratios.** Each of the trigonometric ratios you found has a special name:

The sine ratio $= \dfrac{\text{leg opposite the reference angle}}{\text{hypotenuse}}$ or $\sin = \dfrac{\text{opposite}}{\text{hypotenuse}}$

The cosine ratio $= \dfrac{\text{leg adjacent to the reference angle}}{\text{hypotenuse}}$ or $\cos = \dfrac{\text{adjacent}}{\text{hypotenuse}}$

The tangent ratio $= \dfrac{\text{leg opposite the reference angle}}{\text{leg adjacent to the reference angle}}$ or $\tan = \dfrac{\text{opposite}}{\text{adjacent}}$

Use a graphing calculator to investigate triangle ratios. Since there are several units used to measure angles, first check your calculator's MODE to be sure that it is in the DEGREE mode. If DEGREE is not selected, use the arrow keys to highlight DEGREE. Press ENTER to select (**Figure 8.20**).

FIGURE 8.20.
Checking the MODE.

For example, to find the sine ratio for a 30° angle, press SIN, 30, ENTER. You should get 0.5 (**Figure 8.21**):

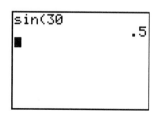

FIGURE 8.21.
Finding the sine.

4. What does sin 30° = 0.5 tell you about the relationship between the length of the leg opposite a 30° angle and the length of the hypotenuse?

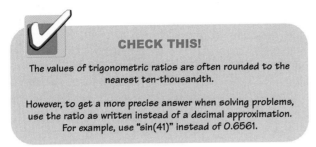

CHECK THIS!

The values of trigonometric ratios are often rounded to the nearest ten-thousandth.

However, to get a more precise answer when solving problems, use the ratio as written instead of a decimal approximation. For example, use "sin(41)" instead of 0.6561.

5. Find sin 32°.

6. Find cos 32°.

7. Find tan 32°.

8. Compare the ratios you found in Section 8.2 for a 55° reference angle with the values that your calculator gives you. Record your data in a table like **Figure 8.22**.

$m\angle A = 55°$	Calculated Ratio	Calculator Value
sin 55°		
cos 55°		
tan 55°		

FIGURE 8.22.
Ratios and values.

9. How do your calculated ratios compare to the calculator values for the sine, cosine, and tangent of $\angle A$?

There are many stories to help remember the trigonometric ratios. One story follows:

Prince Rahib escorted his girlfriend, Princess Rahana, as she rode her camel. He impressed her with his new running shoes and his athletic prowess. The sun beat down on them as they made their way through the burning sands of the desert. Suddenly Rahana's camel lay down and refused to move. All of their efforts to get the camel back on its feet were to no avail. The sun was hot, and they were going to have to walk back to the palace. Princess Rahana was barefoot, so Prince Rahib insisted that she wear his shoes to protect her feet from the hot sands. They hurried, but the Prince's feet were soon blistered. When they rushed onto the palace grounds, Rahana cried, "My hero!" and went to give Rahib a kiss. "Get back!" he cried, "I have no time for you. I just want to SOH-CAH-TOA!"

Trigonometric ratios can be used to solve problems involving angle measures and side lengths. Using trigonometric ratios allows you to take advantage of the power of similar triangles without using a second triangle.

For example, consider this problem.

In **Figure 8.23**, a flagpole's shadow is 23.5 feet long. How tall is the flagpole?

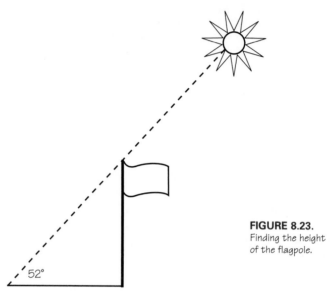

FIGURE 8.23.
Finding the height of the flagpole.

When solving problems with trigonometric ratios, it is helpful to write all known information on the figure and label the hypotenuse, the adjacent leg, and the opposite leg (**Figure 8.24**).

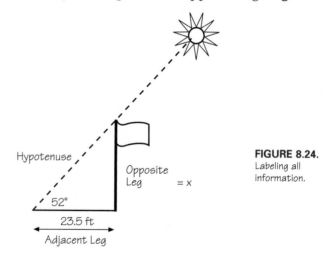

FIGURE 8.24.
Labeling all information.

Next, decide which ratio can be used to solve the problem. In this case, you know the length of the adjacent leg and want to know the length of the opposite leg. Therefore, use the ratio relating the adjacent leg to the opposite leg, which is the **tangent** ratio.

$$\tan = \frac{\text{opposite}}{\text{adjacent}}$$

Now you can substitute the known information into the ratio equation and then solve for x.

$$\tan 52° = \frac{\text{opposite}}{\text{adjacent}}$$

$$\tan 52° = \frac{x}{23.5}$$

$$23.5(\tan 52°) = 23.5\left(\frac{x}{23.5}\right)$$

$$23.5 \tan 52° = x$$

$$30 \approx x$$

So, the height of the flagpole is about 30 feet.

Next, consider this problem:

A right triangle has a 15° angle and a short leg of 18 inches. How long is the hypotenuse?

10. First, draw and label a picture. Be sure to label the reference angle, all known information, the unknown length, and the hypotenuse, adjacent leg, and opposite leg.

11. Which side length is given? Which side length is unknown? Which ratio relates these two?

12. Substitute the known values into the ratio equation and then solve for the unknown.

SECTION 8.3

Assignment

1. Use your calculator to find each of these:

    a) $\sin 76° =$ _____

    b) $\cos 31° =$ _____

    c) $\tan 16° =$ _____

    d) $\sin 60° =$ _____

    e) $\cos 60° =$ _____

    f) $\tan 45° =$ _____

2. Write each ratio in **Figures 8.25 and 8.26** in both words and numbers.

    a)

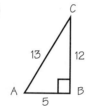

FIGURE 8.25.
Figure for Question 2a.

$\sin \angle A =$ ___ $=$ ___

$\cos \angle A =$ ___ $=$ ___

$\tan \angle A =$ ___ $=$ ___

    b)

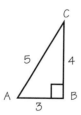

FIGURE 8.26.
Figure for Question 2b.

$\sin \angle C =$ ___ $=$ ___

$\cos \angle C =$ ___ $=$ ___

$\tan \angle C =$ ___ $=$ ___

3. An isosceles right triangle has a hypotenuse that is 10 units long. About how long are its legs?

4. The stairs of the Kukulcan Pyramid (**Figure 8.27**) rise at a 48° angle. When you get to the top, you are 100 feet above the ground (about ten stories). How long is the stairway?

FIGURE 8.27.
Kukulcan Pyramid, Mexico.

In Questions 5 and 6 draw a picture and solve the problem.

5. A right triangle has a 22° angle and a hypotenuse of 10 units. Find the length of each leg.

6. Rigo is holding his kite spool about 3 feet off the ground. The kite is flying at a 55° angle with the horizontal. The kite is directly over his friend Milo's head. If Milo is about 50 feet from Rigo, how high is the kite?

A New Angle on This ...

In Section 8.3 you used sine, cosine, and tangent to find side lengths of triangles.

Consider this problem:

A ski lift rides a 250-meter track up the side of a mountain to the top, which is 160 meters tall. What is the ski lift's angle of elevation?

1. Draw a picture of this situation. Label the known information.

2. Relative to the angle of elevation, label the hypotenuse, adjacent leg, and opposite leg.

3. What does the problem require you to find?

4. How does this problem differ from those you saw in Section 8.3?

5. Which ratio is formed by the lengths of the two known sides?

6. What is the sine ratio for the angle that you want to find?

7. How can you find an angle measure if you know the sine ratio for that angle?

Sometimes mathematicians use tables of trigonometric values to find sine, cosine, and tangent ratios. These tables contain the ratios for certain angles rounded to a certain decimal place. A portion of a trigonometric table is in **Figure 8.28**.

Angle	Sine	Cosine	Tangent
38°	0.616	0.788	0.781
39°	0.629	0.777	0.810
40°	0.643	0.766	0.839
41°	0.656	0.755	0.869
42°	0.669	0.743	0.900

FIGURE 8.28.
Trigonometric table.

8. In the table in Figure 8.28, what angle measure has a sine ratio closest to 0.64? How did you find this value?

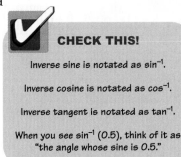

This process of working backwards to find an angle measure whose sine ratio is known is called finding the **inverse sine**. Likewise, you can find the **inverse cosine** or **inverse tangent** if you know an angle's cosine ratio or tangent ratio.

9. The sine ratio of ∠A is 0.64. How would you write this as an equation?

Graphing calculators have built-in features to allow you to compute an inverse trigonometric ratio. You need to press the 2nd or SHIFT key before pressing the SIN, COS, or TAN key to obtain the inverse sine, cosine, or tangent, respectively.

10. Use your graphing calculator to find the angle whose sine ratio is 0.64.

An advantage to using a graphing calculator to find angles is that you do not need to find the decimal value for a ratio before finding the angle measure.

11. Find the measure of ∠M in **Figure 8.29**.

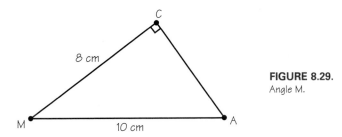

FIGURE 8.29.
Angle M.

Assignment

1. What are the angle measures of a 3-4-5 right triangle?

2. A staircase, which takes up 12.5 feet of horizontal floor space, connects two floors that are 10 feet apart (**Figure 8.30**). What is the angle of elevation of the staircase?

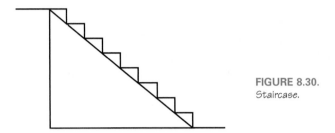

FIGURE 8.30.
Staircase.

3. Go fly a kite! (Draw pictures with your work.)

   a) Shelia is standing 225 feet from a long line of trees 70 feet high. If she is holding the reel of string 4 feet off the ground, how far is it from the reel to the top of the trees?

   b) If Shelia lets out 300 feet of string so that the kite is now directly over a small building 195 feet away, what angle does the string make with the horizontal?

   c) Suppose Shelia moves away from the building so that her kite string makes a 40° angle with the horizontal. If Shelia still has 300 feet of string out, what is the ground distance between Shelia and the building?

4. A right triangle has a hypotenuse of 10 centimeters and one angle that measures 28°. What is the area of the triangle?

The Two Towers

A distance of 2000 feet separates two towers in a state forest. One of the towers is directly west of the other. An observer in the West Tower notices a fire at an angle of 40° from the line of sight between the two towers. An observer in the East Tower notices the same fire due north from her tower (see **Figure 8.31**). Due to the terrain around the towers and through the forest, a fire truck from the West Tower travels at an average speed of 20 miles per hour. A fire truck from the East Tower travels at an average speed of 15 miles per hour. If both leave their towers at the same time, which one will get to the fire first? Justify your answer.

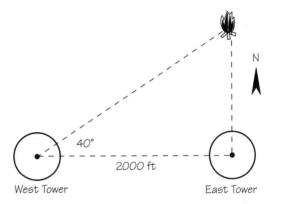

FIGURE 8.31.
Two towers.

Making the Sine Function Connection

You just explored three important ratios in right triangles. Next, you will use a circle to investigate the sine, cosine and tangent ratios for angles larger than 90°.

MEASURING ANGLES IN THE UNIT CIRCLE

Draw x- and y-axes on a sheet of chart paper with grid lines. Place the origin near the paper's center. Tie one end of a piece of string around a pushpin (**be careful**!). Measure off 1 foot of the string then wrap the end of the string around a marker so that the marker is exactly 1 foot from the pushpin. Place the pushpin at the origin of your coordinate plane and draw a large circle with a radius of 1 foot (see **Figure 8.32**).

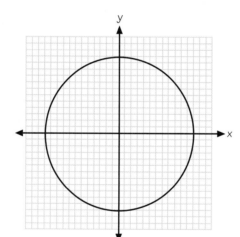

FIGURE.8.32.
Circle with radius of 1 foot.

Label the positive end of the x-axis 0°. Use a protractor to measure a 30° angle from the positive x-axis and draw this radius. Label the point where the radius touches the circle. Now, use a protractor to measure a 60° angle from the positive x-axis and draw this radius. Label the point where the radius touches the circle. Continue this process constructing angles in 30° steps until you reach the other side of the circle (see **Figure 8.33**).

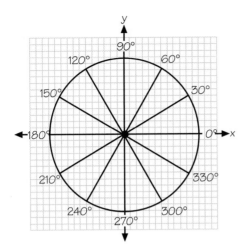

FIGURE 8.33.
Circle with 30° intervals

1. If you extend the 0° line to the other side of the circle, what is the measure of that angle?

2. Did you measure in a clockwise or counterclockwise direction from the positive *x*-axis?

CHECK THIS!

Remember that to convert inches to feet, divide by 12.

3. Create a table like **Figure 8.34**. For each angle you drew, measure the vertical distance in **feet** between the *x*-axis and the endpoint of the radius that lies on the circle as shown in **Figure 8.35**. If the point is above the *x*-axis, record that distance as *positive*. If the point is below the *x*-axis, record that distance as *negative*.

Measure of Angle	Vertical Distance
0°	
30°	
60°	
90°	
120°	
150°	
180°	
210°	
240°	
270°	
300°	
330°	
360°	

FIGURE 8.34.
Data table.

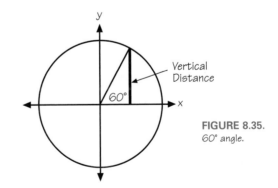

Vertical Distance

FIGURE 8.35.
60° angle.

4. Use your table to make a scatterplot of vertical distance versus angle measure. If you use a graphing calculator, be sure to state your window.

5. Describe the shape of the scatterplot. What would happen if you continued measuring vertical distances for angles beyond 360°, such as 390° or 420°?

Consider the vertical distance that you measured and that is shown in **Figure 8.36**. Look at the right triangle made by the vertical distance, the *x*-axis, and the radius. For the 60° reference angle, label the hypotenuse, adjacent leg, and opposite leg.

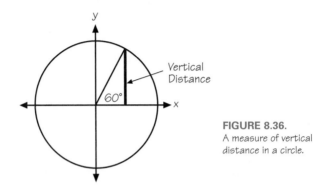

FIGURE 8.36.
A measure of vertical distance in a circle.

6. Which trigonometric ratio involves the opposite side and the hypotenuse?

7. What is the length of the hypotenuse?

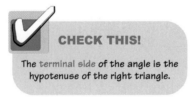

CHECK THIS!

The terminal side of the angle is the hypotenuse of the right triangle.

8. How does the vertical distance compare to the sine ratio of a 60° angle?

9. Make a general statement about the relationship between the sine ratio and the vertical distance in a circle with a radius of 1 unit, or **unit circle**.

Spinning Your Wheels

In 1893, George W. Ferris built the world's first Ferris wheel for the World's Fair in Chicago. In 1985, the Texas Star opened at the Fair Park in Dallas. Today, the Texas Star is the largest Ferris wheel in the United States. From its maximum height of 212 feet, a rider gets a great view of the Dallas skyline.

One Texas Star rider wondered how he could find his height above ground at any given time.

To help solve this problem, first consider a simpler one. You will need something smaller than a Ferris wheel to model the motion of a rider. Instead of an actual Ferris wheel, you will collect data from a hula-hoop or some other round object.

COLLECTING THE DATA

1. What is the diameter of your wheel? What is its circumference?

   Set up the experiment as follows:

   a) Extend the tape measure and tape it to the floor with masking tape. (Later, you will roll your wheel alongside the tape measure.)

   b) Place another piece of tape on the floor perpendicular to the tape measure at its zero end. Place the wheel directly on top of this marker. The wheel should touch the masking tape; its center should be directly above the tape.

   c) Place a dot on the wheel (a sticker or piece of tape will do) at the point where the wheel sits on the tape.

Roll the wheel forward alongside the tape measure. Stop the wheel every 12 inches. When you stop, measure and record two things: the height of the dot above the floor and the total distance the wheel has traveled. (See **Figure 8.37**.) Record your data in Handout 8.1. (See Handout 8.1 for more instructions.) Be sure to save Handout 8.1 because you will need the data in Section 8.8.

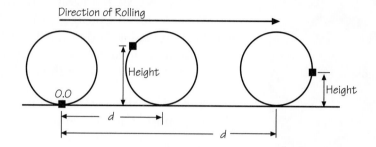

FIGURE 8.37.
Measuring the distance the wheel rolls and height of the dot.

GRAPHING THE MOTION

2. Make a scatterplot of height versus distance. If you use a graphing calculator, be sure to state your window.

3. Sketch your scatterplot on paper. On your plot, sketch a smooth curve (no sharp corners) that you think best represents the relationship between height and distance.

CHECK THIS!

A function that has a graph that repeats itself on intervals of equal length is called periodic.

4. Do you think that the relationship between the height of a marked spot on the wheel and the distance rolled is periodic? Explain.

5. Does your scatterplot appear to be periodic? Explain.

6. What is the height of the center of your wheel above the ground? How do you know?

7. Based on the height of the center of your wheel above the ground and the diameter of the wheel, what would you expect the maximum and minimum heights of your marked spot to be? Explain.

8. From your data, what is the difference between the maximum and minimum heights? How does this compare to what you expected? Explain any differences.

9. How much distance did your wheel cover before your mark on the heights began repeating?

10. What is the horizontal length of one repetition of your graph?

11. How do your answers in Questions 9 and 10 compare?

12. For what distances is the marked spot on the wheel at the top of the wheel? How can you get this answer from your graph?

13. For what distances is the marked spot horizontally level with the center of the wheel? How can you get this answer from your graph?

Assignment

Time (sec)	Height of Right Pedal (in)
0	
0.25	
0.5	
0.75	
1	
1.25	
1.5	
1.75	
2	
2.25	
2.5	
2.75	
3	

FIGURE 8.38.
Height-time data for a bicycle pedal.

Angle Measure	Sine Value

FIGURE 8.39.
Table of sine values.

Suppose you pedal a bicycle so the pedals revolve once a second. Each pedal is 5 inches from the ground at its lowest and 18 inches from the ground at its highest. You start with the right pedal in its lowest position.

1. Copy the table in **Figure 8.38** and complete the height entries.

2. Make a scatterplot of height of the right pedal versus time. Be sure to label the axes.

3. Use your scatterplot to sketch a graph of a function that you think best represents the relationship between height and time.

The graph that you drew for Question 3 oscillates between 5 inches and 18 inches. The up-down pattern repeats itself each second. Any function that repeats itself on intervals of equal length is called **periodic**.

4. Create a table of sine values (rounded to two decimal places) of angles from 0° to 360° in 15 degree steps. (You can copy and fill the table in **Figure 8.39**.)

5. Make a scatterplot of sine versus angle measure.

6. What kind of graph is this?

7. What would you expect to happen if you continued the graph to 720°?

SECTION 8.8

Periodic Motion and Sinusoidal Regression

Carousel horses like those in **Figure 8.40** move up and down as the carousel spins. The graph in **Figure 8.41** represents the height above the floor of a carousel horse's back hooves versus time.

FIGURE 8.40. Carousel horses.

FIGURE 8.41. Height vs. time graph.

CHECK THIS!

An oscillating pattern is one that swings back and forth or varies between two alternate extremes.

1. According to this graph, what happens to the height of the hooves over time?

2. What do the maximum values (the peaks) in the graph represent? What do the minimum values (the troughs) in the graph represent?

3. Where in the graph are the hooves in the middle? Explain.

The centerline of the **oscillation** is a line halfway between the graph's peaks and troughs. The centerline, also called the **axis of oscillation**, is shown in **Figure 8.42** with a dotted line.

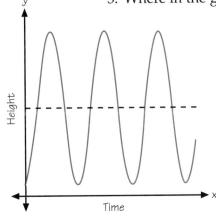

FIGURE 8.42. Axis of oscillation.

In Section 8.7 you rolled a wheel along the floor and measured the height of a marked spot as the wheel rolled.

4. What happened to the height of the center of your wheel as the wheel rolled along the ground?

5. What were the minimum and maximum heights that the marked spot reached?

The height of the marked spot above the floor oscillated between the minimum and maximum values. This oscillation can be described with a **periodic function**. A periodic function is a function in which the value of the dependent variable, in this case the height of a marked spot, oscillates between a minimum value and a maximum value.

Think back to Section 8.6 where you used a large unit circle with a radius of 1 foot to explore the relationship between vertical distance and angle measure (**Figure 8.43**). You noticed that the vertical distance is equal to the sine ratio for any given angle:

$$\sin(x°) = \frac{\text{opposite}}{\text{hypotenuse}}$$

$$\sin(x°) = \frac{\text{vertical distance}}{\text{radius}}$$

$$\sin(x°) = \frac{\text{vertical distance}}{1}$$

$$\sin(x°) = \text{vertical distance}$$

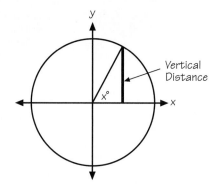

FIGURE 8.43. $x°$ angle.

In Section 8.6 you made a scatterplot of the sine function, $y = \sin(x)$, when you plotted the vertical distance (length of the opposite leg of the right triangle whose hypotenuse is 1 unit) versus angle measure (**Figure 8.44**).

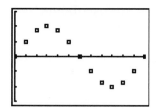

FIGURE 8.44. Scatterplot.

6. Refer back to your data in Handout 8.1. Also, look at the scatterplot that you made of the data. How does this scatterplot compare to the scatterplot of the function $y = \sin(x)$?

One way to model periodic relationships is to use the **sine function** as a parent. You can use your graphing calculator's regression features to find a function in this family. This type of regression is called **sinusoidal regression**.

7. Use your graphing calculator to find a sine function to fit your wheel data in Handout 8.1.

8. Write down your function rule. Round to the nearest ten-thousandth.

CHECK THIS!

When the independent variable is measured in degrees, the calculator's degree mode is used for sinusoidal regression and for related calculations and graphing. Otherwise, the radian mode is used.

Once you have a function rule that models a periodic relationship, you can use the TRACE feature of your calculator to answer questions. Use your function rule to answer these questions.

9. Find the minimum and maximum values of your function. How do they compare to the minimum and maximum values of your data?

10. Subtract the maximum and minimum values. What does this difference represent?

11. Take half of the difference from Question 10. Add this number to your minimum value. How does this value compare to your axis of oscillation?

The **amplitude** is half of the vertical length of the basic repeating shape or half of the difference between the maximum and minimum y-coordinates of the graph.

12. Use the TRACE feature to find the x-values that correspond to consecutive maximum values. Record these x-values.

13. Subtract the two x-values from Question 12. In terms of the cycle of repetition, what does this difference represent? What does this difference represent in terms of the situation?

The shortest horizontal length of a basic repeating shape is called the **period**. On a graph, you can find the period by tracing to get the x-coordinates of consecutive corresponding points, as you did in Question 12.

As with many other functions, the sine family has a general form:

$$y = A \sin(B(x - C)) + D$$

You can use a graphing calculator to see the effect of each control number A, B, C, and D.

14. To understand the impact of A, graph the parent function $y = \sin(x)$ and several others with different values of A. For example, graph $y = \sin(x)$, $y = 2 \sin(x)$, and $y = 3 \sin(x)$. Use your calculator's radian mode. What can you conclude?

15. Conduct a similar investigation into the role of B.

16. Conduct a similar investigation into the role of C.

17. Conduct a similar investigation into the role of D.

Assignment

Use this information for Questions 1–3.

Suppose you are pushing your sister on a swing. If you sketch a graph of her distance from you over time it might look like **Figure 8.45**.

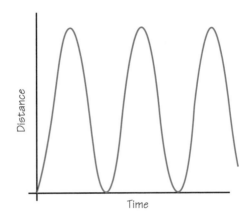

FIGURE 8.45.
Graph of swing distance.

1. What happens to the distance between you and your sister over time?

2. What do the maximum values in the graph represent? What do the minimum values in the graph represent?

3. Where in the graph are the swing's chains perpendicular to the ground? Explain.

Use **Figure 8.46** for Questions 4 and 5.

Figure 8.46 shows graphs of three periodic functions. Each function's graph repeats some basic shape.

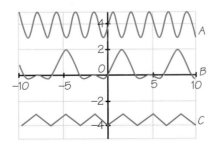

FIGURE 8.46.
Graphs of three periodic functions.

4. Find an approximate period for each graph.

5. What is the amplitude for each graph?

Use **Figure 8.47** for Questions 6–8.

The height of a carousel horse's stirrup changes periodically over time as the carousel spins. The graph of this motion is shown in Figure 8.47.

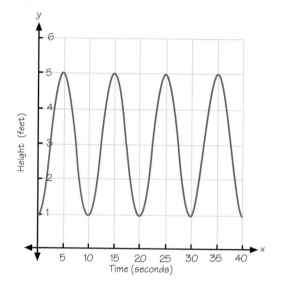

FIGURE 8.47.
The height of the stirrup of a carousel horse versus time.

6. What is the period?

7. What is the maximum height of the stirrup? What is the minimum height?

8. What is the amplitude?

Use this information for Questions 9–13.

In Section 8.7 you drew periodic graphs by plotting the heights of dots on a rolling wheel. Although wheels come in only one shape—a circle—they come in many sizes. If different groups used different sized wheels, then different height-versus-distance graphs would have been drawn.

9. For Section 8.7 Jerry's group used a wheel with a 26-inch diameter. Their height-versus-distance graph is shown in **Figure 8.48**.

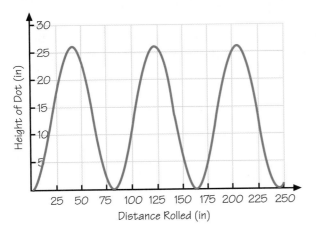

FIGURE 8.48.
Height-versus-distance graph from Jerry's group.

10. Based on the wheel diameter, what are the period and amplitude? Explain how you got your answer. Check that your answer is consistent with the graph in Figure 8.48.

11. What is the axis of oscillation?

Imagine doing Section 8.7 with a wheel that has a one-foot radius (or two-foot diameter).

12. How far would you have to roll the wheel for the dot to make one complete turn? Give both an exact answer (involving π) and a decimal approximation.

13. How far would you have to roll the wheel for the dot to move from the bottom of the wheel to the top of the wheel for the first time? And then the second time? Give both exact and approximate answers.

Use **Figure 8.49** for Questions 14–16.

Assume that the Ferris wheel in Figure 8.49 does one revolution every 30 seconds. The ride lasts about 5 minutes.

14. Thia is the last person to board the Ferris wheel. How far has she traveled after the wheel makes one complete revolution? (To find the circumference of the wheel, use $c = 2\pi r$, where r is the radius and c is its circumference.)

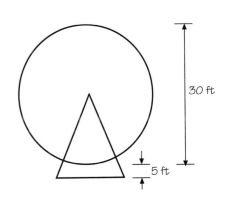

FIGURE 8.49.
Ferris wheel rotating at a constant speed.

15. Plot Thia's height above the ground every 7.5 seconds for the first two minutes of the ride. Then use these data points to draw a graph of a function that you think best represents the relationship between height and time.

16. The graph that you drew for Question 15 is periodic. What is the period? What is the amplitude? What is the axis of oscillation?

It's Just the Swing of the Old Pendulum

In the last three sections you modeled relationships using the sine function as a parent function. These situations involve **periodic motion**, in which motion oscillates, or cycles, back and forth between two values. Most of these situations involve circular motion, where an object is moving along a circle and you measure something, such as height, that changes over time or distance.

Other kinds of motion can be modeled with the sine function. One of these involves the distance between a pendulum and a stationary object over time.

THE PENDULUM STATIONS

There are three pendulum stations. Your teacher will assign student pairs to each station. Each station has a CBR and a pendulum with a length that differs from the other stations.

Place a CBR about half a meter from the maximum swing of the pendulum and at the same height as the pendulum at rest. You might need to use books to adjust the height of the pendulum or the CBR. (See **Figure 8.50**.)

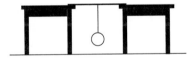

FIGURE 8.50.
Pendulum stations.

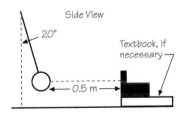

Side View

20°

Textbook, if
necessary

0.5 m

CBR SET UP

1. Connect the CBR to the calculator using the link cable.

2. Run the **RANGER** program on your calculator:

   a) Press **APPS** (**PROGRAM** on some calculators).

   b) Arrow down to **CBL/CBR**. Press **ENTER**.

   c) Press any key.

   d) Arrow down to **RANGER**. Press **ENTER**.

   e) Arrow to **SET UP SAMPLE**. Press **ENTER**.

   f) The Main Menu window should be set up as shown. If it is not, arrow down and press **ENTER** until the correct label is showing.

MAIN MENU	START NOW
REAL TIME:	YES
TIME(S):	15
DISPLAY:	DIST
BEGIN ON:	[ENTER]
SMOOTHING:	NONE
UNITS:	METERS

   g) Arrow up to **START NOW**. When you are ready to collect your data after you release the pendulum, press **ENTER**.

1. Record the length of your pendulum from the meter stick to the bottom of the pendulum.

For the first data collection, pull the pendulum back about 30° from vertical and release it so that it swings back and forth in the motion detector's path. Practice pulling back the pendulum a few times.

COLLECTING THE DATA

You and a partner will make **two** data collections using the CBR at your assigned station:

❖ One collection is with the pendulum in about a 20° swing.

❖ The second collection is with the pendulum in about a 40° swing.

❖ The collections should be made on different calculators.

❖ Each person in the pair is responsible for one of the data collections.

2. Record the angle of the swing for which you are responsible.

3. Before collecting your first set of data, sketch what you think a distance-versus-time graph will look like. Explain why you drew the graph as you did.

HINTS FOR COLLECTING DATA WITH THE CBR

If you do not get good data, there are two likely reasons: The most likely is that the CBR is placed too close to the pendulum. Less likely, is that the pendulum swings a path that is not in line with the CBR, so that data collection is skewed.

Set up your CBR using the directions on page 530. The CBR should be **perpendicular to the floor** and aimed directly at the pendulum. When you are ready to begin collecting data, release the pendulum, get out of the way, and start the CBR.

Collect both sets of data, then answer Questions 4–9.

4. Make a scatterplot of your data. State the window. Sketch the scatterplot.

5. Does the scatterplot match your sketch in Question 3? Why or why not?

6. Mark one period on the sketch. Use the TRACE key to find the length of that period. Be sure to record the units of measure.

> **CHECK THIS!**
>
> A pendulum completes a cycle as it moves from one extreme position to another and back again. The time required for the pendulum to complete a cycle is called the period.

7. Mark the amplitude on your sketch. Use the TRACE key to find the minimum and maximum distances of the pendulum. Use these values to compute the amplitude. Be sure to record the units of measure.

> **CHECK THIS!**
>
> The amplitude is the distance the pendulum moves from at rest to the farthest position away. The amplitude can be found by dividing the total distance the pendulum moves by two.

8. How do your period and amplitude compare with your partner's?

9. How can you tell who used the larger angle from the graph?

For Questions 10–14, compare your findings with other students who collected at the same station.

10. What is similar about the data? What do you think causes the similarities?

11. What is different? What do you think causes the differences?

12. Can you tell from the graphs who pulled the pendulum back the farthest? The least?

13. How do the periods compare?

14. You are about to compare data with students who collected data from stations with pendulum lengths different from yours. How do you think their data will compare to yours?

For Questions 15–17, compare your findings with those of students who collected from different stations.

15. Did their data meet your expectations? Explain.

16. What is similar about your data? What do you think causes the similarities?

17. What is different? What do you think causes the differences?

18. You can adjust a pendulum clock by changing the length of the pendulum. If a clock runs slow, how should you adjust the pendulum? Should it be longer or shorter? Why?

Assignment

1. Over a year, the length of daylight (the number of hours from sunrise to sunset) changes. **Figure 8.51** shows the length of daylight every 30 days from 12/31/97 to 3/26/99 for Boston, Massachusetts.

Date	Day Number	Length (hours)	Date	Day Number	Length (hours)
12/31	0	9.1	8/28	240	13.3
1/30	30	9.9	9/27	270	11.9
2/1	60	11.2	10/27	300	10.6
3/31	90	12.7	11/26	330	9.5
4/30	120	14.0	12/26	360	9.1
5/30	150	15.0	1/25	390	9.7
6/29	180	15.3	2/24	420	11.0
7/29	210	14.6	3/26	450	12.4

FIGURE 8.51.
Data on length of day.

a) Draw a set of axes similar to **Figure 8.52**. Then plot the data from Figure 8.51.

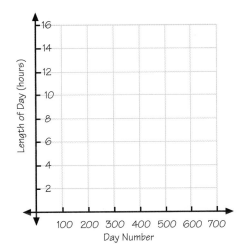

FIGURE 8.52.
Axes for length of day versus day number.

b) Draw a smooth curve through the points on your graph. Extend your graph to show how you think the graph should look over two years. Explain why you think the graph should look this way.

c) What is the period of your graph?

d) What is the graph's amplitude?

e) What is the axis of oscillation?

2. Figure 8.51 gives data on length of daylight for Boston, Massachusetts. A plot of the data indicates that the relationship between length of daylight and day number is periodic.

a) Write a sinusoidal function that models this relationship. Explain how you got your function.

3. Ariana and Terry are turning a long jump rope for Tracy. Ariana and Terry turn the rope once per second. The maximum height is seven feet. At its lowest the rope just touches the ground. Assume the rope is on the ground moving up and away from Tracy at starting time.

a) What is the amplitude for the relationship between the height of the rope and time?

b) What is the period?

The Sun is Shinning Down

Brockport, New York, is a small town on the Erie Canal in western New York state. Winters there can be very cold. Average high temperatures in January are 31°F. Summers, on the other hand, can be quite warm, with average high temperatures in July rising to about 82°F.

One factor that affects temperature is the amount of sunlight received during the day. Scientists can measure and record the amount of sunlight, or solar radiation. They often use the units of calories per square centimeter. **Figure 8.53** contains data from Brockport, New York. The values for each month are the average daily solar radiation for the 15th day of that month.

Month	Solar Radiation	Month	Solar Radiation
January	132	July	495
February	190	August	418
March	274	September	310
April	365	October	210
May	446	November	117
June	500	December	92

FIGURE 8.53.
Solar radiation data.

Based on these data, predict the approximate solar radiation received at the beginning of March. Justify your prediction.

Mathematics and the Sound of Music

Music is played in many forms all over the world. Instruments vary from the didgeridoo of the Australian aborigines (a hollow tree branch that plays only two tones) to large concert organs. People everywhere appreciate good music.

When you pluck a violin or guitar string, it vibrates back and forth. To see this effect, try plucking a stretched rubber band once. The blur you see is the rubber band vibrating back and forth.

The vibrations of a violin string push on nearby air particles. These air particles push on other nearby air particles. Pressure waves are created. Sound waves are pressure waves. The energy from a vibration causes changes in air pressure. The pressure variations create a wave that travels through the air to reach the eardrum, which then vibrates and causes us to hear sound.

Sound waves are longitudinal waves. You can create a longitudinal wave with a Slinky.

❖ Attach a Slinky to a stationary object on a table.

❖ Tie a string to one of the coils of the Slinky.

❖ Stretch the non-stationary end of the Slinky out about 2 meters.

❖ Push and pull back once on the non-stationary end of the Slinky.

1. What happens as the non-stationary end of the Slinky is pushed?

2. How is the movement of the Slinky similar to the movement of sound waves? (See **Figure 8.54**.)

Music is, of course, not the only place where sound waves are used. Sound waves are used in ultrasound, CDs, and cell phones. Not only does sound travel through air, it also travels through liquids, solids, and other gases. You may have tried to make a phone with two cans and string when you were a child. When one person speaks into a can sound travels through the string and the other person can hear the sound in the other can.

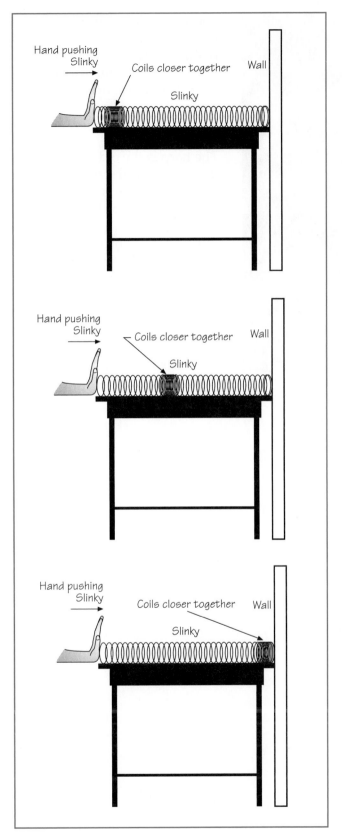

FIGURE 8.54.
The Slinky sound wave experiment.

Sound waves produced by music have a regular pattern. Sound waves produced by noise do not have a regular pattern. You may have heard of singers who can break a glass with their voice. By singing a perfect note the singer's voice causes the glass to vibrate so much that it shatters from the pressure.

The harp is the oldest stringed instrument in the world. Each string vibrates to produce a different note. Some string instruments use a few strings to produce many notes. For example, violins and guitars produce different notes by changing the length of the string. Wind instruments, such as trombones, clarinets, and oboes play different notes by changing the length of the column of air that vibrates inside the instrument. Percussion instruments, such as xylophones and steel drums, create vibrations that ring at a certain pitch when struck.

Do You See the Music?

Have you ever blown across the top of a soda bottle? What happened?

You may have noticed that a sound is produced when you blow across the top of the bottle.

You can use a microphone collection device to measure the pressure variations that cause us to hear sound.

1. Sketch what you think a sound pressure versus time graph might look like.

2. Your teacher will demonstrate how to collect sound pressure data. After the demonstration, record the scatterplot.

3. Label one cycle on the scatterplot.

CHECK THIS!

A cycle is one complete pattern.

4. On the scatterplot, mark and label two points. One point should be at the beginning of a cycle and the other at the end. Record these values in a table like **Figure 8.55** and compute the period. Show your work.

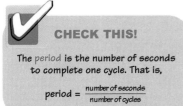

CHECK THIS!

The period is the number of seconds to complete one cycle. That is,

$$\text{period} = \frac{\text{number of seconds}}{\text{number of cycles}}$$

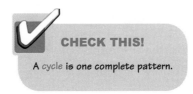

Point	x	y
A		
B		

FIGURE 8.55.
Table for Question 4.

5. Find the frequency of the graph. Explain the meaning of this value.

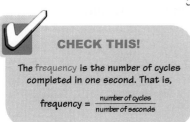

CHECK THIS!

The frequency is the number of cycles completed in one second. That is,

$$\text{frequency} = \frac{\text{number of cycles}}{\text{number of seconds}}$$

A standard unit of sound measure is the Hertz. Hertz means "cycles per second." FM radio waves are given in megahertz. AM radio waves are given in kilohertz. For example, 97.9 FM is 97.9 megahertz. The frequency of a wave that a 97.9 FM station sends out is 97,900,000 cycles per second. Frequencies of radio waves are outside the range of human hearing. Only after a radio receives the waves can we hear the music. The Hertz is named after Heinrich Hertz, the first person to successfully experiment with the transmission and reception of radio waves.

CHECK THIS!

Pitch refers to the highness or lowness of a tone. For example, a low note at the left end of the piano keyboard has a low pitch. A high note at the other end of the keyboard has a high pitch.

6. Repeat the experiment in your group with an assigned water height. Before adding water, blow across the top of the bottle and listen to the sound. Now add the assigned height of water. How does adding water change the sound?

7. Collect new data with water in the bottle. Make a scatterplot.

8. Use the TRACE feature of the calculator to mark and label two points on your scatterplot. Find the period. Show your work.

9. How does the new period compare to the period of the original data?

10. Find the frequency of your new data. Label your answer with appropriate units.

11. How does the frequency of the new data compare to the frequency of the original data?

12. What is the relationship between frequency and pitch?

Complete the following:

13. If the period of a sound vs. time graph decreases, the frequency ____.

14. If the period of a sound vs. time graph increases, the frequency ____.

15. To decrease the frequency of a bottle, you would ____.

16. Consider this: Sound wave A has a shorter period than sound wave B. What can you say about their relative frequencies? Justify your answer.

Assignment

1. Use the data collected by groups in your class to fill in a table like **Figure 8.56**.

Height of Water (inches)	Period (sec/cycle)	Frequency (cycles/sec)

FIGURE 8.56.
Table for Question 1.

2. Make a scatterplot of the relationship between period and height of the water.

3. Which quantity did you use as the independent variable?

4. Which quantity did you use as the dependent variable?

5. What do you notice about a period versus height of water scatterplot?

6. If the height of the water is 3.75 inches, what do you predict the period to be?

7. Make a scatterplot of frequency versus height of water.

8. What do you notice about the frequency versus height of water scatterplot?

9. If you want the frequency of a bottle's note to be 400 cycles per second, about what should the height of the water be?

10. Make a scatterplot of frequency versus period.

11. What do you notice about the relationship between frequency and period? Explain your answer.

Name That Note

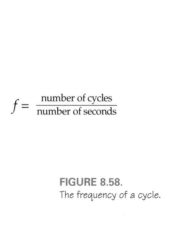

Blowing across the top of a bottle produces sound pressure waves. These waves look like a scatterplot of the data you generate when you roll a wheel or when you swing a pendulum.

Sound waves have properties that determine the note that your ears hear. In Section 8.12 you investigated these properties of sound waves.

The *period* is the number of seconds per cycle. The period can be measured by finding the time it takes to get from one crest of the graph to the next crest (**Figure 8.57**).

$$P = \frac{\text{number of seconds}}{\text{number of cycles}}$$

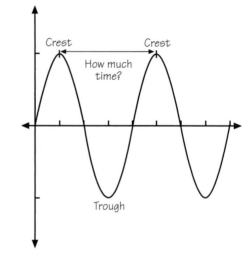

FIGURE 8.57.
The period of a cycle.

The *frequency* is number of cycles per second (**Figure 8.58**).

$$f = \frac{\text{number of cycles}}{\text{number of seconds}}$$

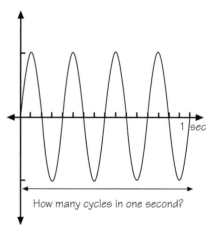

FIGURE 8.58.
The frequency of a cycle.

1. Suppose the period is 0.0048 seconds. Find the frequency.

2. Suppose the frequency is 153.85 cycles per second. Find the period.

Period and frequency have a reciprocal relationship.

Another property of sound is loudness. The larger the amplitude of a sound wave, the more energy the sound wave has and the louder the note (**Figure 8.59**).

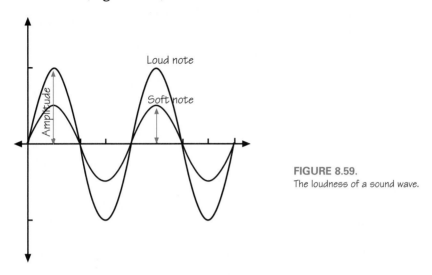

FIGURE 8.59.
The loudness of a sound wave.

In Section 8.8 you investigated properties of periodic functions such as amplitude and period.

In a sinusoidal function for a musical note, $f(t) = A \sin(B(t - C)) + D$, t represents time, and A represents the amplitude or loudness of the note. B is equal to the frequency multiplied by 2π. Assuming that the function includes the point (0, 0), we can use the simpler $f(t) = A \sin Bt$.

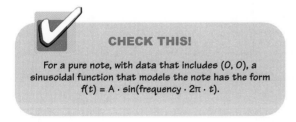

CHECK THIS!

For a pure note, with data that includes (0, 0), a sinusoidal function that models the note has the form
$f(t) = A \cdot \sin(\text{frequency} \cdot 2\pi \cdot t)$.

3. The data in **Figure 8.60** were collected when three tuning forks were struck.

Tuning Fork 1		Tuning Fork 2		Tuning Fork 3	
Time, *t* (seconds)	Sound Pressure, *f(t)*	Time, *t* (seconds)	Sound Pressure, *f(t)*	Time, *t* (seconds)	Sound Pressure, *f(t)*
0.00000	0.00000	0.00000	0.00000	0.00000	0.00000
0.00029	0.00285	0.00022	0.00116	0.00057	0.00285
0.00057	0.00400	0.00057	0.00200	0.00114	0.00400
0.00096	0.00190	0.00093	0.00110	0.00191	0.00190
0.00114	0.00000	0.00114	0.00000	0.00227	0.00000
0.00134	−0.00214	0.00137	−0.00121	0.00287	−0.00295
0.00171	−0.00400	0.00171	−0.00200	0.00341	−0.00400
0.00201	−0.00265	0.00204	−0.00119	0.00402	−0.00265
0.00227	0.00000	0.00227	0.00000	0.00455	0.00000
0.00259	0.00304	0.00252	0.00127	0.00488	0.00180

FIGURE 8.60.
Tuning fork data.

Make a scatterplot of all three data sets. Put all three on the same grid, and use a different color pencil for each set.

4. For each data set find the amplitude, period, frequency, and a function rule. Fill in a table like **Figure 8.61**.

	Tuning Fork 1	Tuning Fork 2	Tuning Fork 3
Amplitude			
Period (seconds/cycle)			
Frequency (cycles/second)			
Function model using sinusoidal regression on graphing calculator			

FIGURE 8.61.
Table for Question 4.

5. Sketch graphs of the sine function that models each tuning fork note on your scatterplot.

6. Which note(s) was (were) played the loudest?

7. Which note(s) has (have) the lowest frequency?

8. Which note(s) has (have) the highest pitch?

Assignment

1. Instead of identifying notes by their frequencies, musicians have given them names. **Figure 8.62** displays the frequencies for the notes in the middle octave of a tuned piano. Make a scatterplot of frequency versus note number.

Name of the Note	Note Number	Frequency
C_4	0	261.63
$C_4^{\#}$	1	277.18
D_4	2	293.66
$D_4^{\#}$	3	311.13
E_4	4	329.63
F_4	5	349.23
$F_4^{\#}$	6	369.99
G_4	7	392.00
$G_4^{\#}$	8	415.30
A_4	9	440.00
$A_4^{\#}$	10	466.16
B_4	11	493.88
C_5	12	523.25

FIGURE 8.62.
Table for Question 1.

2. Do these data represent a function?

3. Find a function rule to model the relationship between frequency and note number.

4. What note do you think was played by tuning fork in Section 8.13?

Match each term with its description.

5. high amplitude   a) a sound wave with a short period

6. low amplitude   b) the number of cycles of a sound wave in one second

7. period   c) a loudly played note

8. frequency   d) a sound wave with a lower frequency

9. low note   e) the length of time for one complete cycle of a periodic function

10. high note   f) a softly played note

11. Joseph collected the data in the scatterplot in **Figure 8.63** from a tuning fork using a CBL. Find the period, the frequency, the amplitude, and a function rule.

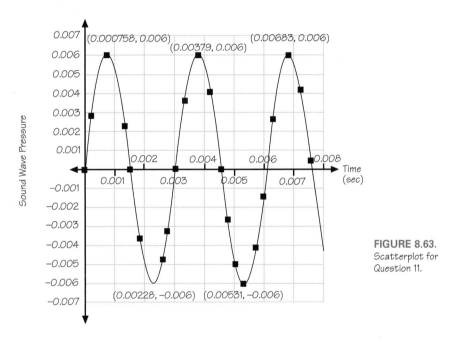

FIGURE 8.63.
Scatterplot for Question 11.

The Big Bottle Band

In Section 8.12 you filled a soda bottle with a certain height of water and found the frequency of the note you played. Your teacher will give you a card with a frequency, and you will tune your bottle to this frequency. Use what you learned in Section 8.12 and 8.13 to help you tune your note. When all of the groups tune their bottles, the class will play a song. As in a bell choir, each person is responsible for playing a certain frequency.

1. My assigned frequency is _____.

2. Record your first graph and find the frequency. Explain how you found the frequency.

3. Do you need to add or remove water from the bottle? Explain your reasoning.

4. Repeat the process until your bottle is tuned. Record the trial number and frequency for each trial in a table like **Figure 8.64**.

Trial Number	Frequency (note)

FIGURE 8.64.
Table for Question 4.

5. Measure the final amount of water you used to tune your bottle.

6. How many trials did it take to tune your bottle?

7. What have you learned about bottle tuning that would help you the next time you need to tune a bottle?

Assignment

1. Group 1 collected data for the bottle band activity. They used the TRACE feature of the calculator to find the two sets of coordinates shown in **Figure 8.65**. The note that Group 1 was supposed to tune to has a frequency of 300. Should they add or remove water to get closer to this frequency?

FIGURE 8.65.
Coordinates for Question 1.

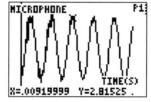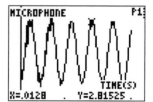

2. Group 2 collected data for the bottle band activity. They used the TRACE feature of the calculator to find the two sets of coordinates shown in **Figure 8.66**. The note that Group 2 was supposed to tune to has a frequency of 400. Should they add or remove water to get closer to this frequency?

FIGURE 8.66.
Coordinates for Question 2.

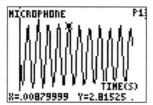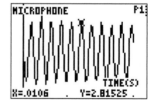

Sound Assessment

1. Describe the graph of a periodic function.

2. What is meant by the period of such a graph?

3. How can you find the period of a graph? Why might someone choose to find the average of several values instead of using just one?

4. What is the frequency of a periodic graph? How is it related to the period?

5. Find the period and frequency of the sound wave from the graph and data in **Figure 8.67**.

FIGURE 8.67.
Figure and table for Question 5.

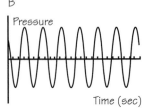

Point	x	y
A	0.0028	−1.917
B	0.0042	0
C	0.0056	1.917
D	0.0069	0
E	0.0081	−1.917
F	0.0095	0
G	0.0109	1.917
H	0.012	0
I	0.013	−1.917
J	0.0148	0
K	0.0159	1.917
L	0.0174	0
M	0.0187	−1.917
N	0.02	0
O	0.0212	1.917
P	0.0227	0
Q	0.0239	−1.917

6. The graphs in **Figure 8.68** are of different tones. Which one represents the tone with the lower pitch? Justify your answer.

FIGURE 8.68.
Graphs for Question 6.

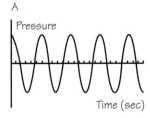

Modeling Project How Does Your Cycle Go?

You have learned that some things can be modeled with periodic functions. Because periodic functions have consistent cycles, they can be used to answer questions such as:

* How high will the tide be next September 15th?

* What will be the demand for air conditioners next April?

* How much coffee should a coffee shop have ready for each hour of the day?

However, in these examples there are things that could interrupt the cycle, such as a change in customers' habits or unusual weather.

For this modeling project you will choose a relationship that you believe is periodic. You can gather information from the Internet, library research, or by interviewing people in your community. You will create a presentation for your class that:

* Describes how your choice is cyclical, both mathematically (using terms like period and amplitude), and in words;

* Discusses the strengths and weaknesses of a model for describing your choice and for making predictions based on your model;

* Discusses the usefulness of your model to people (i.e. how it helps them run a business, protect people from harm, or some other purpose).

Practice Problems

1. In triangle ABC, ∠C is a right angle, c measures 13 cm, and a measures 5 cm.

   a) Make a sketch of the triangle. Label the hypotenuse and the legs that are opposite and adjacent ∠A.

   b) Find the length of b.

   c) List the sine, cosine, and tangent ratios for ∠A.

   d) To the nearest degree, what is the measure of ∠A?

2. Find x and y for the right triangle in **Figure 8.69.**

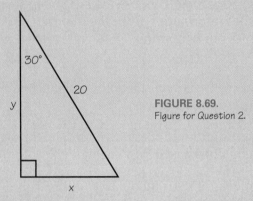

FIGURE 8.69.
Figure for Question 2.

3. Two holes are drilled in a metal plate as shown in **Figure 8.70.**

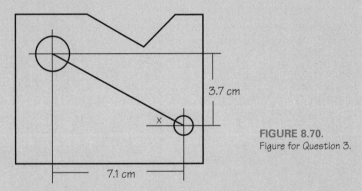

FIGURE 8.70.
Figure for Question 3.

   a) Find the distance between the center of the large hole and the center of the small hole.

   b) Find ∠x.

4. A wheelchair ramp makes an angle of 5° with the ground. It rests on the ground at one end and is 4 feet off the ground at the other end.

   a) How long is the ramp?

   b) Federal disability guidelines say that ramps should rise no more than 1 foot for every 12 feet of horizontal distance. Does this ramp meet the guidelines? Explain.

5. Metals and other materials expand and contract with temperature. As a result, railroad tracks must be laid in sections with small gaps between them to avoid buckling due to expansion. Suppose a track 1 mile long is laid in one continuous piece and that its length increases 0.1% in warm weather. If the track buckles straight up at its center, how far off the ground is the middle of the track? Show your work and give your answer to the nearest foot.

6. **Figure 8.71** is a graph of a sinusoidal function.

   a) What is the period?

   b) What is the amplitude?

   c) Find the equation of the axis of oscillation.

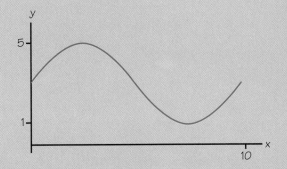

FIGURE 8.71.
Figure for Question 6.

7. A population of gray wolves in Yellowstone Park varies periodically over a year (365 days). A graph of a function that models the relationship between wolf population and time is shown in **Figure 8.72.**

   a) State the graph's period and discuss what it means in this situation.

   b) Give the graph's axis of oscillation and discuss what it means in this situation.

   c) Give the graph's amplitude and discuss what it means in this situation.

   d) When is the wolf population largest, and how many wolves are there at that time?

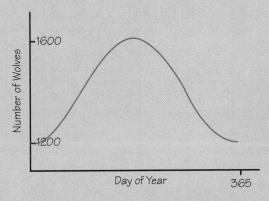

FIGURE 8.72.
Graph for Question 7.

8. A group of students do a rolling wheel experiment. They mark a spot on a wheel, roll the wheel along a tape measure, and collect data on the height of the spot. They model the relationship between height of the spot and distance rolled with a sinusoidal function whose graph is shown in **Figure 8.73.**

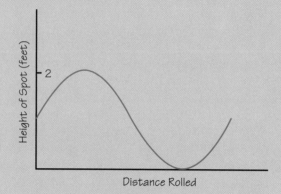

FIGURE 8.73.
Graph for Question 8.

a) What is the diameter of their wheel?

b) What is the wheel's circumference?

9. Ryan is watching his sister Hannah ride a carousel that turns once every 30 seconds. Hannah's distance from Ryan varies between 5 and 25 feet.

a) Sketch a graph of Hannah's distance from Ryan vs. time.

b) Identify your graph's period, amplitude, and axis of oscillation.

10. Tides are examples of periodic events in nature. The table in **Figure 8.74** has data of the depth of water at an ocean beach.

a) Create a scatterplot for these data.

b) Find a sinusoidal regression model for the relationship between depth and time. Graph the model on your scatterplot.

c) Use your model to predict the water's depth 11 hours after data collection began.

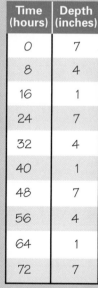

Time (hours)	Depth (inches)
0	7
8	4
16	1
24	7
32	4
40	1
48	7
56	4
64	1
72	7

FIGURE 8.74.
Data table for Question 10.

11. A group of students are trying to tune a bottle to a note that has a frequency of 311 cycles per second. They put water in the bottle and blow across it while collecting sound pressure data with a CBL. They estimate the period of the scatterplot to be 0.003. Should they add or remove water from their bottle?

12. **Figure 8.75** shows graphs of sound pressure waves collected from two sources.

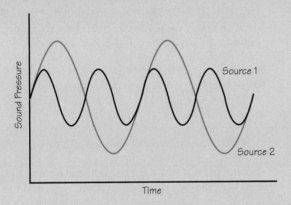

FIGURE 8.75.
Sound pressure graphs.

a) Which source is louder? How do you know?

b) Which source has the highest pitch? How do you know?

Glossary

KEY CONCEPTS

Amplitude (of a periodic function):
Half the vertical height of the basic repeating shape; that is, half of the vertical distance between maximum and minimum values of the periodic graph.

Axis of oscillation (of a periodic function): The center line between a periodic graph's peaks and troughs.

Central angle: An angle whose vertex is the center of a circle.

Cosine (cos) : In a right triangle, the ratio of the length of the side adjacent to an acute angle to the length of the hypotenuse; or $\cos = \frac{\text{adjacent}}{\text{hypotenuse}}$.

Cycle: One complete pattern.

Frequency (sound): The number of cycles a sound wave completes in one second.

Inverse sine: The process of working backwards to find an angle measure whose sine ratio is already known. It is written as $\sin^{-1}$. When you see $\sin^{-1}(0.5)$, it is the angle whose sine is 0.5.

Oscillation or oscillating pattern: Where data points consistently swing back and forth between two alternate extremes.

Period (of a periodic function): The shortest horizontal length of a basic repeating shape.

Periodic function: A function that repeats itself on intervals of a fixed length (equal to the period).

Periodic motion: Motion that follows a predictable back and forth pattern between two extremes. A swinging pendulum is an example of this type of motion.

Pitch: The highness or lowness of a tone.

Sine (sin): In a right triangle, the ratio of the length of the side opposite an acute angle and the length of the hypotenuse; or $\sin = \frac{\text{opposite}}{\text{hypotenuse}}$.

Sine function: A function that describes the relationship between the y-coordinate of a point on the unit circle and a central angle.

Sinusoidal function: Any function that can be expressed in the form $y = A \sin(B(x - C)) + D$, where neither A nor B is zero.

Sinsoidal regression: A method for finding a sinusoidal function for a periodic data set.

Tangent (tan): In a right triangle, the ratio of the length of the side opposite an acute angle and the length of the side adjacent to the acute angle; or $\tan = \frac{\text{opposite}}{\text{adjacent}}$.

Unit circle: Circle with radius one unit and centered at the origin of a rectangular coordinate system.

SPECIAL SECTION

Probability: What Are the Chances?

In 2002 Boeing Corporation found that fuel pumps on some airplanes were faulty. A company in California made the pumps for Boeing. The pumps were installed in large commercial jets during January and April. No serious incidents resulted from the faulty pumps, but all the planes that had pumps installed during those months had to be inspected.

The problem was that wires in some pumps were too close to a rotor. Friction with the rotor might create a spark that could cause a fuel explosion. Only a small percentage of the pumps were actually faulty. So most planes that had these pumps were not at risk. But how much risk was there for any particular plane? There are several important questions that could be asked:

❖ What is the chance that a particular pump had the misplaced wiring?

❖ If a pump had misplaced wiring, what was the chance of an explosion?

❖ Some large planes had several fuel tanks and pumps. What was the chance that one of the pumps on such a plane was faulty?

Answers to questions like these are based on **probability**. In this section you will learn to

❖ find the probability for a single event by doing an experiment.

❖ compute the probability for a combination of events.

❖ determine the probable cause of a failure event.

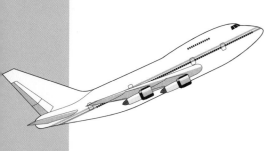

Determining the Probability

Any company that sells a product or service must be concerned about quality. If a product doesn't work well or has flaws, people will not buy it. Many companies have *quality control* departments. Items produced on assembly lines are inspected on a regular schedule by quality control inspectors or machines.

A manufacturing operation cannot produce perfect items 100% of the time. The inspection process determines what fraction of the total are defective and what fraction are not defective (or nondefective). These fractions also describe the probabilities that a single item is defective or not defective.

You may already know that a probability can sometimes be easily predicted. For example, the probability of spinning the number 1 (or any other number) on the spinner in **Figure 1** is $\frac{1}{4}$. This means that, on the average, we expect one-fourth of all spins to result in the number 1. The probability can also be expressed as a decimal, 0.25, and as a percent, 25%.

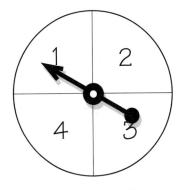

FIGURE 1. *Spinner.*

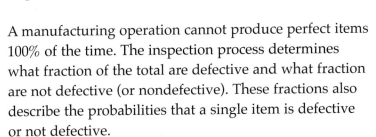

THEORETICAL PROBABILITY

The probability that a particular event will occur when an experiment is performed is found by counting all the different possible outcomes of the experiment:

Probability of an event = $\dfrac{\text{Total number of ways the event can occur}}{\text{Total number of equally likely outcomes}}$.

But sometimes theory and logic are unable to provide an answer. In such cases, only the collection of data can provide the information necessary to determine probabilities.

The Probability of a Defect

Materials: two different brands of jumbo paper clips, graph paper

1. Hold a paper clip, and open it up (see **Figure 2**). Did it break?

FIGURE 2.
An opened paper clip.

Paper clips are not intended to be opened all the way. But suppose that was their purpose. If one breaks, we could say it is defective. A quality control inspection would involve testing several clips to see if they break when opened.

2. A common sample size for inspection is five. Test five clips, and record the number of defective ones.

3. Now repeat your test on many more samples of size five. Record your results in a table similar to **Figure 3.** The third column should contain the total number of defective clips found in all samples up to that point. The fourth column should contain the percent of defective clips up to that point.

FIGURE 3.
Table for Question 3.

Sample Number	Number of Defects in Sample	Cumulative Number of Defects	Cumulative Percent of Defects
1			
2			
3			
etc.			

4. Construct a line graph of the percent of defects versus the total number of clips tested.

As the number of paper clips tested increases, your graph should begin to level off. The height of the graph as it levels off provides an approximate value for the probability that a given clip is defective. This type of probability is called an **experimental probability**.

5. Estimate the probability of a defective paper clip. Express your answer as a percent, a decimal, and a fraction.

6. Now repeat Questions 2–5 for a different brand of paper clip.

PROBABILITY NOTATION

A shorthand notation is often used to denote probabilities. The symbol P(A) stands for "the probability of event A" or "the probability that event A occurs." For example, the probability that a paper clip is defective can be written P(paper clip is defective).

You may have found that the probability of a defective paper clip was different for the two brands you tested. If so, then it is not meaningful to try to identify a single value for P(paper clip is defective). The probability in this case depends on the brand. When the probability of an event depends on whether another event occurs, it is called a **conditional probability**.

CONDITIONAL PROBABILITY

A conditional probability measures the likelihood of an event based on whether another event occurs.

The probability of event A, given that another event B occurs, is symbolized by P(A | B).

EXAMPLE 1

Two machines produce optical lenses for cameras (like the one in **Figure 4**). Of 200 lenses made by Machine 1, six are defective. Five of the 250 lenses produced by Machine 2 are defective.

a) Explain what is meant by P(defective lens | Machine 1), and find its value.

b) Explain what is meant by P(defective lens | Machine 2), and find its value.

FIGURE 4.
A camera lens.

SOLUTION:

a) P(defective lens | Machine 1) means the probability that a lens is defective if it was made by Machine 1. Its value is $\frac{6}{200} = 0.03$, or 3%.

b) P(defective lens | Machine 2) means the probability that a lens is defective if it was made by Machine 2. Its value is $\frac{5}{250} = 0.02$, or 2%.

Two events are called **independent** if the probability of one does not depend on the occurrence of the other. For example, we expect that when a coin is tossed there is a 50% chance that it will land heads up. The type of coin used should not affect the outcome. That is, P(heads | coin is a penny) = P(heads | coin is a dime) = P(heads | coin is a quarter) = 50%, or $\frac{1}{2}$. Whether a coin lands heads or tails is independent of the type of coin used.

THE COMPLEMENT OF AN EVENT

An **event** is a particular outcome of an experiment or process. The **complement** of the event is all *other* possible outcomes. Here are a few examples:

❖ The complement of heads is tails for the tossing of a coin.

❖ The complement of winning a game is tying or losing.

❖ The complement of the door is open is the door is closed.

❖ The complement of red is green or yellow for a traffic light.

❖ The complement of even is odd for the rolling of a six-sided die.

❖ The complement of one or two is three, four, five, or six for the rolling of a six-sided die.

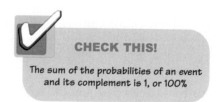

CHECK THIS!

The sum of the probabilities of an event and its complement is 1, or 100%

EXAMPLE 2

A manufacturing process makes circuit boards for computers (like the one in **Figure 5**). The probability that a circuit board made by this process is defective is 0.001. What is the probability that a circuit board is *not* defective?

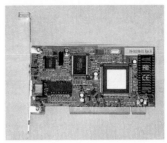

SOLUTION:

A circuit board is either defective or it is not defective. So nondefective is the complement of defective.

P(nondefective) = 1 − P(defective) = 1 − 0.001 = 0.999.

Assignment

1. A parts cabinet in an electronics laboratory has a drawer that contains resistors of various sizes, all mixed together. The drawer's contents include 40 one hundred ohm resistors, 20 two hundred ohm resistors, and 30 five hundred ohm resistors. If a student reaches into the drawer and randomly picks one, what is the probability that it will be a one hundred ohm resistor?

2. The median of a numerical data set is the middle number of the set when it is arranged in numerical order. In a large data set, what is the probability of randomly choosing a value that is below the median?

3. Veteran's Day is one of the few non-religious holidays that is celebrated on its original date (November 11), rather than always being on a Monday. In any arbitrary year, what is the probability that Veteran's Day will be on a weekend?

4. If you randomly guess at the answer to a multiple choice test question for which there are five choices, what is the probability that you will guess correctly?

5. According to a NASA space scientist, a city could be destroyed once every 30,000 years by an asteroid hitting the Earth. If he is correct, what is the probability of a city being destroyed by an asteroid in any year?

6. When the euro coin was issued, a German newspaper reported that a coin-flipping experiment with the Belgian one-euro coin showed that heads came up 140 times out of 250. Because of this, a ban on using euros at the start of soccer matches was considered. According to the experiment, what is the probability that a flipped euro results in heads?

7. Design and conduct an experiment to determine the experimental probability that an ordinary thumbtack will land with its point facing up when dropped on a horizontal surface.

8. When a penny is tossed, we expect that the probabilities of heads and tails are equal, at 50% each. And if the coin is actually tossed many times, the observed fraction of heads (or tails) should approach 0.5. Will the same results hold true if the penny is spun instead of tossed? Is there any reason to think that the fraction of heads might not be 50% in this case? You can experiment to find out.

   a) Use one finger to hold a penny vertically so its edge touches a table. "Flick" the penny with a finger of your other hand so that it spins. Notice which side lands face up. Repeat for a total of at least 100 spins and record your results. How many times did the penny land with heads face up?

   b) Based on your experiment, what is the conditional probability that a penny will land heads up if it is spun?

   c) What is the conditional probability that a penny will land heads up if it is flipped? See if you can verify your answer experimentally.

9. Some materials must be flexible in order to perform correctly under normal use. For example, the steel in a bridge must be flexible enough to bend, but not break, when subjected to stresses. The battery compartment to a calculator has a flexible plastic clasp that can be pressed to open it.

   a) You can perform a flexibility test on a piece of pasta. Press the pasta firmly onto a table top so that 4 inches hang over the edge of the table. Press the free end of the pasta down until it is 3 inches below the edge of the table. Did it break?

   b) Repeat the experiment for many pieces of pasta. How many pieces of pasta can be bent downward 3 inches without breaking?

   c) Now repeat the experiment for similar pasta that has gotten wet but has been allowed to dry out.

   d) What is the conditional probability that a piece of pasta can be bent downward 3 inches without breaking if it has gotten wet?

e) What is the conditional probability that a piece of pasta can be bent downward 3 inches without breaking if it has *not* gotten wet?

10. Air travelers occasionally find that their luggage is missing when they arrive at their destination.

a) In May 2004, there were 199,788 passengers on major airlines in the United States whose checked baggage was missing. The total number of passengers on these airlines was 48,368,609. What was the probability that any one passenger's luggage was lost on a flight?

b) Do you think that a traveler's chance of losing his or her luggage is independent of the choice of airline? What do you need to know in order to decide?

c) **Figure 6** shows May 2004 data on lost luggage for major airlines.

Airline	Number of Passengers	Number of Reports of Lost Luggage
JetBlue Airways	971,841	2115
Hawaiian Airlines	460,080	1304
Airtran Airways	1,136,947	3265
America West Airlines	1,741,982	5122
Southwest Airlines	7,321,279	22,473
Alaska Airlines	1,217,055	3801
Delta Air Lines	7,407,135	23,597
ATA Airlines	839,193	2708
Continental Airlines	2,786,245	9497
United Airlines	5,408,451	19,302
US Airways	3,202,200	12,883
Northwest Airlines	3,393,330	13,852
American Airlines	6,399,856	27,199
Expressjet Airlines	1,067,563	5815
Skywest Airlines	1,069,487	8487
American Eagle Airlines	1,267,581	10,618
Comair	1,110,230	9874
Atlantic Southeast Airlines	877,237	9601
Atlantic Coast Airlines	690,917	8275

FIGURE 6.
Data are from the July 2004 "Air Travel Consumer Report" of the Office of Aviation Enforcement and Proceedings, Aviation Consumer Protection Division.

Find the conditional probabilities of losing luggage for travel on JetBlue Airways and for travel on Atlantic Coast Airlines. Round your answers to four decimal places.

d) Do you think whether an airline passenger lost luggage in May 2004 was independent of whether he or she flew on JetBlue or Atlantic Coast? Explain.

e) Do you think whether an airline passenger lost luggage in May 2004 was independent of whether he or she flew on Delta or ATA? Explain.

11. In 2000, for every 1000 licensed 16-year-old girls in the U.S., 175 got in car accidents. That same year, 210 of every 1000 licensed 16-year-old boys got in accidents.

a) Find the conditional probability that a licensed 16-year-old got into a car accident in 2000 if the driver was a girl.

b) Find the conditional probability that a licensed 16-year-old got into a car accident in 2000 if the driver was a boy.

12. A study of poor African-American children who grew up in Ypsilanti, Michigan in the 1960s looked at the effects of attending preschool. The study found that 71% of the children who went to preschool earned a high school diploma or GED. Of those who didn't go to preschool, 54% earned a high school diploma or GED. Use conditional probability notation to express these results.

13. The following probability statements involve late flights from U.S. airports during 2002. Write a sentence to explain each one.

a) P(late departure | Detroit) = 0.20

b) P(late departure | Boston) = 0.14

14. United Airlines had the best on-time arrival record of the major U.S. airlines in 2002. The probability of a United flight being on time was 0.84. What was the probability that a United flight was *not* on time in 2002?

15. An injection molder is used to make plastic bicycle reflectors. If the probability of a defective reflector is 0.13, what is the probability that a reflector is not defective?

16. Many companies use automated telephone voice systems to provide their customers with account information. A study was made of such systems at Massachusetts utility companies in 2003. For one gas company, the probability that the automated system provided a caller with the desired information was 0.932. What was the probability that the system failed to provide the desired information?

17. Write your results from Part 1 using conditional probability notation.

Questions 18–21 involve the concept of product *reliability*. The reliability of a product is defined as the probability that it will perform its intended function correctly for a specified time interval, under normal conditions of use.

18. A control valve for a water feed line has a one-year reliability of 0.92. What is the probability that the valve will fail sometime during the year?

19. The probability that a hair-dryer's blower motor will fail during its first year of use is 21%. What is the one-year reliability for the blower motor?

20. The probability that a typical thermocouple will fail during 1000 hours of use is 1%. What is the 1000-hour reliability of such a thermocouple?

21. The 100-hour reliability of a steam turbine is 0.996. What is the probability that the steam turbine will fail within 100 hours?

Airline Pump Failures

The airplane fuel pump problem mentioned at the beginning of this section required inspection of 30,000 pumps. The initial study suggested that only 3% of the pumps may have had faulty wiring. So there was a 3%, or 0.03, probability that any one pump was faulty.

But many planes had more than one pump. Large planes like the Boeing 747 have fuel tanks in the fuselage and also in the wings, and sometimes in the tail section. And each tank may have several pumps. In fact, one plane can have as many as 17 or 18 fuel pumps.

If the probability of one pump being faulty is known, how can we find the probability that more than one pump in a plane might be faulty? If a plane has several pumps, what are the chances that all are faulty, or that none are faulty? To explore these questions, we will first consider a situation in which a plane has two pumps. And to make it easier to visualize, we will assume a high defect rate.

How Many Faulty Pumps?

Materials: graph paper, ruler

Imagine that a pump manufacturer has such poor quality control that one-fourth of its pumps are defective. Also imagine that a certain type of airplane is fitted with two fuel pumps. Such a plane might have no defective pumps. But it's possible that just one or even both pumps might be defective. You can use an area model to see the combined probabilities.

1. Draw a square to represent a total probability of 1, or 100%. Probabilities for all possible pump combinations will be contained in the square. Draw a vertical line that divides the square into two rectangles. One rectangle should contain $\frac{1}{4}$ of the total area. The other should contain $\frac{3}{4}$ of the total area. These rectangles represent the probabilities that Pump #1 is either defective or OK. Write appropriate labels along the top edge of the square. (Include the probability values.)

2. Now draw a horizontal line that divides the left edge of the square in a similar way. Write labels along the left edge to show the probabilities that Pump #2 is either defective or OK.

3. The square is now divided into four regions. The smallest one represents the event that both pumps are defective. Label each region according to the combination of pump conditions it represents.

4. Compute the areas of the regions and complete **Figure 7.**

FIGURE 7.
Table for Question 4.

State of the System	Probability
Both pumps defective	
Pump #1 defective, Pump #2 OK	
Pump #1 OK, Pump #2 defective	
Both pumps OK	

5. The probabilities in the table are called *joint probabilities*. Each of them is a combination of the probabilities of two events. In each case, the state of the system depends on two separate events *both* occurring. Look at the way you computed each joint probability. Then write a rule for finding the probability that two events both occur.

6. Suppose an airline flies 80 aircraft of this type. Use your answer to Question 4 to help you complete **Figure 8.** Fill in the number of airplanes in each category.

		Fuel Pump #1	
		OK	Defective
Fuel Pump #2	OK		
	Defective		

FIGURE 8.
Table for Question 6.

7. How many of the 80 planes have only one defective fuel pump?

8. What is the probability that a plane has exactly one defective pump?

9. Explain how you could use the information in your answer to Question 4 to find the probability that a plane has exactly one defective pump.

INDEPENDENT AND DEPENDENT EVENTS

In Part 2 you found that the joint probability that both fuel pumps fail could be found by multiplying the failure probabilities for the individual pumps. In fact, even for a plane that contains more than two fuel pumps, the probability that all of them fail is equal to the product of the individual failure probabilities.

But this assumes that fuel pump failures are independent events. That is, the probability that one of the pumps fails does not depend on whether the other pump fails. For example, if failure of Pump #1 made Pump #2 more (or less) likely to fail, the events would be *dependent*. In such a case, a conditional probability would have to be used for the failure probability of Pump #2. And the probability that both pumps fail would be found from

P(both pumps fail) = P(Pump #1 fails) × P(Pump #2 fails | Pump #1 fails).

EXAMPLE 3

According to the Asthma and Allergy Foundation of America, one-fifth of all Americans have allergies of some kind. But if a person has one parent with allergies, that person has a one-third chance of being allergic.

a) What is the probability that two randomly selected Americans have allergies?

b) What is the probability that four randomly selected Americans all have allergies?

c) What is the probability that a mother and child both have allergies?

SOLUTION:

a) If the two people are randomly selected, then we can assume that whether each has allergies or not are independent events. That is, the probability that one person is allergic does not depend on whether the other is allergic. So the joint probability that they both have allergies is

P(first person has allergies) × P(second person has allergies) = $\left(\frac{1}{5}\right)\left(\frac{1}{5}\right) = \frac{1}{25}$ or 4%.

b) Assume independent events:

P(four people have allergies) = $\left(\frac{1}{5}\right)\left(\frac{1}{5}\right)\left(\frac{1}{5}\right)\left(\frac{1}{5}\right) = \frac{1}{625}$ or 0.16%.

c) These are dependent events. A child is more likely to have allergies if his or her mother has allergies.

P(mother and child have allergies)

= P(mother has allergies) × P(child has allergies if mother has allergies)

= $\left(\frac{1}{5}\right)\left(\frac{1}{3}\right)$

= $\frac{1}{15}$ or about 6.7%.

EXAMPLE 4

Studies have been made of human error rates for simple tasks. For example, a computer technician will insert a printed circuit board incorrectly about 0.4% of the time. If a technician has to replace 20 circuit boards, what is the probability that at least one of the 20 will be incorrectly inserted?

SOLUTION:

The events "at least one of 20 incorrectly inserted" and "all 20 correctly inserted" are complementary. And if there is a 0.4% chance of incorrect insertion, there is a 99.6% chance of correct insertion. So

P(at least one of 20 incorrectly inserted) = 1 – P(all 20 correctly inserted)

$= 1 - P(\text{one correctly inserted})^{20}$

$= 1 - (0.996)^{20}$

$\approx 1 - 0.923$

$= 0.077$ or 7.7%.

In this solution it is assumed that correct insertions of the 20 boards are all independent.

ADDING PROBABILITIES

In Question 9 of Part 2, you found that the chance of only one pump being defective is equal to the sum of two probabilities. In many cases, the joint probability that one or the other of two events occurs is equal to the sum of their individual probabilities. But this is only true if the events *cannot both occur*. Such events are called **mutually exclusive** events.

In general, the probability that any one of many events occurs is equal to the sum of their individual probabilities, provided they are all mutually exclusive.

Number of Days	Probability
5	0.02
6	0.04
7	0.11
8	0.22
9	0.25
10	0.19
11	0.13
12	0.04

FIGURE 9.
Slate roof installation.

EXAMPLE 5

A project scheduler for a roofing company has data on the number of days it takes to install a slate roof on a certain type of building. (See **Figure 9.**)

What is the probability that such a roof will take more than 9 days to install?

SOLUTION:

Only one length of time is possible for a particular installation. For example, it is impossible that a particular roof installation will take both 10 days *and* 11 days. All of the possibilities in the table are mutually exclusive.

P(more than 9 days) = P(10 days) + P(11 days) + P(12 days)

= 0.19 + 0.13 + 0.04

= 0.36

CHECK THIS!

Rules for Combining Probabilities

P(A and B) = P(A) × P(B | A) But if events A and B are independent, P(A and B) = P(A) × P(B). In general, to find the joint probability that any number of independent events all occur, multiply their individual probabilities.

P(A or B) = P(A) + P(B) only if events A and B are mutually exclusive. In general, to find the probability that any one of a number of mutually exclusive events occurs, add their individual probabilities.

Assignment

1. **Figure 10** shows data on the performance of two lathes in a machine shop.

	Total Number of Pieces Produced	Number of Defective Pieces Produced
Lathe #1	500	40
Lathe #2	300	30

FIGURE 10.
Lathe data.

a) Find the probability that a randomly selected piece was produced on Lathe #1.

b) Find the probability that a piece produced on Lathe #1 was defective.

c) Find the probability that a randomly selected piece was produced on Lathe #1 and was also defective.

d) Repeat a–c for pieces produced on Lathe #2.

e) Are the choice of lathe and the occurrence of a defective piece independent or dependent?

2. **Figure 11** shows data on items made on two assembly lines in a manufacturing plant.

	Total Number of Pieces Produced	Number of Defective Pieces Produced
Line #1	7520	188
Line #2	10,680	267

FIGURE 11.
Assembly line data.

a) Find the probability that a randomly selected item was produced on Line #1.

b) Find the probability that an item produced on Line #1 was defective.

c) Find the probability that a randomly selected item was produced on Line #1 and was also defective.

d) Repeat a–c for items produced on Line #2.

e) Are the choice of assembly line and the occurrence of a defective item independent or dependent?

3. On January 28, 1986, the space shuttle *Challenger* exploded over Florida. Many authorities believe this tragedy could have been prevented if closer attention had been paid to the laws of probability. The rocket that carried the shuttle aloft was separated into sections that were sealed by large rubber O-rings. Experts believe that at least one of the O-rings leaked burning gases that caused the rocket to explode. Studies after the tragedy found that the probability a single O-ring would work was 0.977. If the six O-rings were truly independent of one another, what is the probability that all six would function properly during the mission?

4. A particular structural beam in a building has a 0.02 probability of failure when subjected to a load of a certain critical size. If the probability of such a load occurring during the design life of the building is one in 100, what is the probability of beam failure during the building's design life?

For Questions 5–9, assume that events are independent.

5. In 2002, 85% of Massachusetts high school students passed the MCAS mathematics test on their first attempt. Of four randomly selected test-takers, what is the probability that all passed the test on their first attempt?

6. One type of computer modem has a one-year reliability of 98%. What is the probability that five such modems will all work correctly for one year?

7. A radio receiver consists of five assemblies. The 100-hour reliability of each is given in **Figure 12.**

Each component assembly must function correctly for the receiver to work. (This kind of system is called a "series" system.) What is the 100-hour reliability of the receiver?

8. A simple electronic circuit consists of 4 transistors, 3 capacitors, 5 resistors, and 1 switch. The reliabilities for each type of component are listed in **Figure 13.**

All components must operate for the circuit to work. What is the 1000-hour reliability for the circuit?

Assembly	Reliability
Power supply	0.892
rf stage	0.943
Mixer	0.961
if stage	0.907
Power output	0.939

FIGURE 12.
Data for Question 7.

Component	1000-hr Reliability
Transistor	0.9990
Capacitor	0.9998
Resistor	0.9950
Switch	0.9980

FIGURE 13.
Data for Question 8.

9. Four electronic components are wired in series. Their 100-hour reliabilities are 0.98, 0.995, 0.96, and 0.975. What is the 100-hour reliability of the system; that is, what is the probability that all components will work for 100 hours?

10. In Part 1 you compared the quality of two brands of paper clips. For that type of test, suppose 20% of the Brand A paper clips and 45% of Brand B are defective. Also suppose 100 Brand A clips and 150 Brand B clips are mixed together. A randomly selected paper clip is tested.

    a) What is the probability that the selected paper clip is Brand A and it breaks?

    b) What is the probability that the selected paper clip is Brand B and it breaks?

    c) What is the total probability that the selected paper clip breaks?

11. **Figure 14** shows the U.S. car production shares of major companies for the first half of 2004.

    a) What is the probability that a randomly selected car produced in the U.S. in the first half of 2004 was made by Toyota?

    b) What is the probability that a randomly selected car produced in the U.S. in the first half of 2004 was made either by Ford or General Motors?

    c) Are the events "made by Ford" and "made by General Motors" mutually exclusive? Explain.

Company	Percent of U.S. Cars
Daimler Chrysler	8.7
Ford	18.6
General Motors	28.7
Honda	13.0
Nissan	7.9
Toyota	10.6
Other	12.5

FIGURE 14.
Data are from the Automotive News Website.

12. **Figure 15** shows data on U.S. traffic fatalities for 2002.

    a) What is the probability that a randomly selected traffic fatality was either under 16 or over 54 years old?

    b) Are the events "person is under 16" and "person is over 54" mutually exclusive? Explain.

    c) Are the events "person is under 25 years old" and "person is from 16 to 24 years old" mutually exclusive? Explain.

Age Group	Percent of Fatalities
Under 16	6
16–24	25
25–54	45
Over 54	24

FIGURE 15.
Data are from the National Highway Transportation Safety Administration.

13. A quality control inspector checks electrical transformers for defects. **Figure 16** shows the probabilities of different numbers of defects.

Number of Defects Observed	Probability
0	0.03
1	0.07
2	0.18
3	0.23
4	0.16
5	0.14
6	0.09
7	0.06
8	0.04

FIGURE 16.
Data for Question 13.

a) What is the probability that a transformer has fewer than three defects?

b) What is the probability that a transformer has at least five defects?

14. The probabilities of a randomly selected blood sample in the New York Blood Center being type O, type A, type B, or type AB are 0.45, 0.40, 0.10, and 0.05, respectively. What is the probability that a sample is either type A or type B?

15. A student takes a ten-question multiple choice quiz. Each question has five choices.

a) If the student guesses each answer, what is the probability that the student will get all the questions wrong?

b) What is the probability that the student will get all the questions correct?

16. A company that makes refrigerator ice-making units has a 4% defect rate. What is the probability that a quality control inspector might examine a sample of five randomly selected units and find no defectives in the sample?

17. Two jeeps are used on a military mission that is expected to take 150 hours. The motor in each jeep has a 150-hour reliability of 0.94. The mission can succeed provided at least one of the jeeps completes the mission. What is the probability that the motor in at least one of the two jeeps lasts for the entire mission?

18. Hinges are stamped from a punch press in a process that has a 10% defect rate. What is the probability that a random sample of five hinges contains no defective hinges?

19. Work studies suggest that there is a probability of 0.002 that a worker will incorrectly calibrate a dial. Assume that a certain technician has to calibrate a dial on a meter at the start of each workday. In a month that has 21 workdays, what is the probability that the dial will be incorrectly calibrated at least once?

20. An air traffic control system has three identical central processing units (CPUs). Each unit has a 5% probability of failure in a year. The system will function normally unless all three CPUs fail. What is the one-year reliability of the system?

21. A McDonnell Douglas MD-11 airplane has 17 fuel pumps. Assume each pump has a 3% probability of being defective.

    a) What is the probability that all of the fuel pumps on one MD-11 are defective?

    b) What is the probability that at least one of the pumps is defective?

22. The fuel pumps used by Boeing were thought to have a 3% chance of having faulty wiring. Consider a plane with two such pumps.

    a) What is the probability that both pumps are faulty?

    b) What is the probability that neither pump is faulty?

    c) What is the probability that exactly one of the pumps is faulty?

23. Consider an airplane with three fuel pumps. Assume each has a 3% chance of having faulty wiring.

    a) What is the probability that all three pumps are faulty?

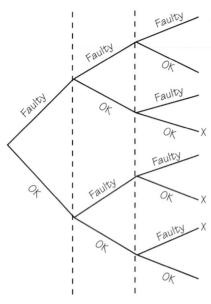

First pump Second pump Third pump

FIGURE 17.
Tree diagram.

b) What is the probability that none of the pumps are faulty?

c) Finding the probability that only one of the pumps is faulty is a little more involved than in the case of two pumps. The various combinations that include one faulty pump and two OK pumps must be counted. A tree diagram can be of help in the counting. **Figure 17** shows a tree in which each stage of branches represents one of the pumps.

Each path through the tree that is marked with an X represents a case where one pump is faulty, with probability 3%, and two pumps are OK, with probability 97%. Find the total probability that exactly one pump is faulty.

d) Use the tree diagram in Figure 17 to help you find the probability that exactly two of the three pumps are faulty.

24. When a basketball player is fouled, he or she shoots free throws. Over the years, there have been several different formats for awarding free throws. Use your knowledge of probability rules to determine the probability that an 80% free throw shooter will make two free throws in each of these formats. (Assume that free throws are independent.)

a) A player fouled in the act of shooting is awarded two free throws.

b) A player is allowed one free throw, and then can try another only if the first one is made. (This is called a one-and-one situation, and is currently not used above the college level.) It will be helpful if you first list all of the possible outcomes for this case.

c) A player is given three chances to make two shots. If the first two are made, there is no third shot. (This format was used in the NBA for a few years and then discarded.) A tree diagram may be helpful.

What is the Cause?

Charles De Gaulle International Airport is near Paris, France. A section of a new terminal at the airport collapsed in May 2004. Several people were killed, and the terminal was closed.

It is often very difficult to find the cause of such a major disaster. But a knowledge of probability can sometimes be helpful.

New Cracks Stop Search At Terminal After Collapse
(*New York Times*, May 25, 2004)

PARIS, May 24 – Cracking noises and new fissures at an airport terminal at Charles De Gaulle International Airport here Monday interrupted the search for survivors in a section of the building that collapsed early Sunday.

With more cracks discovered at the terminal on Monday, airport officials suggested that much of the $900 million structure may have to come down, forcing the airport to shift as many as 25,000 passengers a day to other terminals for months, if not years, to come.

While it is too early to say whether a design flaw, engineering error or construction mistake caused the collapse, architects and engineers asked whether the fashion for increasingly innovative buildings has strained the limits of what a group of specialized construction companies can safely build together.

Critics warn that the increasingly extreme designs rely on complex engineering that creates many opportunities for critical errors in implementation. About 400 contractors were involved in the construction of Terminal 2E.

The collapsed terminal's design was hailed as groundbreaking when it opened last year, using technology developed for concrete-lined tunnels adapted to be free-standing. The graceful elliptical shell was built in prefabricated 12-foot sections by the French firm GTM Construction and installed on reinforced concrete pylons by the company Hervé.

Each 60-ton section rested on four pylons, and at least seven of the sections collapsed.

One engineer who worked on the project told French television that the pylons had shown signs of weakening before the accident, supporting a hypothesis that the pylons were not sufficient to bear the load placed upon them, or that shifting soil may have played a part in the collapse. Guy Nordenson, a structural engineer in New York who helped evaluate the causes of the World Trade Center collapse, said that in most building collapses there is a "cascade of problems" that could come from design or construction errors or omissions.

"Often what appears to be at the beginning the cause that triggered the problem is, in the end, not the real one," said Roberto Leon, a professor of structural engineering at the Georgia Institute of Technology. In most cases, he said, the underlying cause of a building collapse comes down to improper construction or quality control rather than the initial design.

Probable Cause

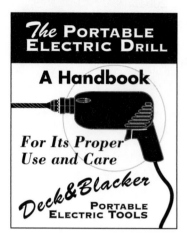

Materials: none

If a piece of equipment is misused, it could wear out sooner than it should. Suppose you lend a one-year-old electric drill to a friend, and it is returned with a broken motor. Maybe the drill was old and the motor was ready to go. But your friend might have overloaded the motor by trying to drill into something that was too hard for the drill to penetrate. How can you know what caused the failure?

In order to analyze the drill's failure, you would need information about the drill and the way people use drills. Suppose that research on a sample of 200 drills shows that:

❖ Under normal use, only 5% of drill motors wear out during the first year.

❖ Of drills that are overloaded, 30% of their motors wear out during the first year.

❖ 10% of users overload their drills.

1. A tree diagram can be used to picture this situation. The box at the left contains the sample size. The upper branch in **Figure 18** represents overloaded drills. The lower branch represents drills that weren't overloaded.

   Fill in the boxes with the number of drills in each category.

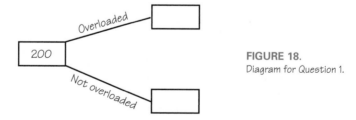

FIGURE 18.
Diagram for Question 1.

2. Consider only the overloaded drills. **Figure 19** shows a second stage of tree branches. In this stage, the branching factor is whether or not a drill fails within a year. The upper branch represents the overloaded drills that fail. The lower

branch represents the overloaded drills that don't fail. Use the fact that 30% of overloaded drills fail to fill in the boxes at the end of the branches.

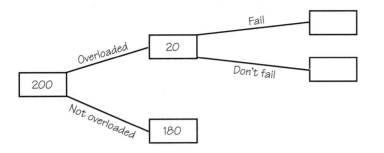

FIGURE 19.
Diagram for Question 2.

3. **Figure 20** completes the second stage of branches. Fill in the boxes at the ends of these branches. They represent the results for drills that are *not* overloaded.

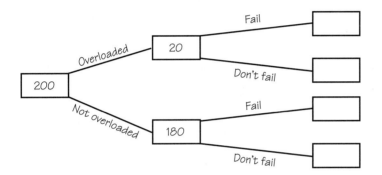

FIGURE 20.
Diagram for Question 3.

4. Altogether, how many drills fail?

5. How many of the drills that fail were overloaded?

6. If a certain drill has failed, what is the probability that it was overloaded?

7. Your tree can be redrawn as a probability tree. Each branch is labeled with the probability of a drill being on that branch. For example, the upper branch in Stage 1 is labeled P(O), for the probability that a drill is overloaded. The top branch in Stage 2 is P(F | O), for the conditional probability that a drill fails if it was overloaded (see **Figure 21**). Label the remaining branches, along with the probability values.

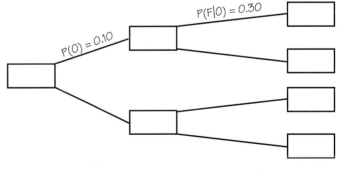

FIGURE 21.
Diagram for Question 7.

8. The tree contains four different paths. Each one begins at the box on the left and ends on one of the boxes on the right. The probability that a drill follows the top path is a joint probability. It is the probability that a drill is overloaded *and* it fails. This probability is found from the product $P(O) \times P(O|F)$. What is this probability?

9. Find the joint probabilities for each of the four paths. Write each value next to the appropriate box.

10. You can use your answer to Question 9 to calculate the probability that a failure was caused by an overload. First, find the total probability of failure by adding the probabilities at the ends of paths that include failure.

11. If your drill was returned broken, then you know it followed one of these two paths. To find the probability that it was overloaded, divide the probability of following the overloaded and failed path by your answer to Question 10.

Notice that there are two ways to use a tree diagram that give the same result. The probability that the drill failure was due to an overload was found using only probabilities and also by using a fixed initial sample size. In fact, *any* sample size could have been placed in the leftmost box. The result would have been the same.

CONSTRUCTING TREE DIAGRAMS

When tree diagrams are used to investigate the probable cause of an accident or other failure, a few guidelines should be followed:

❖ The tree branches out from left to right;

❖ The branching factor for Stage 1 is based on the conditions preceding the failure;

❖ The branching factor for Stage 2 is whether or not failure occurred;

❖ The sum of the probabilities for all branches that leave any branch point is 1;

❖ The sum of the probabilities for all complete paths is 1.

EXAMPLE 6

A production line in a parts factory operates around the clock. There are three shifts of workers: day, evening, and night. The day shift makes 40% of the parts. Each of the other two shifts makes 30% of the parts. Only 1% of the parts made on the day shift are defective. The evening shift has a 2% defect rate, and the night shift produces 5% defects.

Suppose the parts from all three shifts are mixed together.

a) If a randomly selected part is defective, what is the probability that it is from the night shift?

b) If a randomly selected part is not defective, what is the probability that it is from the night shift?

SOLUTION:

a) The tree diagram in **Figure 22** represents this situation.

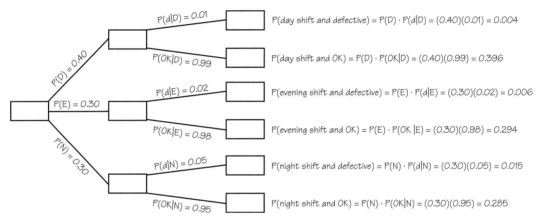

FIGURE 22.
Tree diagram.

In the figure, P(D), P(E), and P(N) represent the probabilities that a part is made by the day shift, evening shift, or night shift, respectively. P(d | D) represents the conditional probability that a part is defective if it is made by the day shift, etc.

If a part is known to be defective, then it must have followed one of the three paths that are circled in **Figure 23.**

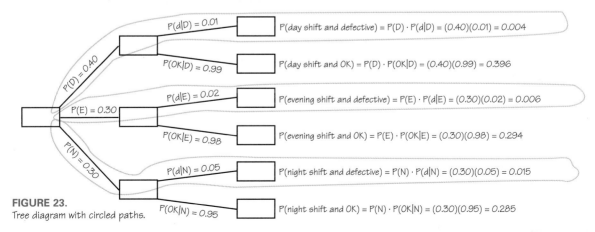

FIGURE 23.
Tree diagram with circled paths.

The total probability that a part is defective is 0.004 + 0.006 + 0.015 = 0.025. The amount contributed to this total by the night shift is 0.015. So the fraction of defective parts that were made on the night shift is

$$\frac{P(\text{night shift and defective})}{P(\text{defective})} = \frac{0.015}{0.025} = 0.60, \text{ or } 60\%.$$

The probability that a particular defective part was made on the night shift is 60%.

The same result could be found by assuming a sample size of, say, 2000 parts (see **Figure 24**).

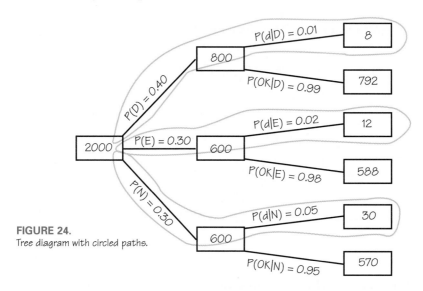

FIGURE 24.
Tree diagram with circled paths.

The probability that a particular defective part was made on the night shift is

$$\frac{\text{Number of defectives produced on the night shift}}{\text{Total number of defectives produced}} = \frac{30}{8+12+30} = \frac{3}{5} \text{ or } 0.60.$$

b) The other three paths represent nondefective parts. The probability that a nondefective part came from the night shift is

$$\frac{P(\text{night shift and OK})}{P(\text{OK})} = \frac{0.285}{0.396+0.294+0.285} \approx 0.29, \text{ or } 29\%.$$

Assignment

1. A particular brand of electric hedge clippers is intended for use on branches no larger than $\frac{1}{4}$-inch diameter. They carry a one-year warranty. If used as intended, their one-year reliability is 92%. If used on branches larger than $\frac{1}{4}$-inch, there is a 60% probability of failure. Only 20% of owners use the clippers on branches larger than $\frac{1}{4}$-inch diameter.

    a) Draw a tree diagram for this situation. Use a sample size of 400. Label probabilities on the branches, and fill in all boxes.

    b) Of 400 sets of clippers, what is the total number of failures expected in a year?

    c) What is the probability that a particular set of failed hedge clippers was used on branches larger than $\frac{1}{4}$-inch?

2. A parts bin contains 500 switches. One-fourth of the switches are from Vendor A, and the rest are from Vendor B. The switches are mixed randomly. The defect rate for Vendor A is 16%, and the defect rate for Vendor B is 5%.

    a) Draw a tree diagram for this situation.

    b) If a particular switch from the bin is defective, what is the probability that it came from Vendor A?

3. Suppose that, on the average, you do your math homework four out of every five nights, or 80% of the time. Also suppose your math teacher gives you a short quiz daily. If you do your homework the night before a quiz, the probability that you get a good score on the quiz is 5/6. If you don't do your homework, the probability that you get a good score on the quiz is only 1/3.

    a) Draw a tree diagram for this situation for a school year of 180 days.

    b) Your math teacher grades one of your quizzes, and finds that you got a good score. What is the probability that you did your homework last night?

4. A welding machine is used for welding the parts of chairs. If a weld is properly made, it will fail under stress with a probability of 0.002. But 3% of the welds are improper, and as a result are weak. A weak weld will fail when stressed with probability 0.30.

   a) Draw a tree diagram for this situation.

   b) If a weld fails under stress, what is the probability that it was improperly made?

5. The mortar (a cement and water mixture) that is used to bond bricks in a wall, staircase, or chimney slowly wears away. When this happens, the bricks are repointed, with fresh mortar applied after loose mortar is removed. Assume that:

   ❖ Mortar that is properly mixed and cured fails within one year with a probability of 0.03.

   ❖ Mortar that is mixed with the wrong proportions fails within one year with a probability of 0.75.

   ❖ Mortar that is improperly cured fails within one year with a probability of 0.50.

   ❖ Of all home owners that attempt brick repointing themselves,

      ❖ 50% do it correctly;

      ❖ 20% don't mix the mortar with the right proportions;

      ❖ 30% don't properly cure the mortar.

   a) Draw a tree diagram for this situation.

   b) If the mortar fails within a year, what is the probability that it was mixed in the wrong proportions?

   c) If the mortar fails within a year, what is the probability that it was improperly cured?

6. Tests for medical conditions and drug usage are usually not 100% accurate. And even if a test has a high accuracy, the results can be misleading.

One test is often used to test for a type of cancer that occurs only in men. It is approximately 80% accurate in identifying the presence of cancer. In other words, if a man has cancer, there is a 0.80 probability that he will test positive. This is called a true positive test. On the other hand, if he doesn't have cancer, the test will be negative (and correct) about 97% of the time.

a) If a test on a healthy person turns out positive for a disease, the result is called a false positive. What is the probability that a healthy man will receive a false positive for this test?

b) Of all the men who take the test, 4% have cancer. What is the probability that a man who takes the test has cancer and has a true positive test?

c) Draw a tree diagram for this situation for a sample of 1000 men. The branching factor for Stage 1 should be whether or not a man has cancer. The branching factor for Stage 2 should be whether the test is positive or negative.

d) What is the probability that a man who takes the test doesn't have cancer and tests positive (a false positive)?

e) Out of every 1000 men who take the test, how many will have a true positive?

f) Out of every 1000 men who take the test, how many will have a false positive?

g) What is the probability that a man who tests positive for cancer on this test is healthy?

7. Some areas of the United States have levels of arsenic in the drinking water that pose health risks. Suppose that a test for arsenic in the water supply correctly reports the presence of arsenic with probability 0.95, and correctly reports the absence of arsenic with probability 0.82. In reality, only about 4% of the drinking water in the United States is thought to contain arsenic.

If a water sample tests positive, what is the probability that it actually contains arsenic?

a) Complete a tree diagram for this situation, if a random sample of 5000 water supplies is tested.

b) Estimate the fraction of all tests that are positive.

c) Is the percentage of positive test results a good estimate for the percentage of water supplies in the United States that contain arsenic?

d) What is the probability that a water sample that tests positive contains arsenic?

8. A 2001 federal study of emissions testing equipment in Arizona found that many results were not correct. Test results for all cars in the study were compared with actual emissions levels as determined by more thorough measurements. **Figure 25** shows results similar to those found in the study.

		Actual Emissions Condition	
		Met standards	Didn't meet standards
Test	Passed	627	24
	Failed	384	18

a) Draw a tree diagram for this situation.

b) If a car passed the test, what was the probability that it met emissions standards?

c) If a car failed the test, what was the probability that it did not meet emissions standards?

9. In criminal trials in the United States, a person is innocent until proven guilty. Assume that a guilty verdict is like a positive test result. Also assume an acquittal (not guilty verdict) is like a negative test result.

a) For a criminal trial, what is a false positive?

b) What is a false negative?

c) For a criminal trial, which do you think is worse: a false positive or a false negative?

Practice Problems

These practice problems are taken from state tests. Many of these tests decide whether or not a student graduates from high school. Your experience in this module should help you do these problems.

1. Graduate students are selected for grants from a pool of applicants. Half of the students in the pool are selected and half are rejected. Because some students decline grants, the students who are rejected have a second chance. The probabilities of being selected and rejected are shown in the diagram below.

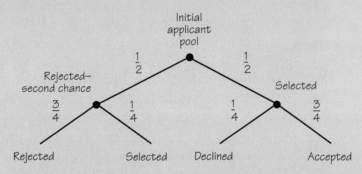

What is the probability that a student will be rejected in the first round and selected in the second round?

A. 1/8

B. 1/4

C. 3/8

D. 7/8

2. The two spinners at right are used for a math game. Each arrow has an equal chance of landing on any one number on its spinner. If both arrows land on even numbers, one point is given. Which of these fractions is the probability of getting two even numbers?

A. 5/25

B. 6/25

C. 10/25

D. 19/25

3. Cindy tossed a coin onto the playing board shown below.

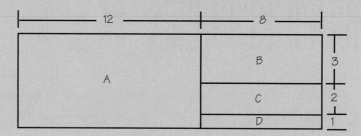

What is the probability the coin will land in region C?

4. A quality control engineer for a toy company tested 800 video games and found 3 defective. The company plans to make 500,000 video games this year. Based on the findings, how many can be expected to be defective?

A. 16

B. 20

C. 1875

D. 2000

5. Tara plays a game using 2 bags of game pieces. One bag has 6 blue game pieces and 6 red game pieces. The other bag has ten game pieces numbered 1 through 10. On her turn, Tara must draw one game piece from each bag. What is the probability that she draws a red piece and an even-numbered piece?

6. To get home from work, Curtis must get on one of the three highways that leave the city. He then has a choice of four roads that lead to his house. In the diagram below, each letter represents a highway, and each number represents a road.

		Highway		
		A	B	C
	1	A1	B1	C1
Road	2	A2	B2	C2
	3	A3	B3	C3
	4	A4	B4	C4

If Curtis randomly chooses a route to travel home, what is the probability that he will travel Highway B and Road 4?

A. 1/16

B. 1/12

C. 1/4

D. 1/3

7. What is the probability that the spinner will NOT stop on red if it is spun one time?

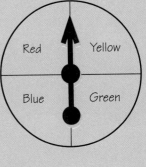

A. 1/4

B. 1/3

C. 3/4

D. 4/3

8. Mrs. Penner has 7 keys for her school on her key ring. All of the keys are different but look alike. One of these keys is for the bathroom and two are for her classroom. If she picks one key at random, what is the probability that the key is *not* for the bathroom or her classroom?

9. Jared has a white cube and a red cube. The surfaces of each cube are numbered with a unique number from 1 to 6. If Jared tosses the cubes, what is the probability he will get a 4 on the white cube and an even number on the red cube?

A. 1/12

B. 1/3

C. 1/2

D. 2/3

10. Pedro designed a board game with a spinner that is divided into four equal sections: 1, 2, 3, and "Lose a Turn." He wanted to determine how many times a player would spin "Lose a Turn" in 10 spins. He conducted a simulation where one trial consisted of 10 spins. The table at right shows his results for 50 trials.

Based on these results, what is the probability that a player will spin "Lose a Turn" 3 or more times in 10 spins?

A. 7/50

B. 15/50

C. 22/50

D. 43/50

Simulation Results

Number of "Lose a Turn" Spins	Frequency
0	5
1	12
2	11
3	15
4	4
5	2
6	0
7	1
8	0
9	0
10	0

11. Lani had a box that contained

  ❖ 1 blue marble;

  ❖ 1 green marble;

  ❖ 1 purple marble;

  ❖ 1 yellow marble; and

  ❖ 2 red marbles.

  Lani removed one marble without looking, and she recorded the result. She placed the marble back in the box and repeated the procedure one more time. What is the probability that Lani removed a red marble followed by a blue marble?

  A. 1/36

  B. 1/18

  C. 1/3

  D. 1/2

12. Weatherpersons predict tomorrow's weather based on what has happened in the past. During the past 50 years, there have been 380 days that have been just like today, and of those, 200 have been followed by a clear day. Which of these is the approximate probability of a clear day tomorrow?

  A. 13%

  B. 34%

  C. 53%

  D. 66%

13. Anita and Ajay play a game of chess. The probability that Anita will win is 48%, and the probability that Ajay will win is 42%. What is the probability that this game will end in a tie?

  A. 6%

  B. 10%

  C. 52%

  D. 58%

14. Jeremiah is doing an experiment in his math class. He flips four pennies in the air. What is most likely to happen?

    A. Two of the pennies will be heads and two will be tails.

    B. Three of the pennies will be heads and one will be tails.

    C. All four pennies will be heads.

    D. None of the pennies will be heads.

15. When Joe bowls, he gets a strike (knocks down all the pins) 60% of the time. How many times more likely is it for Joe to bowl *at least* three strikes out of four tries as it is for him to bowl zero strikes out of four tries? Round your answer to the *nearest whole number*.

16. The rectangular garden shown at right has a rectangular, brick walkway.

    What is the probability that a seed tossed randomly into the garden will land on the walkway?

    A. 1/6

    B. 1/5

    C. 1/4

    D. 1/3

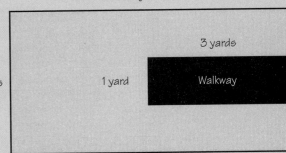

17. If a number is selected at random from the set {−3, −2, −1, 0, 1, 2, 3}, what is the probability that it is odd?

18. A computer simulated tossing 3 coins 400 times. The results are shown at right.

    a) Calculate the experimental probability of tossing 2 heads and a tail as shown by this simulation.

    b) Find the theoretical probability of tossing 2 heads and a tail.

    c) Compare the two probabilities and explain any differences.

HHH	41
HHT	54
HTH	48
THH	57
TTH	50
THT	53
HTT	45
TTT	52

19. What is the probability of a team losing a game if the probability of the team winning a game is 0.735?

20. A bag contained four green balls, three red balls, and two purple balls. Jason removed one purple ball from the bag and did not put the ball back in the bag. He then randomly removed another ball from the bag. What is the probability that the second ball Jason removed was purple?

A. 1/36

B. 1/9

C. 1/8

D. 2/9

21. A spinner has spaces labeled X, Y, and Z. The spinner was spun 20 times. The results are shown in the table below.

Letter	Number of Times Spinner Landed on Letter
X	4
Y	3
Z	13

According to these data, what is the experimental probability that the spinner will land on Y?

A. 3/20

B. 4/20

C. 13/20

D. 17/20

22. The table below shows information about the students in Ms. Murphy's algebra class.

	Male	Female
Rides the bus	6	8
Does not ride the bus	9	5

What is the probability that a randomly selected male student does <u>not</u> ride the bus to school?

A. 9/28

B. 9/14

C. 15/28

D. 9/15

23. The probability that Caitlin gets an A on a math test is 0.4. Find the probability that she earns an A on exactly two of three math tests.

24. Janet is playing a game using the two spinners shown below. She will spin the arrow on each spinner once and will move a specified number of steps forward or backward according to the results.

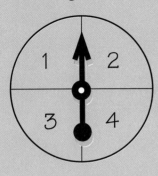

Spinner 1   Spinner 2

What is the probability that Janet will have to move backward **less than** 4 steps?

A. 1/8

B. 3/8

C. 1/2

D. 3/4

25. For quality control, a light bulb company conducted a random sampling of their light bulbs. The results are shown below.

Number of Defective Bulbs	Number of Nondefective Bulbs
4	96

The light bulb company makes 6000 light bulbs in a day. Based on this sample, how may defective light bulbs can the company expect to make in a day?

A. 240

B. 250

C. 1500

D. 2400

26. Captain Crunch contains one of 6 different prizes in each box of cereal. The prizes are distributed evenly and randomly. If two boxes of the cereal are purchased, what is the probability that the prizes are different?

    A. 2/3

    B. 5/6

    C. 6/5

    D. 1/6

27. Mark wants to compare theoretical and experimental probabilities. He plans to roll two cubes with faces numbered 1 through 6 and add the numbers showing on the top face of each cube.

    a) What is the **theoretical** probability of rolling a sum that is less than or equal to 4?

    b) Mark performed 100 trials of this experiment and recorded the results in the table below. Based on Mark's data, what is the **experimental** probability of rolling a sum that is less than or equal to 4?

Sum	2	3	4	5	6	7	8	9	10	11	12
Number of Occurrences	2	4	6	10	14	17	13	14	11	5	4

    c) The experimental probability can be used as an estimate of the theoretical probability. How could Mark alter his experiment so that the experimental probability is likely to be closer to the theoretical probability?

INDEX

Symbols and Numerics

Δ (delta), **25**
16% model, 107

A

acceleration, 282, 333
  due to gravity, 282, 330
accumulation, defined, 492
accuracy of predictions, checking,
  102–105, 128
adding probabilities, 573–574
addition method. *See* linear
  combination method
addition of matrices, 155–156, 192
additive inverse property, 151, 192
adjacent leg of a right triangle,
  499, 501
adjusted capitalized cost, 474
adjusted rate mortgages (ARMs),
  481, 486, 492
Age of Enlightenment, 4
air volume and pressure. *See*
  Boyle's Law
allowances, 423–424, 492
altitude of triangles, 227–228
amortization, 473–476, 492
  formula, defined, 492
amplitude of oscillations, 523, 531,
  543, 556
angle of elevation, 496–497
angles in the unit circle, 514–516
annual percentage rate (APR),
  469, **464,** 471, 492
  car loans, 464
  home loans, 481
annually compounded interest,
  444, 452
annuities, 455–458, 492
apartment rentals, 482, 486
apparent height, 227
approximating data with trend
  lines, 75–77
APR (annual percentage rate), 492
  car loans, 464
  home loans, 481
Archimedes, 3, 6
Archimedes' Principle, 3
area of triangles, 228
areas of regions, 367–370
ARMs (adjusted rate mortgages),
  481, 486, 492
arrow diagrams, 295

art transformations, 197–198,
  202–204, 254, 341
  algebraic representation of,
  207–211
assets, 455
Astrodome, 166
average, 128. *See* measures of
  central tendency
axes, 12, 67
axis of oscillation, 521–522, 556

B

balance (financial), 418, 492
bank interest. *See also* leasing vs.
  purchasing
  adjusted rate mortgages
  (ARMs), 481, 486, 492
  amortization, 473–476, 492
  compound interest, 444–448,
  451–452, 492
  fixed-rate loans, 480, 486, 492
  mixed growth, 455–458, 493
  simple interest, 417, 425, 439,
  444, 493
base (exponents), 347, 396, 413
  changing, formula for, 402, 407
best fit, line of, 103–104
binomial perfect square,
  completing, 298
binomials, defined, 341
body height, estimating, 74
  femur length and, 89–90, 92
  forearm length and, 93
  tibia length and, 91–92
bone length. *See* body height,
  estimating
Boyle, Robert, 4, 54
Boyle's Law, 4, 54–56, 67
Brown, Dan, 148
Brown, Warren, 131
Burns, Karen, 71
buying vs. leasing
  cars, 464–470, 485–486
  homes, 480–482, 486

C

Caesar, Julius, 147
Cake Love, 131
capitalized cost, defined, 492
car purchases
  amortization, 473–476, 492
  buying vs. leasing, 464–470,
  485–486
  depreciation, 461–463
central angle, defined, 556

central tendency, measures of,
  76–77
change, rate of, 11, 25, 31, 37, 68, 78
change of base formula, 402, 407
checking accounts, 439
clockwise rotations, 205, 210
codes and coded messages
  breaking with known
  frequencies, 150
  Internet information, 149
  matrices for, 150–151
  shift transformations, 147–148
coefficient matrix, 161, 192
coefficients, defined, 341
columns in a matrix, 150
combining probabilities, 573
common logarithms, 396
common multiplier (common
  ratio), 355–356, 413
complement of an event, 562–563
completing the square, 296–300,
  341
components of a matrix, 154
compound interest, 444–448,
  451–452, 492
compounding period, defined,
  492
conditional probability, 561–562
congruent polygons, 226
constant matrix, 162, 192
constant of proportionality, 11, 21,
  26, 67
Consumer Price Index (CPI),
  441–443, 451, 492
continuous compounding,
  defined, 492
continuous graphs, 12, 67
conventional fixed-rate loans,
  480, 486
convergence and vanishing
  points, 232–237, **253**–254
coordinate plane, 12, 67
coordinates, 12, 67
  symmetries and isometries,
  207–211
corner stones, 236–238
correlation between variables,
  11–13, 43, 67–68
  inversely proportional
  relationships, 50, 67
  linear. *See* linear relationships
  positive and negative, 13, 43,
  67–68, 98–99, 128
  proportional relationships, 26,
  68, 83

S

salary, defined, 493
sales tax, 475, 486
saving for retirement, 437
savings formula, 458, 493
savings plan, defined, 493
savings vehicles, 455
scalar multiplication, 193
scalars, 156
scale, 254
scale factor, 228, 254
scale of objects, 227
scaling, 225
scatterplots, 9, 12, 68
  linear form of, 99
  "versus" in, 25
second differences, 262, 268,
  341, 354
  writing functions with, 270
secret keys, 148
secret messages. *See* codes and
  coded messages
semiannual, defined, 493
sequences, defined, 493
sequences, geometric, 457, 492
series, geometric, 457, 492
Seurat, George, 217
shift transformations, 147–148
SI system of units, 13
sight lines, 225–228, 253
  convergence and vanishing
    points, 233–237
similar triangles, 225–226, 254, 497
simple interest, 417, 425, 439, 444,
  493
sine function (sin), defined, 556
sine ratio, 504–507, 556
  inverse of, 511, 556
sinusoidal functions, 521–524,
  543–545, 556
**sinusoidal regression, defined,
  556**
16% model, 107
size, perspective and, 218–221
slides, 197
slope-intercept form (*y = mx + b***),
  32,** 38, 68, 84, 90, 128, 261, 263
  y-intercepts, 32, 37, 68, 84, 356,
    363
slope of a line, 31, 33, 37, 68, 78
social security program, 493
social security tax, 429–430, 433,
  435

solving quadratic functions
  with factoring, 288–290, 303
  with inverse operations,
    295–300, 303
  with Quadratic Formula,
    305–308
solving systems of equations
  by graphing, 135–136
  with linear combinations,
    139–140, 192
  with matrices, 160–163
  by substitution, 140–141, **193**
  with tables, 135
solving systems of inequalities,
  167, 173–175, 178–181, 193
sound intensity, 406
sound waves, 542–545
  musical vibrations, 536–538, 540
speed, 260
  acceleration due to gravity,
    282, 330
  change in (acceleration), 282,
    333
sphere, volume of, 15
spins, 198
springs, 4, 23. *See also* Hooke's Law
square, completing, 296–300, 341
square root, defined, 341
square root functions, 341
SSE (sum of squared residual
  errors), 103–104, 128
St. Cyr Cipher strip, 148
  solving, 150
stadiums, domed, 166
steepness. *See* slope of a line
strong relationships, 97, 128
subscripts, 454, 493
substitution method, 140–141, 193
subtraction of matrices, 155, 193
successive quotients, 356, 363, 368
sum of squared residual errors
  (SSE), 103–104, 128
symmetry, 200–201
  algebraic representation of,
    207–211
systems of equations, 193
  solving graphically, 135–136
  solving using a table, 135
  solving with linear combination
    method, 139–140, 192
  solving with matrices, 160–163
  solving with substitution
    method, 140–141, 193

systems of inequalities, 167,
  173–175, 178–181, **193**

T

tables of trigonometric values, 510
tangent (tan), defined, 556
tangent ratio, 504–507
  inverse of, 511
tax withholding table, 421–422
taxes, 420–424, 493
  social security and medicare
    taxes, 429–430
  withholding, 421–422, 428–429,
    493
terminal side, 516
theoretical probability, 559
Torricelli, Evangelista, 55
trade-in allowance, defined, 493
trading in an old vehicle, 468, 493
transformations, 197–198, 202–204,
  254, 341
  algebraic representation of,
    207–211
translations, 197, 203, 254, 341
  algebraic representation of,
    207–208
  of quadratic functions, 273–277
tree diagrams, 583–586
trend lines, 75–77
triangles, 496–502
  altitude of, 227–228
  area of, 228
  inverse trigonometric ratios, 511
  similar, 225–226, 254, 497
  trigonometric ratios, 504–507,
    510–511, 514–516
trigonometric ratios, 504–507,
  510–511
  angles in the unit circle, 514–516
  inverses of, 511
Trotter, Mildred, 89
TVM Solver, 466, 493
two-point perspective, 232–237,
  254

U

unit circle, 514–516, 556
units of measurement, 10, 68
US Treasury bill index, 493

V

Vallejo, Cerapio, 197
vanishing point, 233–237, 254
variable matrix, 161, 193
variable relationships, 11–13, 43, 67–68
  inversely proportional relationships, 50, 67
  linear. *See* linear relationships
  positive and negative, 13, 43, 67–68, 98–99, 128
  proportional relationships, 26, 68, 83
  strong and weak relationships, 97, 128
variables, 10, 68
  dependent and independent, 10, 12, 48, 67
  input and output variables, 138
  proportional, 13
  rate of change, 11, 25, 31, 37, 68, 78
vehicle purchases
  amortization, 473–476, 492
  buying vs. leasing, 464–470, 485–486
  depreciation, 461–463
"versus," in graphs, 24, 68
vertex form of quadratic functions, 295–298, 303, 310, 341
vertex of a parabola, 264, 341
vertical compression/stretch of exponential functions, 356, 363
vertical shift of lines. *See* y-intercepts
vibrations of music, 536–538
volume, defined, 68
volume of sphere, 15
volume–pressure relationship. *See* Boyle's Law

W

wages. *See* earnings
water displacement, 6, 8–10
waves. *See* periodic functions
weak relationships, 97, 128
withholding allowances, 423–424, 492
withholding tax, 421–422, 428–429, 493

X

x-axis, reflection across, 209, 211
***x*-intercepts, 288,** 341

Y

y-axis, reflection across, 208
***y*-intercepts, 32,** 37, 68, 84
  exponential functions, 356, 363
$y = a(x - h)^2 + k$ family, 295–298, 303, 310, 341
$y = a(x - r_1)(x - r_2)$ family, 288, 303, 310, 341
$y = ax^2 + bx + c$ family, 264, 288, 303, 310, 341. *See also* quadratic functions
$y = b^x$ parent function, 355. *See also* exponential functions
$y = mx$ family. *See* direct variation function
$y = mx + b$ family (slope-intercept form), 32, 38, 68, 84, 90, 128, 261, 263. *See also* y-intercepts
$y = x$ parent function, 263, 322
  reflections across. *See* inverses
$y = x^2$ parent function, 264

Z

Zero-Product Property, 289, 303, 342
zeros of quadratic functions, 288, 342

ACKNOWLEDGMENTS

Project Leadership

Jo Ann Wheeler
Region 4 ESC, Houston, TX

David Eschberger
Region 4 ESC, Houston, TX

Solomon Garfunkel
COMAP Inc., Lexington, MA

Roland Cheyney
COMAP Inc., Lexington, MA

Lead Authors

Gary Cosenza
Region 4 ESC, Houston, TX

Paul Gray
Region 4 ESC, Houston, TX

Julie Horn
Region 4 ESC, Houston, TX

Authors

Sharon Benson
Region 4 ESC, Houston, TX

Roland Cheyney
COMAP, Lexington, MA

David Eschberger
Region 4 ESC, Houston, TX

Jo Ann Wheeler
Region 4 ESC, Houston, TX

Publisher

Craig Bleyer

Marketing Manager

Cindi WeissGoldner

Director of High School Sales and Marketing

Mike Saltzman

Senior Media Editor

Roland Cheyney

Associate Editor

Brendan Cady

COMAP Staff

Laurie Aragón, Rafael Aragón, Roland Cheyney, Kevin Darcy, Gary Feldman, Gary Froelich, Solomon Garfunkel, Daiva Kiliulis, Tim McLean, Sheila Sconiers, Anne Sterling, John Tomicek, George Ward, Pauline Wright

Index Editor

Seth Maislin
Focus Publishing Services, Arlington, MA

Illustrations

Lianne Dunn, George Ward

Photo Research

COMAP Production

PHOTO/ILLUSTRATION CREDITS

Rafael Aragón: 572

David Barber: 112, 120, 147, 411, 564

Jackson Barber: 195

Bridgeman Art Library International LTD: 216, 217, 239, 244

Vin Catania Studios: 45, 220, 221, 234, 237

Corbis Images: 62, 70, 82, 94, 96, 100, 114, 117, 119, 125, 127, 130, 337, 455, 538

Dover Publications Inc.: 76, 147, 352

Lianne Dunn: 41, 71, 74, 88, 93, 112, 122, 123, 417, 441, 470, 477, 478

EyeWire: 109, 439, 461, 464, 529

Mark Finkensteadt: 131

Gary Froelich: 200, 201

Hewlett-Packard: 258

Image Bank: 233, 243

Daiva Kiliulis: 123, 124, 468, 480

NASA: 282

National Anthropological Archives: 89

Novica.com: 197

Purdue University: 71

Shutterstock: 2, 6, 8, 14, 16, 18, 20, 22, 23, 30, 35, 42, 47, 57, 59, 79, 98, 132, 137, 142, 143, 145, 149, 154, 163, 164, 165, 166, 171, 172, 176, 177, 182, 184, 186, 187, 190, 191, 194, 206, 248, 256, 257, 261, 266, 280, 285, 294, 295, 302, 314, 323, 332, 340, 344, 345, 347, 351, 356, 359, 361, 362, 376, 381, 395, 400, 405, 406, 414, 415, 426, 431, 444, 459, 487, 494, 495, 510, 513, 517, 521, 525, 536, 542

Syi Tong: 215

Susan Van Etten: 106, 107, 266, 418, 460, 473, 489

George Ward: 215, 534, 546, 548, 565, 582

F. G. Wilson: 559